# Environmental Science and Engineering

## Environmental Science



**Series Editors**

Ulrich Förstner, Technical University of Hamburg-Harburg, Hamburg, Germany

Wim H. Rulkens, Department of Environmental Technology, Wageningen, The Netherlands

Wim Salomons, Institute for Environmental Studies, University of Amsterdam, Haren, The Netherlands

The protection of our environment is one of the most important challenges facing today's society. At the focus of efforts to solve environmental problems are strategies to determine the actual damage, to manage problems in a viable manner, and to provide technical protection. Similar to the companion subseries Environmental Engineering, Environmental Science reports the newest results of research. The subjects covered include: air pollution; water and soil pollution; renaturation of rivers; lakes and wet areas; biological ecological; and geochemical evaluation of larger regions undergoing rehabilitation; avoidance of environmental damage. The newest research results are presented in concise presentations written in easy to understand language, ready to be put into practice.

More information about this subseries at http://www.springer.com/series/3234

Prashanthi Devi Marimuthu ·
Rajakumar Sundaram · Aravind Jeyaseelan ·
Thamaraiselvi Kaliannan
Editors

# Bioremediation and Green Technologies

## Sustainable Approaches to Mitigate Environmental Impacts

Springer

*Editors*
Prashanthi Devi Marimuthu
Department of Environmental Science and Management
Bharathidasan University
Tiruchirappalli, Tamil Nadu, India

Aravind Jeyaseelan
Department of Civil Engineering
Dhirajlal Gandhi College of Technology
Salem, Tamil Nadu, India

Rajakumar Sundaram
Department of Marine Biotechnology
Bharathidasan University
Tiruchirappalli, Tamil Nadu, India

Thamaraiselvi Kaliannan
Department of Environmental Biotechnology
Bharathidasan University
Tiruchirappalli, Tamil Nadu, India

ISSN 1863-5520 ISSN 1863-5539 (electronic)
Environmental Science and Engineering
ISSN 1431-6250 ISSN 2661-8222 (electronic)
Environmental Science
ISBN 978-3-030-64124-5 ISBN 978-3-030-64122-1 (eBook)
https://doi.org/10.1007/978-3-030-64122-1



This Springer imprint is published by the registered company Springer Nature Switzerland AG
The registered company address is: Gewerbestrasse 11, 6330 Cham, Switzerland

*Dedicated*
*to*
*Our Loved Ones*

# Preface

In recent years, the environment is contaminated to a substantial extent from both natural and anthropogenic sources. The prevalence and persistence of a wide range of toxic pollutants in the soils, sediments and waters are of major concern. Though lot of physical and chemical methods are available, they are not economical, safe and eco-friendly. Hence, alternate viable methods using microbes and plants for cleaning up the pollutants are employed. Due to the regulations of strict laws for the protection of the environment, alternate clean technologies has become an important area of research in academia, government and industry. Bioremediation and clean technologies is an area of promising research over the recent years giving scope to many biotechnologists and microbiologists to engineer technologies for the restoration of contaminated zones. This book is a compilation of research conducted over varied geographical areas contaminated by regional commercial activities. The editors have sincerely worked over three years to carefully select and compile the contributed manuscripts, edit and revise them to the scope of the book. Several engineered solutions are present in the market to degrade contaminants at small and large scales accordingly; however, a long-standing technology is yet to be standardized. In this book, the impact of contamination on various environmental resources is documented, and several solutions researched through microbial, biotechnological methods and biochemical methods are put forth. Contaminant removal and degradation by the use of simple natural methods and conversion to innovative alternate have been discussed. The contributed articles suggest novel approaches to combat the ever increasing contamination and pollution problem. Though this book suggests alternative from regional studies, the findings can be adopted globally paving way for new research areas.

Tiruchirappalli, India — Prashanthi Devi Marimuthu
Tiruchirappalli, India — Rajakumar Sundaram
Salem, India — Aravind Jeyaseelan
Tiruchirappalli, India — Thamaraiselvi Kaliannan

**Acknowledgements** The editors acknowledge the hard work and efforts by the contributors and the support by our research scholars in revising and compiling the book manuscript.

# Contents

**Impact of Contaminants on Environmental Health**

**Impact of Water Quality on the Anatomical and Histochemical Characteristics of *Eichhornia Crassipes*** . . . . . . . . . . . . . . . . . . . . . . . . . 3
D. MadhuBala, J. Narmathasri, V. Priyadarshini, S. Saranya, S. Sneka, and K. S. TamilSelvi

**Insights into the Status of Heavy Metal Resistant Rhizobacterial Communities in the Heavy Metal Contaminated Sites** . . . . . . . . . . . . . . 13
Karthikeyan KirupaSree, Vijay Karuppiah, Sathiamoorthi Thangavelu, and Kavitha Thangavel

**Microbiome Analysis of Pesticide Contaminated Soils Its Impact on the Microbial Isolates** . . . . . . . . . . . . . . . . . . . . . . . . . . . . . . . . . 35
S. Sowmiya Sri, V. Suji, and G. Sophia Reena

**Impact of Sugar Mill Effluent on the Quality of Groundwater in Cuddalore District, Tamil Nadu, India** . . . . . . . . . . . . . . . . . . . . . . . . 41
A. Arivoli, X. Agnello J. Naveen, and Sathiamoorthi Thangavelu

**Bioremediation Studies of Contaminated Sites**

**Biosorption of Ni(II) Using Seeds of *Mirabilis Jalapa*** . . . . . . . . . . . . . . . 55
V. Anitha, A. Balakumaran, V. V. Ramprakash, A. V. Vinothini, P. Muthukumaran, and Aravind Jeyaseelan

**Bioremediation of Heavy Metals and Toxic Chemicals from Muttukadu Lake, Chennai by Biosurfactant and Biomass Treatment Strategies** . . . . . . . . . . . . . . . . . . . . . . . . . . . . . . . . . . . . . . 67
C. Elizabeth Rani, V. Balaji Ayyadurai, and K. K. Kavitha

**Degradation Efficacy of *Pinus radiata* Don Needle Leaf, Twig and Bark by Wood Degrading Fungi in Forest Ecosystem** . . . . . . . . . . . 87
M. N. Abubacker and M. Prince

**Biodegradation of Lead from Accumulated Municipal Solid Waste Using Bacterial Consortium for Effective Biomining** . . . . . . . . . . . . . . . 103
Mihigo Felix, Manikandan Arjunan, Kavya Siddeshwar, Rajakumar Sundaram, and Prashanthi Devi Marimuthu

**Bioremediation Approaches for Degradation of Textile Dye Effluents** . . . . . . . . . . . . . . . . . . . . . . . . . . . . . . . . . . . . . . . . 113
S. Lakshmana Prabu, A. Umamaheswari, and A. Puratchikody

**Bioremediation of Dye Effluent Using Dodonae Viscosa Seed Powder** . . . . . . . . . . . . . . . . . . . . . . . . . . . . . . . . . . . . . . . . 135
B. Sivasathya, A. Kalaicelvi, and K. K. Kavitha

**Optimization of Culture Conditions for Enhanced Decolorization of Azo Dyes by *Aspergillus flavus* Isolated from Dye Contaminated Soil** . . . . . . . . . . . . . . . . . . . . . . . . . . . . . . . . . . 143
M. Dexilin, B. Gowri Manogari, and Thamaraiselvi Kaliannan

**Potential Utilization of a Weed *Prosopis Juliflora* Leaf Extract Nanoparticle for Dye Degradation and Antibacterial Activity** . . . . . . . . 155
S. Sahithya and C. Krishnaveni

**Experimentation and Evaluation of Biodegradative Kinetic Parameters of Nitrate Removal from Drinking Water** . . . . . . . . . . . . . . 163
Rajakumar Sundaram, P. M. Ayyasamy, S. Umadevi, Y. C. Song, and P. Lakshmanaperumalsamy

**Clean Technology and Promising Microbial Technological Findings for Environmental Mitigation**

**Economic and Environmental Benefits of Cleaner Technology in Industrial Pollution Control: Case Study of Select Sugar Industry in Tamil Nadu** . . . . . . . . . . . . . . . . . . . . . . . . . . . . . . . . . . . . . . . 185
X. Agnello J. Naveen, S. Boopathi, A. Arivoli, and A. Kannan

**Case Experience on WIN WIN Approach Towards Industrial Pollution Control-Economic and Environmental Efficacy of Cleaner Technology in Select Cement Industry in Tamil Nadu** . . . . . . . . . . . . . 209
X. Agnello J. Naveen, S. Boopathi, and A. Arivoli

**Analysis of Microbial Isolate from Contaminated Street Food with a Potential to Degrade Food Waste** . . . . . . . . . . . . . . . . . . . . . . . 235
S. Swathi Priya, C. Shanmuga Priya, and J. Maria Shyla

**Screening, Isolation and Molecular Characterization of Biosurfactants Producing *Serratia Marcescens* from Petrochemical Exposed Site** . . . . . 243
V. Vijayalekshmi, P. Muthukumaran, and Aravind Jeyaseelan

**Alternative Approaches and Suggested Studies for Prospective Environmental Remediation**

**Sustainable Packaging from Waste Material: A Review on Innovative Solutions for Cleaner Environment** . . . . . 259
Paneti Rajesh and V. Subhashini

**Influence of Sugar Mill Effluent on Germination and Growth of Fenugreek (*Trigonella foenumgraecum*) and Mustard (*Brassica nigra*)** . . . . . 271
A. Arivoli, Sathiamoorthi Thangavelu, and X. Agnello J. Naveen

**Quorum Sensing Inhibitors as an Alternate to Antibiotic Against Biotic Pressure Induced Bacterial Contamination in Aquaculture** . . . . . 283
Arumugam Veera Ravi, Sivasubramanian Santhakumari, Issac Abraham Sybiya Vasantha Packiavathy, and Jeyaraj Godfred Ponraj

**Ecofriendly Approach for Bioethanol Production from Microalgae** . . . . 301
Pandian Prabakaran, Virumandi Pradeepa, Nagasundaram Rashiya, Sundaram Ravikumar, Sathiamoorthi Thangavelu, and Gopal Selvakumar

**Novel QR Code Tagging System for Campus Vegetation to Promote Ecorestoration** . . . . . 319
V. E. Chahana, S. Abirami, S. Karpagavalli, and C. Arunpriya

# Impact of Contaminants on Environmental Health

# Impact of Water Quality on the Anatomical and Histochemical Characteristics of *Eichhornia Crassipes*

**D. MadhuBala, J. Narmathasri, V. Priyadarshini, S. Saranya, S. Sneka, and K. S. TamilSelvi**

**Abstract** Water hyacinth, a worst aquatic weed and also a right candidate for pollutant removal, was collected from four different locations. Our work aimed to study the influence of water quality over the anatomical and histochemical features of water hyacinth from four water bodies. A significant difference in the size of the plant, size of the different cells in leaf, petiole and root was observed between the locations. Histochemicals were localized in the hypodermal, cortical and in some cases, stellar region of parts of water hyacinth. Water quality parameters such as pH, temperature, chloride and sulphate content of the water samples varied between locations.

**Keywords** *Eichhornia crassipes* · Aquatic weed · Water quality · Anatomical and histochemical features · Stomata

## 1 Introduction

*Eichhornia crassipes* (Pontederiaceae) is a free-floating perennial aquatic herb growing in tropical and sub-tropical regions. Its common name is Water hyacinth, and it was often a highly problematic invasive species. The primary factors that influence the growth of aquatic plants are nutrients, depth of water, bottom type, and water clarity of the water bodies. These factors vary from lake to lake, and influenced by many factors, including the surrounding land that contributes water (from rainfall or groundwater) to the lake. Studies revealed that water hyacinth could grow prolific in wastewater and can efficiently accumulate heavy metals (Yahya 1990; Vesk et al. 1999; Ali and Soltan 1999; Soltan and Rashed 2003). The objectives of the current investigation were to analyze the anatomical features, stomatal variations, histochemical and to evaluate the influence of water quality over anatomical and histochemical characteristics of *E. crassipes*.

---

D. MadhuBala · J. Narmathasri · V. Priyadarshini · S. Saranya · S. Sneka · K. S. TamilSelvi (✉)
Department of Botany, PSGR Krishnammal College for Women, Coimbatore, Tamil Nadu 641004, India
e-mail: tamilselviks@psgrkcw.ac.in



M. Prashanthi et al. (eds.), *Bioremediation and Green Technologies*, Environmental Science and Engineering, https://doi.org/10.1007/978-3-030-64122-1_1

*E. crassipes*, today has become a prominent and malicious invasive species. When left unchecked and undisturbed, it spreads over the entire lakes and ponds; thus, drastically impeding water movement and obstruct sunlight, preventing its reach to native aquatic flora, triggering their death and decay, culminating in the depletion of the dissolved oxygen in water, galvanize the demise of other aquatic fauna. The plants serve as a convenient niche for mosquito breeding, the usual vectors of many dangerous diseases. Water hyacinth is also responsible for posing similar environmental problems in many countries and ends as an unmanageable problem. if adequate control measures are not employed they encroach waterbodies that has been exposed to anthropogenic effects; unbalance the existing lifecycles in the eutrophicated lakes that acted as a sink of nutrient runoffs (Patel 2012). Because of *E. crassipes* invasiveness, there has been established report of trails of biological control weevils (*Neochetina bruchi* Hustache and *Neochetina eichhorniae*), and moth *Niphograpta albiguttalis* (Richard and John 1981). *Neochetina eichhorniae* makes a lasting impact on controlling and eradicating this pest by reducing its mass, enabling the plant to produce fewer offspring. A semi-aquatic grasshopper, tested in South Africa as a supplementary control agent (Christiane and Hendrik 2008).

In an in-depth scientific investigation, *E. crassipes* has been established for its competency to liberate 60–80% nitrogen and 69% of potassium from water. In another similar study, *E. crassipes* roots were found to evacuate particulate matter and nitrogen from a natural eutrophicated wetland. The plant absorbs heavy metals and assorted toxins from polluted water sources (Sheffield 1967). The roots *Eichhornia crassipes* take-up contaminants naturally; lead, mercury, strontium-90, including organic compounds in higher concentrations capacity, thus the water hyacinths can be grown explicitly for wastewater remediation (Abid et al. 2014). Hence in this current work, *E. crassipes* was explored as an environmental indicator of pollution in water bodies.

## 2 Materials and Methods

*E. crassipes* collected from four different places which includes a river—Yakra in Kerala, and three lakes, namely Singanallur, Sulur and Ukkadam in Coimbatore, Tamil Nadu. Anatomical studies of water hyacinth done using saffranin and tryphan blue stains. Histochemical localization for starch, protein, alkaloid, lignin was performed on the plant using corresponding dyes. Cell measurements were also done using image analyzer software. Chloride and sulphate content of the water sample were estimated using standard methods (APHA 2005). pH and temperature of the water samples were measured using pH meter and thermometer, respectively.

# 3 Results and Discussion

## *3.1 Anatomical Studies of Water Hyacinth*

### 3.1.1 Leaf

Leaf possess predominant upper and lower epidermal layers. Epidermis was composed of single-layered parenchyma cells arranged without any spaces. Epidermis was surrounded by thin cuticle. Bulliform cells were seen in the upper epidermis. Ground tissue contains uniform palisade cells. Palisade layer was found on both the upper and lower side beneath the epidermis. Aerenchyma cells were found in the mesophyll layer. Vascular bundles were found randomly in the mesophyll of the leaf. Bundles were of two types (smaller and larger). Xylem tissues consist of tracheids, vessels, fibres and parenchyma cells. Phloem was composed of sieve tubes and companion cells. Phloem fibres and parenchyma cells were also present. Each bundle has bundle sheath made up of sclerenchyma cells. Sclereids were found in cells facing airspaces. Water cavity or protoxylem lacuna was found in the mesophyll of the leaf. Stomata were found in the upper epidermis, but in some cases the leaf is amphistomatic. The subsidiary cells were parallel and exhibited paracytic stomata. Subsidiary cells were surrounded by many epidermal cells. Guard cells contain chloroplast and thus help in photosynthesis. Stomata help in controlling transpiration, thereby aid in leaf cooling, metabolite fluxes, and long-distance signalling as well as acting as a barrier to harmful pollutants such as ozone and pathogens (Meidner and Mansfield 1968; Mansfield and Majernik 1970). In the leaf of Singanallur sample, vascular bundles were surrounded by 7–8 cells. In leaf samples collected from Sulur, two or more vascular bundles were found attached together like a single bundle (Table 1).

### 3.1.2 Petiole

Epidermis of petiole was made of single-layered parenchyma cells (Mahmood et al. 2005). Cuticle was absent. Large air spaces were present in the hypodermis of petiole. Sclereids were observed arising from the aerenchyma cells projecting into air spaces. Few raphide crystals were also observed in petiole. Inclusions were found in the hypodermis. Scattered vascular bundles were found smaller and large in number towards the epidermis. Xylem tissues consist of tracheids, vessels, fibres and parenchyma cells. Phloem was composed of sieve tubes and companion cells. Phloem fibres and parenchyma cells were also present. Each bundle has bundle cap made up of sclerenchyma cells. The hexagonal air spaces were surrounded by bands of single-layered parenchyma cells. Dutta (1986) reported that petiole was a pseudobulb with many air chambers for facility of floating. Pith was absent. Petiole of this plant collected from Singanallur, Yakra and Sulur exhibited the presence of numerous sclereids and those gathered from Ukkadam samples possessed numerous raphides (Table 2).

**Table 1** Size of different types of cells in the leaves of water hyacinth

| Location | Diameter in μm* | | | | | | | | |
|---|---|---|---|---|---|---|---|---|---|
| | Upper epidermal cells | Lower epidermal cells | Xylem cells | Phloem cells | Vascular bundle | Storage cells | Sclereids | Parenchyma cells | Air cavity |
| Singanalur | 16.36 | 12.09 | 27.16 | 16.21 | 125.32 | 68.32 | 11.74 | 39.34 | 204.94 |
| Sulur | 15.30 | 9.82 | 5.41 | 3.86 | 29.58 | 32.61 | 37.87 | 51.61 | 32.21 |
| Ukkadam | 14.73 | 10 | 2.44 | 5.46 | 44.52 | 54.47 | 46.54 | 49.86 | 35.57 |
| Yakra | 18.89 | 13.51 | 7.07 | 7.11 | 73.46 | 31.16 | 70.75 | 25.59 | 69.33 |

*Mean of six replicates

**Table 2** Size of different types of cells in petiole of water hyacinth

| Location | Diameter in μm* | | | | | | | |
|---|---|---|---|---|---|---|---|---|
| | Epidermis | Hypodermis | Air space | Vascular bundle | Endodermis | Sclereids | Storage cells | Inclusions |
| Singanalur | 23.20 | 26.48 | 177.16 | 139.58 | 56.61 | 113.52 | 197.71 | 33.68 |
| Sulur | 5.44 | 6.72 | 80.91 | 52.60 | 21.73 | 41.08 | 52.35 | 7.65 |
| Ukkadam | 10.57 | 15.23 | 46.11 | 36.17 | 19.63 | 26.34 | 43.26 | 7.88 |
| Yakra | 13.79 | 23.49 | 184.99 | 99.36 | 44.72 | 93.67 | 189.15 | 18.22 |

*Mean of six replicates

### 3.1.3 Root

Epiblema of root was single-layered and composed of tightly arranged parenchyma cells. The epidermal regions of the roots perform absorption functions, unlike the other plants where it was responsible for protection (Warrier and Saroja 2008). Root cap was found. Younger roots were white in colour, and the matured roots were brown or violet in colour. Cortex was found beneath the epiblema. Cortex was composed of parenchyma cells and differentiated into outer, middle and inner cortex. Middle cortex forms the diaphragm. Inner cortex consists of 4–5 layers, and it varied from place to place. Cortex composed of air spaces. Each air space has trabaculae or diaphragm, and it was linear in shape. Endodermis and pericycle were single-layered and composed of parenchyma cells. Vascular region consisted of vascular bundles and pith (Table 3). Vascular bundles were found arranged radially in the stellar region. Xylem tissues include tracheids, vessels, fibres and parenchyma cells. Phloem was composed of sieve tubes and companion cells. Phloem fibres and parenchyma cells were also present. Xylem and phloem were arranged in an alternate manner. Each bundle has bundle cap made up of sclerenchyma cells. Roots of samples collected from all the five locations exhibited similar structures.

The epidermis of leaf, root and petiole consists of a single layer of rectangular cells which is characteristically a constant feature of this species. The epidermis is covered with a very thin cuticle, and thin cellulose walls of epidermal cells facilitate steady absorption from surrounding water. The most characteristic anatomical feature of this plant is the presence of gas-filled chambers and passages in roots, leaves and petiole. Air chambers are large, usually regular (circular to hexagonal) intercellular spaces. These chambers provide a sort of internal atmosphere for the plant. In these spaces oxygen emitted during photosynthesis is stored and used again in respiration. No Casparian strips found on endodermis, this finding is confirmed by earlier works of Barnabas (1996).

## *3.2 Histochemical Studies*

Histochemical localization in different sections of *E. crassipes* were analyzed, and the occurrence of orgastic content or secondary metabolites such as starch, protein, tannin, fat and alkaloids in stem, petiole and leaf were observed (Table 4). Starch was present in the hypodermis and around the vascular bundles in petiole, leaf and root of the water hyacinth. Protein was present in the ground tissue of root, petiole and leaf of water hyacinth. Alkaloids were found in petiole, leaf and root of water hyacinth. Lignin was found in roots and petiole of water hyacinth. It was rarely present in leaves.

**Table 3** Size of different types of cells in the root of water hyacinth

| Location | Diameter in µm* | | | | | | | | | |
|---|---|---|---|---|---|---|---|---|---|---|
| | Epidermis | Hypodermis | Outer cortex | Middle cortex | Inner cortex | Vascular bundle | Endodermis | Air cavity | Cortical space | Pericycle |
| Singanallur | 3.47 | 3.02 | 16.78 | 102.82 | 7.10 | 105.32 | 5.68 | 21.98 | 23.68 | 5.42 |
| Sulur | 10.61 | 9.19 | 68.91 | 65.75 | 6.16 | 80.04 | 4.39 | 15.60 | 9.30 | 2.10 |
| Ukkadam | 4.80 | 3.98 | 14.37 | 118.55 | 7.41 | 104.05 | 3.93 | 15.82 | 27.17 | 2.69 |
| Yakra | 2.13 | 8.36 | 16.90 | 51.69 | 6.49 | 60.97 | 4.00 | 16.92 | 15.09 | 2.62 |

*Mean of six replicates

**Table 4** Histochemical studies of *E. crassipes*

| Sample | Histochemicals | Root | Leaf | Petiole |
|---|---|---|---|---|
| Singanallur lake | Starch | + | + | + |
| | Alkaloids | – | – | + |
| | Protein | + | – | – |
| | Lignin | + | – | + |
| Sulur lake | Starch | – | + | + |
| | Alkaloids | + | + | + |
| | Protein | + | + | + |
| | Lignin | + | + | + |
| Ukkadam lake | Starch | + | + | + |
| | Alkaloids | + | + | + |
| | Protein | + | + | + |
| | Lignin | + | – | + |
| Yakra river | Starch | – | + | + |
| | Alkaloids | + | + | + |
| | Protein | + | + | + |
| | Lignin | + | + | + |

## 3.3 *Impact of Water Quality on the Anatomical Features of* E. crassipes

Petiole of the *E. crassipes* sample from Singanallur and Sulur lakes possessed maximum number of sclereids which was due to high pH, sulphates, chlorides and temperature of water (Table 1). Sclereids were present in the petiole of samples from other locations but in minimum numbers. Petiole of sample from Ukkadam lake showed the presence of raphides because sulphate concentration in the water was very low.

Due to high pH, temperature, chloride and sulphate level in the water of Sulur lake, size of xylem cells, phloem cells, storage cells, air cavity and vascular bundles in the leaves were diminished whereas the size of such cells were found larger in Singanallur leaf samples though they contain approximately equal level of pH, temperature, chloride as that of Sulur water. In the petiole of samples from Singanallur, size of cells in all regions was larger than other location samples (Table 2). In the petiole of Ukkadam samples, size of xylem cells, phloem cells, vascular bundle and air cavity were small due to low sulphate concentration whereas in the roots of Ukkadam samples such low level expressed large vascular bundles. Air cavity and inclusions in the petiole of Yakra samples were found larger than other location samples which might be due to low chloride level in the water whereas such low level induced smaller air cavity and inclusions in the roots of Yakra samples (Table 3).

## 4 Conclusion

*E. crassipes* is a potential aquatic weed belonging to the family Pontederiaceae. The present investigation mainly focused on the study of the anatomical and histochemical analysis of petiole, leaf and root of water hyacinth, from different polluted lakes. In anatomical studies of water hyacinth, different structures like sclereids and raphides were observed. Stomatal studies resulted that water hyacinth possessed paracytic type of stomata. Histochemical studies of water hyacinth revealed the presence of starch, protein, alkaloids and lignin in plant parts by the application of corresponding stains. Water quality parameters like pH, temperature, chlorides and sulphates of lakes and rivers were estimated, and such parameters had some influence over the anatomy of water hyacinth.

**Acknowledgements** The authors would like to thank Department of Biotechnology, Government of India, for the financial assistance provided under the STAR STATUS scheme and Foldscope project.

## References

Abid A, Gill S, Khan F, Ghauri N (2014) Phytoremediation systems for the recovery of nutrients from eutrophic waters. In: Eutrophication: causes, consequences and control, vol 2. pp 239–248

Ali MM, Soltan ME (1999) Heavy metals in aquatic macrophytes, water and hydrosoils from the river Nile. Egypt J Union Arab Biol 9:99–115

APHA (2005) Standard methods for the examination of water and wastewater. 21st ed. Am Publ Health Assoc. Washington, DC. Part, 8000:94–100

Barnabas AD (1996) Casparian band-like structures in the root hypodermis of some aquatic angiosperms. Aquat Bot 55:217–225

Christiane A, Hendrik D (2008) Global diversity of true and pygmy grasshoppers (Acridomorpha, Orthoptera) in freshwater. Hydrobiologia 595:535–543

Dutta AC (1986) A class book of botany, XVII. Oxford University Press, New Delhi, pp 309–312

Mahmood Q, Zheng P, Siddiqi MR, Islam E, Azim MR, Hayat Y (2005) Anatomical studies on water hyacinth (*Eichhornia crassipes* (Mart.) Solms) under the influence of textile wastewater. J Zhejiang Univ Sci A 6B:991–998

Mansfield TA, Majernik O (1970) Can stomata play a part in protecting plants against air pollutants. Environ Pollut 1:149–154

Meidner H, Mansfield TA (1968) Physiology of stomata. Blackie, Glasgow, UK

Patel S (2012) Threats, management and envisaged utilisations of aquatic weed *Eichhornia crassipes*: an overview. Rev Environ Sci Biotechnol 11:249–259

Richard GA, John SD (1981) Suppressing water hyacinth with an imported weevil. La Agric 24:4–5

Sheffield CW (1967) Water hyacinth for nutrient removal. J Aquat Plant Manage 6:27–30

Soltan ME, Rashed MN (2003) Laboratory study on the survival of water hyacinth under several conditions of heavy metal concentrations. Adv Environ Res 7(2):82–91

Vesk PA, Nockold CE, Aaway WG (1999) Metal localization in water hyacinth roots from an urban wetland. Plant Cell Environ 22:149–158

Warrier RR, Saroja S (2008) Histochemical studies on water hyacinth with particular reference to water pollution. Int J Integr Biol 2:96–99

Yahya MN (1990) The absorption of metal ions by *Eichhornia crassipes*. Chem Speciat Bioavailab 2:82–91

# Insights into the Status of Heavy Metal Resistant Rhizobacterial Communities in the Heavy Metal Contaminated Sites

**Karthikeyan KirupaSree, Vijay Karuppiah, Sathiamoorthi Thangavelu, and Kavitha Thangavel**

**Abstract** Anthropogenic activities viz. modern agricultural practices, industrialization and mining have long term detrimental effects on environment. All these factors lead to the increase in heavy metal concentration in both hydrosphere and lithosphere. The extreme use of chemical fertilizers in agricultural field possesses a major threat to human and animal health and also causes various environmental hazards. The utilization of the chemical fertilizers can be reduced through biological aspects. Rhizosphere bacterial community is majorly involved in the plant growth and colonized in root zones of the plants with enhanced symbiotic relationship with plant community. These bacteria support the plant growth at normal and stressed conditions. These naturally occurring bacteria will pave a way to minimize the use of chemical fertilizers and hence in reducing the risk hazards. Plant-soil-root ecosystem is an important interface between soil and plants and also plays a significant role in the biosorption of heavy metals from contaminated soils. The rhizobacteria dwelling in this soil are known to affect heavy metal mobility and availability to the growing plant through the release of chelating agents, acidification, phosphate solubilization and redox changes, therefore having tremendous potentials to enhance the bioremediation processes. Bioremoval strategies with appropriate heavy metal-adapted rhizobacteria have received considerable importance. This chapter aims to reveal the sources of heavy metals and its effects on various life forms with special emphasis on PGPR assisted mechanisms for bioremoval of heavy metals from heavy metal implicated sites in the environment.

**Keywords** Heavy metals · Chelating agents · Bioremoval · Plant growth promoting rhizobacteria · Chemical fertilizers

K. KirupaSree · V. Karuppiah · S. Thangavelu · K. Thangavel (✉)
Department of Microbiology, Science Campus, Alagappa University, Karaikudi, Tamil Nadu 630003, India
e-mail: kavithat@alagappauniversity.ac.in



M. Prashanthi et al. (eds.), *Bioremediation and Green Technologies*, Environmental Science and Engineering, https://doi.org/10.1007/978-3-030-64122-1_2

# 1 Introduction

Various industries such as cosmetics, textiles, tanneries emerge worldwide to fulfill the various requirements of enlarging population. These advancements have put an increasing burden on the environment by releasing large quantities of hazardous wastes, heavy metals, metalloids and chemicals that lead to serious problems in an ecosystem (Ayansina and Olubukola 2017). Heavy metals are naturally occurring elements that have a high atomic weight and a density at least 5 times greater than that of water (Tchounwou et al. 2012). The increase in heavy metal concentration in the environment at alarming rate may directly or indirectly affect plants, microbes, animals and human beings. Enormous amount of heavy metals are used in various fungicides and chemical fertilizers, wastewater irrigation and sewage sludge which in turn contaminates water resources as well as agricultural soils (Akcil et al. 2015). Copper conjugated pesticides are very expensive and formulated to have fungicidal and bactericidal actions (CIPAC 1992). Copper-(II) ion ($Cu^{2+}$) enters the fungal spores during germination and accumulates until a high concentration is achieved to kill the spores. However, antifungal activity is restricted to prevent the spore germination. Hence a prophylactic mode of fungicidal action is observed with copper based fungicides. Unfortunately, the deposition does happen on the crop before fungal spores begin to germinate indicating the essentiality of environment risk assessments. A similar mechanism is postulated for antibacterial action of copper based pesticides in day-to-day agriculture by US-EPA (United States Environmental Protection Agency) in the year 2000. Examples of copper based fungicides include GalbinAr, Efdalbakirox, Moltifen and Bromix. Waheed and Nahed (2017) have analyzed the presence of Arsenic (As), Cadmium (Cd) and Lead (Pb) as impurities and copper as conjugates in the above listed copper based fungicidal formulates before and after storage at 54 °C for 21 days.

Microbes play an important role in substance turnover of heavy metal contamination which will clean up the metal contaminated sites (Spain and Alm 2003). If the heavy metal is fewer in concentration it may act as active elements in plants and microorganisms. For example, Copper (Cu), manganese (Mn), molybdenum (Mo), nickel (Ni) and zinc (Zn) are actually micronutrients which mean these heavy metals are needed at very low quantities for the normal growth of plants. Iron (Fe) is not generally considered as a heavy metal because it is essential for growth and metabolism of both plants and animals at its optimum level. However when the above listed heavy metals are presented at supra-optimum levels (i.e., above 0.1%) they are toxic to plants and rhizosphere microorganisms (Nies 1999).To circumvent the metal stress, bacteria progress through many types of mechanisms to overcome the uptake of heavy metal ions. The mechanisms include efflux of metal ions outside the cell, binding and accumulation of the metal ions inside the cell and reduction of heavy metal ions to the less toxic states (Nies 1999). Cations of the heavy metals bind to glutathione in gram negative bacteria, which would result in bisglutathionato complexes. Bisglutathionato complexes in turn react with the molecular oxygen forming the Oxidized bisglutathione (Kachur et al. 1998). Reduced forms of heavy metals, which are less toxic, are released from the biosglutothionato complex into the environment. More heavy metal is uptaken by Gram positive bacteria due to presence of glycoproteins. Less heavy metal uptake by Gram negative bacteria is observed due to phospholipids and Lipopolysaccharides (LPS) (Das et al. 2008).

Plant growth promoting rhizobacteria (PGPR) play a vital role in agriculture mainly in the host plant adaptation to metal contaminated area by activation of several physiological changes in the plant cell metabolism. This improves tolerance of growing plants towards high amount of heavy metal implications (Conrath et al. 2006).The plant growth promoting bacteria assist the phytoremediation process through mechanisms that support plant growth including nitrogen metabolisms, synthesis and secretion of phytohormones such as indole-3-acetic acid (IAA), Cytokinins, acetonin and 2,3 butanediol, and organic acids as well as defense molecules such as siderophores, I—aminocyclopropane—I–carboxylate deaminase (ACC) (Khan et al. 2009; Taghavi et al. 2009).

## 2 Various Methods for Removal of Heavy Metals

A variety of physical, chemical and biological methods were in use to remove metals from the environment which are as follows (Joo et al. 2010; Pagnanelli et al. 2010):

### *2.1 Physical Methods*

Reverse osmosis, Membrane technology, Evaporation recovery, Filtration and Ion exchange.

### *2.2 Chemical Methods*

Electrochemical treatment, chemical precipitation and oxidation/reduction reactions.

Though the above mentioned methods were not the initial choice as they are costly, ineffective, and labor-intensive or the treatment process lacks selectivity (Chen 2008; Tang 2008), the study carried out by Talos 2009, on bioremediation or biosorption—based remediation techniques concluded that the natural processes are found to be cost effective and in that line biological methods to remove heavy metals stands first and employs microorganisms such as bacteria, fungi and algae. The present chapter describes the types of heavy metals, chances of environmental release and its effect on earth and life forms. In addition it specially emphasizes the role of Bacteria and Rhizobacterial communities on heavy metal removal and detoxification and safe release into the environment.

# 3 Effects of Heavy Metal Accumulation on the Earth

Environment is polluted by various ways and heavy metals are the main source for pollution which affects many biological systems in the world as heavy metals does not undergo biodegradation. According to the Biological functions, these metals has been classified as: (i) toxic metalloids and metals with undefined biological functions, (ii) fundamental metals with known biological functions, (iii) non-essential and nontoxic metal with no biological functions (Pepper et al. 2015). Heavy metals ions are chemical moieties which would influence its negative effects through diverse mechanisms such as protein damage, DNA damage and oxidative damage through the production of reactive oxygen species etc., (Assal et al. 2017). Most of the heavy metals are toxic to the environment in that Pb, Co, Cd stands in the first line and are differentiated from other pollutants; those are not biodegraded however these heavy metals can be accumulated in living organisms resulting in various diseases and disorders even at lower concentrations (Pehlivan et al. 2009). Soil is the backbone of agriculture and the worst effects are caused due to the accumulation of heavy metals in plants. Higher organisms that consume heavy metal accumulated plants face numerous health issues. This may also have a negative crash on balance of soil micro flora, plant growth and in the ground cover at the level of ground water purity (Roy et al. 2005). The toxicity of heavy metal against the plant and plant associated microbes are described in the table as well as the most of the PGPR which are involved in the phytoremediation has shown in the Table 1.

## *3.1 Arsenic*

Arsenic is moderately distributed in natural waters which are related with geological sources. Likewise in various locations of anthropogenic inputs, such as the use of pesticides, insecticides as well as the combustion of fossil fuels are the enormously important additional sources. Oxidation states III and V of arsenic occurs in natural waters, in the form of arsenic acid ($H_3AsO_5$) and its salts, arsenous acid ($H_3AsO_3$) and its salts, respectively (Sawyer et al. 2003). Arsenic contamination in groundwater may create major problem during irrigations. For example, it accumulates in plant tissues including grains and contaminates food chain (Verma et al. 2016). In recent, study has been carried out to inspect the molecular mechanisms and physiology of arsenic accumulation, toxicity, detoxification and tolerance in various plants (Kumar et al. 2015) some of the plant growth promoting rhizobacteria which is resistant to the arsenic is shown in the Table 1.

**Table 1** Toxicity of heavy metals towards plants and plant-associated microbes under heavy metal stress

| Metal | Source | Effects on plants | Effects on microbes | Host plant | PGPR involved in phytoremediation | References |
|---|---|---|---|---|---|---|
| Antimony | Soil erosion, volcanic eruption, coal combustion, mining, smelting | Decrease synthesis of some metabolites, growth inhibition, inhibit chlorophyll synthesis | Inhibit enzyme activities, reduced growth rate | Not yet reported | Not yet reported | Blais et al. (2008) and An et al. (2009) |
| Beryllium | Coal and oil combustion, volcanic dust | Inhibits seed germination | Chromosomal aberration, mutation | Not yet reported | Not yet reported | Finnegan et al. (2012) and Bissen et al. (2003) |
| Arsenic | Atmospheric deposition, mining, pesticides, rock sedimentation, smelting | Damage cell membrane, inhibition of growth, inhibits roots extension and proliferation, interferes with critical metabolic processes, loss of fertility, yield and fruit production, oxidative stress, physiological disorders | Deactivation of enzymes | *Brassica nigra*, *Brassica juncea* | *Microbacterium sp. CE3R2*, *Curtobacterium sp. NM1R1*, *Staphylococcus arlettae* | Román-Ponce et al. (2017), Srivastava et al. (2013) |
| Chromium | Dyeing, electroplating, paints production, steel fabrication, tanning, textile | Chlorosis, delayed, senescence, wilting, biochemical lesions, reduced biosynthesis germination, stunted growth, oxidative stress | Elongation of lag phase, growth inhibition, inhibition of oxygen uptake | *Cicer arietinum* | *Mesorhizobium* sp. RC3 | Wani and Khan (2010) |

(continued)

**Table 1** (continued)

| Metal | Source | Effects on plants | Effects on microbes | Host plant | PGPR involved in phytoremediation | References |
|---|---|---|---|---|---|---|
| Cadmium | Fertilizer, mining, pesticide, plastic, refining, welding | Chlorosis, decrease in plant nutrient content, growth inhibition, reduced seed germination | Damage nucleic acid, denature protein, inhibit cell division and transcription, inhibits carbon and nitrogen mineralization | *Brassica Juncea*, *Brassica napus* | *Variovox paradoxus*, *Rhodococcus* sp., *Xanthomonas* sp. RJ3, *Azomonas* sp. RJ4, *Pseudomonas* sp. RJ10, *Bacillus* sp. RJ31 | Belimov et al. (2005), Sheng and Xia (2006) |
| Copper | Copper polishing, mining, paint, plating, printing operations | Chlorosis, oxidative stress, retard growth | Disrupt cellular function, inhibit enzyme activities | *Brassica Juncea*, *Cicer arietinum*, *Brassica nigra* | *Bacillus* sp. PSB10, *Achromobacter xylosoxidans* Strain Ax10, *Bacillus* sp. PSB10, *Microbacterium* sp. CE3R2, *Curtobacterium* sp. NM1R1 | Ma et al. (2009), Wani and Khan (2010), Román-Ponce et al. (2017) |
| Mercury | Batteries, coal combustion, geothermal activities, mining, paint industries, paper industry, volcanic eruption, weathering of rocks | Affects antioxidative system, affects photosynthesis, enhance lipid peroxidation, induced genotoxic effect, inhibit plant growth, yield, nutrient uptake and homeostasis, oxidative stress | Decrease population size, denature protein, disrupt cell membrane, inhibits enzyme function | *Triticumaestivum* | *Enterobacterludwigii*, *Klebsiella pneumoniae* | Gontia-Mishra et al. (2016) |

(continued)

**Table 1** (continued)

| Metal | Source | Effects on plants | Effects on microbes | Host plant | PGPR involved in phytoremediation | References |
|---|---|---|---|---|---|---|
| Lead | Coal combustion, electroplating, manufacturing of batteries, mining, paint, pigments | Affects photosynthesis and growth, chlorosis, inhibit enzyme activities and seed germination, oxidative stress | Denatures nucleic acid and protein, inhibits enzymes activities and transcription | *Brassica Juncea,* | *Azotobacter chroococcum* HKN-5, *Bacillus megaterium* HKP-1, *Bacillus mucillaginosus* HKK | Wu et al. (2006) |
| Nickel | Electroplating, non-ferrous metal, paints, por | Decrease chlorophyll content, inhibit | Enzyme activities and growth, reduced | *Oryzasativa, Brassica juncea* | *Bacillus licheniformis* NCCP-59, *Bacillus subtilis* SJ-101 | Jamil et al. (2014), Zaidi et al. (2006) |
| Thallium | Cement production, combustion of fossil fuels, metal smelting, oil refining | Inhibits enzyme activities, reduced growth | Damages DNA, inhibits enzyme activities and growth | Not yet reported | Not yet reported | Blais et al. (2008) and Babula et al. (2008) |
| Selenium | Coal combustion, mining | Alteration of protein properties, reduction of plant biomass | Inhibits growth rate | Not yet reported | Not yet reported | Dixit et al. (2015) and Germ et al. (2007) |
| Zinc | Brass manufacturing, mining, oil refinery, plumbing | Affects photosynthesis, inhibits growth rate, reduced chlorophyll content, germination rate and plant biomass | Death, decrease in biomass, inhibits growth | *Brassica juncea* | *Bacillus megaterium* HKP-1, *Bacillus mucillaginosus* HKK | Wu et al. (2006) |

## 3.2 Lead

Lead can affect the various life forms through contamination from old lead plumbing pipes, dusts, fuels from various industrial sites and also in the old orchard sites in production where arsenate and lead is mostly used (Tangahu et al. 2011). Long term exposure to lead is found to be intensely toxic to both animals as well as plants which is non-biodegradable and has several harmful effects on organic systems including soil properties i.e., pH, organic carbon, amorphous iron, aluminium oxides (FEAL), and cation exchange capacity. (Bradham et al. 2006; Pehlivan et al. 2009). Several bacteria, such as *Arthrobacter* spp., *Bacillus megaterium, Pseudomonas marginalis, Citrobacter freundii, Staphylococcus aureus*, and *E. coli* have been found to be resistant to lead (Das et al. 2016).

## 3.3 Cadmium

Cadmium is highly soluble in water therefore it is easily up taken by plants which results in phytotoxicity followed by entry through the pathways into the food chain causing serious harmful effects to human beings (Buchet et al. 1990). This heavy metal has been classified into carcinogenic to humans by The International Agency for Research on Cancer (IARC 1993). Still at low concentrations, cadmium alters some enzyme activities including those enzymes involved in Calvin cycle, $CO_2$ fixation, Carbohydrate and Phosphorus metabolisms (Gill and Tuteja 2011). This may result in the underdeveloped growth, alterations in chloroplast ultrastructure, Chlorosis, leaf epinasty, inhibition of photosynthesis and pollen germination, alterations in nitrogen (N) and sulphur (S) metabolism and disruption of the antioxidant machinery (Gill and Tuteja 2011). In a report by Roane et al. (2001), *Pseudomonas* strain H1 and *Bacillus* strain H9 showed an intracellular mechanism of cadmium sequestration (36%) for reducing cadmium toxicity. These strains showed the production of exopolymers (EPS) which accumulated cadmium and reduced soluble cadmium levels by 22% and 11%, respectively.

## 3.4 Chromium

Seventh most rich metal on earth is chromium and exists in two stable states in the environment: they are trivalent $Cr^{3+}$ and hexavalent $Cr^{6+}$. Source of Chromium contamination is due to the use of Cr in many industries such as leather tanning, metal plating, and other metallurgical procedures. The insufficient disposal of their wastes may give rise to concentrations above the natural values (Wuana 2011). Chromium inhibits sulphate membrane transport and causes oxidative damage in bacteria. Against chromium toxicity, microbes involve in two mechanisms. The first and foremost is chromate efflux from the cells, and another one is the reduction of toxic form $Cr^{6+}$ into less toxic form $Cr^{3+}$. The efflux protein of chromate is encoded by chrAgene, which has homologs in eubacteria, archaea, and also in some eukaryotes (Das et al. 2016). Nies et al. (1998) defined about two chromate efflux pumps with six transmembrane segments but Diaz-Magana et al. (2009) described that in *E. coli*, these two pumps are not separate ones, however together form a heterodimer of 12 segments.

## 3.5 Copper

Copper is third oldest abundantly used metal having wide applications in wires, architecture, motors and medicines. Consuming more amount of copper in the body causes Copperiedus, leading to the production of Reactive Oxygen Species (ROS), which have the potential to damage proteins, DNA and lipids. Two oxidation states Cu(I) and Cu(II) involve in copper cycles and dislocate iron (Fe) from available Fe–S clusters in dehydratases and other iron sulphur proteins (Macomber and Imlay 2009). Plasmid pRJ1004-mediated copper resistance in bacteria was first reported by Tetaz and Luke (1983). Later on Bender and Cooksey (1986) identified native pPT23D plasmids in *Pseudomonas syringae* helping tomato from copper toxicity. However, the removal of cytoplasmic copper is involved by ATPase-driven copper efflux system. Periplasmic copper handling, metallochaperones, multicopper oxidases and Resistance-Nodulation-Division (RND) systems are involved in this process (Bondarczuk and Piotrowska-Seget 2013).

## 3.6 Mercury

Compared to other toxic metal pollutants, mercury is one of the most toxic elements in the earth which has severe health concerns. It is strong in sediments, soils, water and atmosphere. Through anthropogenic activities mercury enters into the environment. Mercury resistant bacteria have two operons. One of the operons is narrow-spectrum mer operon and another one is broad spectrum mer operon (Matsui et al. 2016). Consensus sequence GMTCAAC is present in the mercury binding site. MerP scavenges inorganic mercury ions and transports them to the MerT protein (Hamlett et al. 1992). merC expression in *Arabidopsis thaliana* and *Nicotiana tabacum* has led to their doubling ability to accumulate mercury (Sasaki et al. 2006).

### 3.7 Nickel

Nickel is most abundantly disturbed in the environment which exists as five stable isotopes $^{58}Ni$, $^{60}Ni$, $^{61}Ni$, $^{62}Ni$, $^{64}Ni$. Nickel has vital role in the biochemistry of microbes and plants. Most of the enzyme contains nickel active site such as ureases, hydrogenases, superoxide dismutases, and glyoxalase enzyme contains in the form of Ni–Fe clusters or use nickel as a co-factor. Nickel is highly toxic to the humans and animals because it's potential to cross the placenta and affect the developing fetus. Nickel resistance in bacteria is generally mediated by metal efflux pumps, this resistance mechanism is well studied in *Cupriavidus metallidurans* CH34 formerly called *Ralstonia metallidurans* by Grass et al. (2000). It has been reported that in the presence of CnrCBA transenvelope efflux pump encoded by the cnrYHXCBAT gene system, Ni is expelled out of the cell (Maillard et al. 2015). Through cnrY and cnrC the cnr promoter is initiated while the nickel enters into the periplasm, transcription occurs. The CnrCBA encodes a highly efficient pump which is activated only in micromolar concentrations of nickel. These gene products form an efflux pump to efflux excess nickel outside the cell. Certain standard levels of heavy metals allowed to be present in mg per Kg of soil varies with countries as below in Table 2.

## 4 Bioremediation of Heavy Metals by Microorganisms

Bioremediation is the process of removing heavy metals through biological aspects which also involves microorganisms. These microbes decrease the heavy metal ion toxicity by immobilizing, uptake, mobilizing and transformation of heavy metals (Hassan et al. 2017). Numerous symbiotic PGPR resides in plant roots and also as free-living bacteria in the rhizosphere soil that positively alters plant growth and increases the productivity by the production of growth regulators, through supplying and facilitating nutrient uptake from soil (Nadeem et al. 2014). Most of the studies

**Table 2** Permissible level of heavy metals

| Heavy metal (mg/kg) | Austria standard | The European Union standard | Indian standard |
|---|---|---|---|
| Al | – | – | – |
| As | 50 | – | – |
| Cd | 5 | 3 | 3–6 |
| Co | 50 | – | – |
| Cr | 100 | 150 | – |
| Cu | 100 | 140 | 135–270 |
| Ni | 100 | 75 | 75–150 |
| Pb | 100 | 300 | 250–500 |
| Zn | 100 | 300 | – |

have reported that PGPR act as potential elicitors for heavy metal tolerance as well as abiotic stress tolerance (Dary et al. 2010; Tiwari et al. 2016, 2017). These PGPR bind with the bioavailablemetals by forming complexes with siderophores such as Desferroxamines, Dihydroxybenzoic acids and Rhizoferrins (Dimpka et al. 2016), particular metabolites like heme in *Bacillus japonicum* by ferrochelatases, metallothionein cation-binding proteins (Chandrangsu 2017) and bacterial heavy metal transporters such as Pb(II)/Cd(II)/Zn(II)-transporting ATPase in *E. coli* (Rajkumar et al. 2010; Ahemad 2012). The agriculturally important microorganisms evolved various mechanisms to overcome heavy metal stress which includes (a) transport of metals across cytoplasmic membrane; (b) biosorption and bioaccumulation to the cell walls; (c) metal entrapment in the extracellular capsules; (d) heavy metals precipitation; and (e) metal detoxification via oxidation–reduction reactions (Zubair et al. 2016). The harmful effects of heavy metals are reduced through various microbes of Heavy-metal-tolerant PGPR including *Bacillus, Pseudomonas, Streptomyces* and *Methylobacterium* which has the potentials to improve the growth and yield of the crops (Sessitsch et al. 2013).

Plant growth promoting bacteria are involved in the biosorption of heavy metals in which siderophores and IAA are accountable for metal uptake with which indirect defence mechanisms are activated (Spaepen and Vanderleyden 2011). Siderophores reduce the abiotic stresses forced on plants by making stable complexes with toxic heavy metals of environmental concern such as Cd, Cu, Cr, Pb and Zn (Rajkumar et al. 2010).

## *4.1 Role of Microbes in Detoxification of Heavy Metal*

In order to survive at high concentration of heavy metals, bacteria need to develop different mechanisms to confer resistances to these heavy metals. There is no general mechanism for resistance in bacteria towards all heavy metal ions. Though it is well known that both living and dead cells are capable of metal accumulation but there are differences in the mechanism involved.

There are four possible known mechanisms postulated in bacterial heavy metal resistances. They are as follows:

- The first mechanism is by keeping the toxic ion out of the cell by altering a membrane transport system involved in initial cellular accumulation.
- The second mechanism is the intracellular or extracellular sequestration by specific metal-ion binding components (analogous to the phytochelatins in plants and the metallothioneins of eukaryotes, but generally binding occurs at the level of the cell wall in bacteria). Extracellular accumulation/precipitation may be facilitated by using viable microorganisms. However, cell surface sorption or complexation can occur within alive or dead microorganisms, while intracellular accumulation requires microbial activity (Macek et al. 2011).

- The third mechanism includes plasmid mediated bacterial metal ion resistance, which involves highly specific anion or cation efflux systems encoded by resistance genes associated with plasmids.
- The fourth and widely known mechanism involves detoxification of the toxic anion or cation by enzymatically converting it from a more toxic to a less toxic form. This mechanism does essentially occur in detoxification of inorganic and organomercurials.

Some other major mechanisms of microbial metal transformations between soluble and insoluble metal species include chemolithotrophic leaching, chemoorganotrophic leaching, rock and mineral bioweathering and biodeterioration, biocorrosion, redox mobilization, methylation, complexation (For example, complex formation with microbial products such as metallothionein like proteins and extracellular polymers (EPS) in case of soluble metal species whereas for insoluble metal species the process includes biosorption, accumulation, biomineral formation, redox immobilization, metal sorption to biogenic minerals and formation of metalloid nanoparticlesare well notable processes (Tunali 2006)). The key factors in controlling these mechanisms include:

- The nature of the biomass i.e. living or non-living;
- The type of biological ligands available for heavy metal sequestration;
- The chemical, stereochemical and coordination characteristics of the targeted metals and metalloid species (Remoudaki et al. 1999).
- The hysic-chemical characteristics of the metal solution such as pH, and presence of competing co-ions (Esposito 2002).

## 4.2 Interaction Between the Heavy Metals and Microbes

Numerous mechanisms have been executed by bacteria to detoxify and resist heavy metals by which the metal ions bind to the cell surface and incorporate electrostatic interactions, covalent binding, Van der Waals forces, redox interactions and extracellular precipitation or by means of all these as combined processes (Blanco et al. 2000). In general the response of bacteria may fall into two categories: (1) mechanisms dependent on activation by specific metals, and (2) general mechanisms, which convey resistance but do not depend on metal stress for their activation. Uptake of heavy metals and detoxification through Heavy metal tolerant—plant growth promoting microbes is largely responsible for the Bioaccumulation. There are two methods involved in bioaccumulation of Heavy metals: one is passive uptake which is also known as biosorption, a metabolism independent accumulation of heavy metals by inactive non-living biomass or living cells/biological materials. Another one is active uptake that occurs only in alive cells, it requires energy and metabolism for the exchange of metals (Gutierrez-Corona et al. 2016). One or a blend of different processes involved in the biosorption includes coordination, complexation, chelation, ion exchange, entrapment and micro precipitation (Pokethitiyook and Poolpak 2016).

Biosorption of heavy metals from the contaminated site by means of microbes are associated within the cell wall and functional groups like –SH, –OH, and –COOH and other biomolecules have more affinity for heavy metals. Metal-binding peptides and chelators involved in metal binding such as metallothioneins (MT) and glutathione—derived peptides (PC). These PC and MT are secreted by rhizosphere fungi, bacteria and also by plants in response to heavy metals stress (Miransari 2011).

## 4.3 Mechanisms Behind the Utility of Heavy Metals by Microorganisms

Heavy metal ions are trapped by the cellular structure of microorganisms and consequently attached to the binding sites of the cell wall (Malik 2004). This method is known as passive uptake or biosorption, and is exclusively independent of the metabolic cycles. The amount of metals biosorbed depends on the kinetic equilibrium and composition of the metals at the cellular surface. The mechanism involves several process including electrostatic interaction, ion exchange and surface complexation. Absorption of heavy metal is carried out by fragments of cells and tissues, or living cells or by dead biomass as passive uptake by surface complexation on the cell wall and other outer layers (Fomina et al. 2014). In cellular metabolic cycles, these heavy metal ions may pass across the cell membrane and the process is known as bioaccumulation or active uptake.

## 4.4 Bioaccumulation or Active Uptake of Heavy Metals in Bacteria

In gram positive bacteria peptidoglycan layer contains alanine, glutamic acid, meso-diaminopimelic acid, polymers of glycerol and teichoic acid whereas in gram negative bacteria, cell wall contains enzymes, lipoproteins, glycoproteins, lipopolysaccharides and phospholipids which concerned as the active site for metal binding (Lesmana et al. 2009; Fomina et al. 2014; Gupta et al. 2015). The process involved in the bioaccumulation of heavy metals in the living cell is dependent on a variety of physical, chemical and biological mechanisms. The intercellular and extracellular process plays a partial and ill-defined role in biosorption as well (Fomina et al. 2014). The microbes which accumulate heavy metals have the tolerance capacity to one or more metals at higher concentrations, and it has the ability to change the toxic forms to harmless forms thereby reducing the toxic levels of the metals in the environment (Fig. 1). Simultaneously these organisms do retain the heavy metals contained in the cells (Mosa et al. 2016).

Microorganisms hide many kinds of metal-binding metabolites and produce extracellular polymeric substances like polysaccharides and associated components such

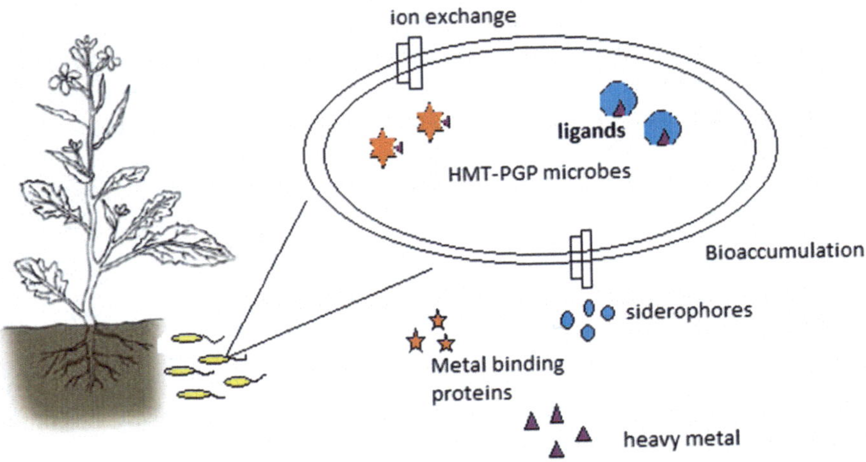


**Fig. 1** Bioaccumulation/active uptake of heavy metals by plant growth promoting rhizosphere bacteria

as capsules, slimes and sheaths, and biofilms for with heavy metals (Fomina et al. 2014).

## *4.5 Biosorption or Passive Uptake in Bacteria*

Among microorganisms, bacteria constitute of being the most abundant, versatile, most diverse creature on this planet earth (Norberg et al. 1984; Abbas et al. 2014). They are fundamentally classified on the basis of their morphology as rod, cocci or spirillum (Wang et al. 2009; Abbas et al. 2014). A bacterium has relatively simple morphology consisting of cell wall, cell membrane, capsule, slime layer and internal structures, such as ribosomes, mesosomes etc. Slime layer contains functional groups like carboxyl, amino, phosphate or sulfate for metal chelation (Abbas et al. 2014). Cell wall in general, is dependable for surface binding sites and the binding strength for different metal ions depend on different binding mechanisms. There exist many intracellular and extracellular events in bacteria involving microbial bioremediation of toxic pollutants from the environment. In response to the presence of toxic metals in the environment, resistant bacteria synthesize many intracellular and extracellular enzymes to remove/degrade the toxic form of metals to non-toxic/less toxic forms (Fig. 2). Various bacterial species belonging to the genus such as *Bacillus, Pseudomonas, Escherichia* exhibit biosorption property because of their small size and ability to grow in different environmental conditions (Vasudevan et al. 2001; Vijayaraghavan et al. 2008; Kinoshita et al. 2013).

Gram positive bacteria are comprised of thick peptidoglycan layer connected by amino acid bridges, also known to contain polyalcohols and teichoic acids. On the

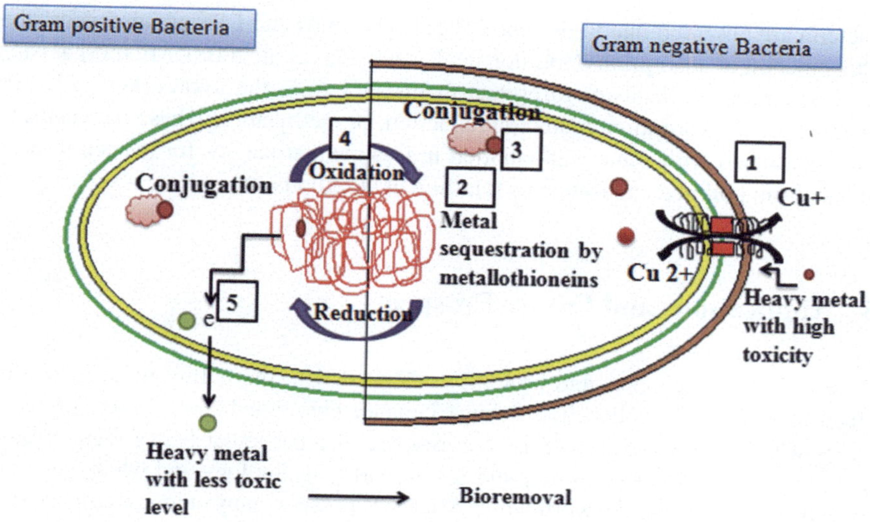


**Fig. 2** Bioremoval of heavy metals from the environment by resistant bacteria through passive uptake in response to the presence of toxic metals in the environment. Here, resistant bacteria synthesize many intracellular and extracellular enzymes to remove/detoxify the toxic form of heavy metals to non-toxic/less toxic forms. These processes include (1) Entry of heavy metal with high toxicity (2) Metal sequestration by metallothioneins, (3) Conjugate formation with organic compounds/precipitation, (4) Oxidation–reduction of metals, (5) Metal efflux by metal transporters followed by bioremoval through microbial products such as biosurfactants or EPS

whole, Gram positive bacterial cell wall comprised of 90% peptidoglycan. Some teichoic acids are linked to lipids of lipid bilayer forming lipoteichoic acid. These lipoteichoic acids again forma part of cytoplasmic membrane. They constitute linkage of peptidoglycan to cytoplasmic membrane. This results in cross linking of peptidoglycan forming a grid like structure. These teichoic acids are responsible for negative charge on cell wall due to presence of phosphodiester bonds between teichoic acid monomers (Abbas et al. 2014). Hence, affinity towards the heavy metals is more favoured in cell wall of gram positive bacteria as these heavy metals are cationic in nature. On the other hand, Gram negative bacterial cell wall contains an additional outer membrane composed of phospholipids and lipopolysaccharides. Gram negative cell wall contains 10–20% peptidoglycan. The negative charge on the Gram negative bacteria is due to lipopolysaccharides, teichoic acids, teichuronic acids. Extracellular polysaccharides also exhibit the properties of metal binding. Bacterial cell wall encountering the metal ion is the first component of biosorption. The metal ions get attached to the functional groups (amine, carboxyl, hydroxyl, phosphate, sulfate, and amine) present on the cell wall (Abdi 2015). The general metal uptake process involves binding of metal ions to reactive groups present on bacterial cell wall followed by internalization of metal ions inside the cell (Abbas et al. 2014). More metal uptake is carried out by Gram positive bacteria due to presence of glycoproteins which facilitates the internalization. However, fewer metal uptakes by Gram negative bacteria are observed due to presence of phospholipids and LPS which favours the phenomenon of adsorption (Gourdon et al. 1990; Das et al. 2008). Microorganisms have advanced mechanisms to protect themselves from the toxic doses of heavy metals such as adsorption, oxidation/reduction, or methylation. These mechanisms can be adopted with some manipulation in treatment strategies for the removal of metals from polluted environments (Hashim et al. 2011).

## 5 Applications and Future Prospects

Detection of novel genes and proteins associated with the ability of eco-friendly clean-up will be a great benefit to achieve enhanced bioremediation. To identify new genes which may be expressed in the presence of a particular heavy metal pollutant, gene expression studies employing microarray technology and whole genome sequencing assays shall be worthwhile. Microbes possess many unique characteristic features such as biofilm formation, biosurfactant productions, secondary metabolites synthesis, and many more to withstand the adverse conditions in the environment. These properties of metal resistant bacteria may be harnessed for their enhanced utilization in bioremediation. Recently, multispecies biofilm communities have been explored for their metal tolerance and bio-mineralization properties (Golby et al. 2014). Though bacteria develop metal-resistant phenotypes and genotypes as a mode of adaptation in the contaminated environments, the gene pool of these resistant bacteria can be a choice of research thirst in near future to have a proper insight into the molecular genetics approaches for an enhanced heavy metal bioremediation so as to save the environment.

## 6 Conclusion

This chapter revealed both the active (Bioaccumulation) and passive (Biosorption) mechanisms of bioremoval of toxic heavy metals by Heavy metal tolerant-plant growth promoting rhizobacteria. These communities are widely involved in the removal of heavy metals from the polluted environment. At this juncture, Bacterial siderophores gain importance because of their capability to interact with the heavy metals like Fe, Ni, Cd, Cu, and Zn. In which the metal uptake is concerned through two special proteins. They are metallo-proteins or metal-binding proteins and peptides. The use of PGPR for plant growth improvement is presently receiving substantial worldwide concentration and the latest successive PGPR researches have exhibit luminous prospects for bioremediation of polluted soil environments. In addition, the rapid development of molecular biological methods is bringing valuable advantages to identify and enhance the rhizobacterial traits involved in heavy metal bioremoval from the areas under heavy metal stress.

**Acknowledgements** Authors gratefully acknowledging the RUSA—Phase 2.0 grant sanctioned vide Letter No. F. 24-51/2014-U, Policy (TN Multi-Gen), Department of Education, Govt. of India, dated 09.10.2018.

## References

Abbas SH, Ismail IM, Mostafa TM, Sulaymon AH (2014) Biosorption of heavy metals: a review. J Chem Sci Technol 3(4):74–102

Abdi O, Kazemi M (2015) A review study of biosorption of heavy metals and comparison between different biosorbents. J Mater Environ Sci 6(5):1386–1399

Ahemad M (2012). Implications of bacterial resistance against heavy metals in bioremediation: a review. J Inst Integr Omics Appl Biotechnol (IIOAB) 3(3)

Akcil A, Erust C, Ozdemiroglu S, Fonti V, Beolchini F (2015) A review of approaches and techniques used in aquatic contaminated sediments: metal removal and stabilization by chemical and biotechnological processes. J Cleaner Prod 86:24–36

An YJ, Kim M (2009) Effect of antimony on the microbial growth and the activities of soil enzymes. Chemosphere 74(5):654–659

Belimov, AA, Hontzeas N, Safronova VI, Demchinskaya SV, Piluzza G, Bullitta S, Glick BR (2005) Cadmium-tolerant plant growth-promoting bacteria associated with the roots of Indian mustard (Brassica juncea L. Czern.). Soil Biol Biochem 37(2):241–250

Bender CL, Cooksey DA (1986) Indigenous plasmids in Pseudomonas syringae pv. tomato: conjugative transfer and role in copper resistance. J Bacteriol 165(2):534–541

Bissen M, Frimmel FH (2003) Arsenic—a review. Part I: occurrence, toxicity, speciation, mobility. Acta Hydrochim Hydrobiol 31(1):9–18

Blais JF, Djedidi Z, Cheikh RB, Tyagi RD, Mercier G (2008) Metals precipitation from effluents. Pract Periodical Hazard Toxic Radioactive Waste Manage 12(3):35–149

Blanco A, Sampedro MA, Sanz B, Llama MJ, Serra JL (2000). Immobilization of non-viable cyanobacteria and their use for heavy metal adsorption from water. Environ Biotechnol Cleaner Bioprocess 135–151

Bondarczuk K, Piotrowska-Seget Z (2013) Molecular basis of active copper resistance mechanisms in gram-negative bacteria. Cell Biol Toxicol 29:397–405

Bradham KD, Dayton EA, Basta NT, Schroder J, Payton M, Lanno RP (2006) Effect of soil properties on lead bioavailability and toxicity to earthworms. Environ Toxicol Chem 25:769–775

Buchet JP, Lauwerys R, Roels H, Bernard A, Bruaux P, Claeys F, Lijnen P (1990) Renal effects of cadmium body burden of the general population. Lancet 336(8717):699–702

Chandrangsu P, Rensing C, Helmann JD (2017) Metal homeostasis and resistance in bacteria. Nat Rev Microbiol 15(6):338

Chen G, Zeng G, Tang L, Du C, Jiang X, Huang G, Liu H, Shen G (2008) Cadmium removal from simulated wastewater to biomass byproduct of Lentinusedodes. Bioresour Technol 99(15):7034–7040

Conrath U, Beckers GJ, Flors V, García-Agustín P, Jakab G, Mauch F, Pugin A (2006) Priming: getting ready for battle. Mol Plant Microbe Interact 19(10):1062–1071

Dary M, Chamber-Pérez MA, Palomares AJ, Pajuelo E (2010) "In situ" phytostabilisation of heavy metal polluted soils using Lupinus luteus inoculated with metal resistant plant-growth promoting rhizobacteria. J Hazard Mater 177(1–3):323–330

Das N, Vimala R, Karthika P (2008) Biosorption of heavy metals—an overview. Indian J Biotechnol 2008(7):159–169

Das S, Dash HR, Chakraborty J (2016) Genetic basis and importance of metal resistant genes in bacteria for bioremediation of contaminated environments with toxic metal pollutants. Appl Microbiol Biotechnol 100(7):2967–2984

Diaz-Magana A, Aguilar-Barajas E, Moreno-Sánchez R, Ramírez- Díaz MI, Riveros-Rosas H, Vargas E, Cervantes C (2009) Short-chain chromate ion transporter proteins from Bacillus subtilis confer chromate resistance in *Escherichia coli*. J Bacteriol 191:5441–5445
Dimkpa C (2016) Microbial siderophores: production, detection and application in agriculture and environment. Endocytobiosis Cell Res 27(2):7–16
Dixit R, Malaviya D, Pandiyan K, Singh U, Sahu A, Shukla R, Paul D (2015) Bioremediation of heavy metals from soil and aquatic environment: an overview of principles and criteria of fundamental processes. Sustainability 7(2):2189–2212
Dubey S, Shri M, Misra P, Lakhwani D, Bag SK, Asif MH (2014) Heavy metals induce oxidative stress and genome-wide modulation in transcriptome of rice root. Funct Integr Genomics 14:401–417
El Assal FM, Abdel-Meguid ZA (2017) Impact of Heavy Metal Pollution on Procambarus clarkii (Crustacea: Decapoda), from Egypt. Int J Waste Resour 7(3):270
Esposito A, Pagnanelli F, Megliò F (2002) pH-related equilibria models for biosorption in single metal systems. Chem Eng Sci 57(3):307–313
Finnegan P, Chen W (2012) Arsenic toxicity: the effects on plant metabolism. Front Physiol 3:182
Fomina M, Gadd GM (2014) Biosorption: current perspectives on concept, definition and application. Bioresour Technol 160:3–14
Germ M, Kreft I, Stibilj V, Urbanc-Berčič O (2007) Combined effects of selenium and drought on photosynthesis and mitochondrial respiration in potato. Plant Physiol Biochem 45(2):162–167
Gill SS, Tuteja N (2011) Cadmium stress tolerance in crop plants: probing the role of sulfur. Plant Signal Behav 6:215–222
Golby S, Ceri H, Marques LLR, Turner RJ (2014) Mixed-species biofilms cultured from an oil sand tailings pond canbiomineralize metals. Microb Ecol 68:70–80
Gontia-Mishra I, Sapre S, Sharma A, Tiwari S (2016) Alleviation of mercury toxicity in wheat by the interaction of mercury-tolerant plant growth promoting rhizobacteria. J Plant Growth Regul 35:1000–1012
Gourdon R, Bhende S, Rus E, Sofer SS (1990) Comparison of cadmium biosorption by gram-positive and gram-negative bacteria from activated sludge. Biotech Lett 12(11):839–842
Grass G, Grobe C, Nies DH (2000) Regulation of the cnr cobalt and nickel resistance determinant from Ralstonia sp. strain CH34. J Bacteriol 182:1390–1398
Gupta VK, Nayak A, Agarwal S (2015) Bioadsorbents for remediation of heavy metals: current status and their future prospects. Environ Eng Res 20(1):1–18
Gutierrez-Corona JF, Romo-Rodriguez P, Santos-Escobar F, Espino-Saldana AE, Hernandez-Escoto H (2016) Microbial interactions with chromium: basic biological processes and applications in environmental biotechnology. World J Microbiol Biotechnol 32:191
Hamlett NV, Landale EC, Davis BH, Summers AO (1992) Roles of the Tn21 merT, merP, and merC gene products in mercury resistance and mercury binding. J Bacteriol 174(20):6377–6385
Hashim MA, Mukhopadhyay S, Sahu JN, Sengupta B (2011) Remediation technologies for heavy metal contaminated groundwater. J Environ Manag 92(10): 2355–2388
Hassan TU, Bano A, Naz I (2017) Alleviation of heavy metals toxicity by the application of plant growth promoting rhizobacteria and effects on wheat grown in saline sodic field. Int J Phytorem 19(6):522–529
Jamil M, Zeb S, Anees M, Roohi A, Ahmed I, ur Rehman S, Rha ES (2014) Role of Bacillus licheniformis in phytoremediation of nickel contaminated soil cultivated with rice. Int J Phytorem 16(6):554–571
Joo JH, Hassan SH, Oh SE (2010) Comparative study of biosorption of Zn2+ by Pseudomonas aeruginosa and Bacillus cereus. Int Biodeterior Biodegradation 64(8):734–741
Kachur AV, Koch CJ, Biaglow JE (1998) Mechanism of copper-catalyzed oxidation of glutathione. Free Radical Res 28(3):259–269
Khan MS, Zaidi A, Wani PA, Oves M (2009) Role of plant growth promoting rhizobacteria in the remediation of metal contaminated soils. Environ Chem Lett 7(1):1–19

Kinoshita H, Sohma Y, Ohtake F, Ishida M, Kawai Y, Kitazawa H, Kimura K (2013) Biosorption of heavy metals by lactic acid bacteria and identification of mercury binding protein. Res Microbiol 164(7):701–709

Kumar S, Dubey RS, Tripathi RD, Chakrabarty D, Trivedi PK (2015) Omics and biotechnology of arsenic stress and detoxification in plants: current updates and prospective. Environ Int 74:221–230

Lesmana SO, Febriana N, Soetaredjo FE, Sunarso J, Ismadji S (2009) Studies on potential applications of biomass for the separation of heavy metals from water and wastewater. Biochem Eng J 44:19–41

Ma Y, Rajkumar M, Freitas H (2009) Inoculation of plant growth promoting bacterium Achromobacter xylosoxidans strain Ax10 for the improvement of copper phytoextraction by Brassica juncea. J Environ Manag 90(2):831–837

Macek T, Mackova M (2011) Potential of biosorption technology. In: Microbial biosorption of metals. Springer, Dordrecht, pp 7–17

Macomber L, Imlay JA (2009) The iron-sulfur clusters of dehydratases are primary intracellular targets of copper toxicity. Proc Natl Acad Sci 106(20):8344–8349

Maillard AP, Künnemann S, Grobe C, Volbeda A, Schleuder G, Petit-Härtlein I, Covès J (2015) Response of CnrX from Cupriavidus metallidurans CH34 to nickel binding. Metallomics 7(4):622–631

Malik A (2004) Metal bioremediation through growing cells. Environ Int 30(2):261–278

Matsui K, Yoshinami S, Narita M, Chien MF, Phung LT, Silver S, Endo G (2016) Mercury resistance transposons in Bacilli strains from different geographical regions. FEMS Microbiol Lett 363(5)

Miransari M (2011) Hyperaccumulators, arbuscular mycorrhizal fungi and stress of heavy metals. Biotechnol Adv 29(6):645–653

Mosa A, El-Ghamry A, Trüby P, Omar M, Gao B, Elnaggar A, Li Y (2016) Chemo-mechanical modification of cottonwood for Pb2+ removal from aqueous solutions: sorption mechanisms and potential application as biofilter in drip-irrigation. Chemosphere 161:1–9

Nadeem SM, Ahmad M, Zahir ZA, Javaid A, Ashraf M (2014) The role of mycorrhizae and plant growth promoting rhizobacteria (PGPR) in improving crop productivity under stressful environments. Biotechnol Adv 32:429–448. https://doi.org/10.1016/j.biotechadv.2013.12.005

Nies DH, Koch S, Wachi S, Peitzsch N, Saier MH (1998) CHR, a novel family of prokaryotic proton motive force-driven transporters probably containing chromate/sulfateantiporters. J Bacteriol 180:5799–5802

Nies DH (1999) Microbial heavy-metal resistance. Appl Microbiol Biotechnol 51(6):730–750

Norberg AB, Persson H (1984) Accumulation of heavy-metal ions by Zoogloea ramigera. Biotechnol Bioeng 26(3):239–246

Pagnanelli F, Viggi CC, Toro L (2010) Isolation and quantification of cadmium removal mechanisms in batch reactors inoculated by sulphate reducing bacteria: biosorption versus bioprecipitation. Bioresour Technol 101(9):2981–2987

Pehlivan E, Özkan AM, Dinç S, Parlayici Ş (2009) Adsorption of $Cu^{2+}$ and $Pb^{2+}$ ion on dolomite powder. J Hazard Mater 167(1–3):1044–1049

Pepper IL, Gerba CP, Gentry TJ (2015) Environmental Microbiology, 3rd edn. Academic, San Diego USA

Pokethitiyook P, Poolpak T (2016) Biosorption of heavy metal from aqueous solutions. In: Phytoremediation, pp. 113–141. Springer, Cham

Rajkumar M, Ae N, Prasad MNV, Freitas H (2010) Potential of siderophore-producing bacteria for improving heavy metal phytoextraction. Trends Biotechnol 28(3):142–149

Remoudaki E, Tsezos M, Hatzikioseyian A, Karakoussis V (1999) Mechanism of palladium biosorption by microbial biomass. The effects of metal ionic speciation and solution co-ions. Process Metall 9:449–462

Román-Ponce B, Reza-vázquez DM, Gutiérrez-paredes S, María de jesús DE, Maldonado-hernández J, Bahena-osorio Y (2017) Plant growth-promoting traits in rhizobacteria of heavy

metal-resistant plants and their effects on *Brassica nigra* seed germination. Pedosphere 27:511–526. https://doi.org/10.1016/S1002-0160(17)60347-3
Roy S, Labelle S, Mehta P, Mihoc A, Fortin N, Masson C, Olsen C (2005) Phytoremediation of heavy metal and PAH-contaminated brownfield sites. Plant Soil 272(1):277–290
Sasaki Y, Hayakawa T, Inoue C, Miyazaki A, Silver S, Kusano T (2006) Generation of mercury-hyperaccumulating plants through transgenic expression of the bacterial mercury membrane transport protein MerC. Transgenic Res 15(5):615
Sawyer RH, Washington LD, Salvatore BA, Glenn TC, Knapp LW (2003) Origin of archosaurian integumentary appendages: the bristles of the wild Turkey beard express feather-type β keratins. J Exp Zool Part B: Mol Dev Evol 297(1):27–34
Sessitsch A, Kuffner M, Kidd P, Vangronsveld J, Wenzel WW, Fallmann K et al (2013) The role of plant-associated bacteria in the mobilization and phytoextraction of trace elements in contaminated soils. Soil Biol Biochem 60:182–194. https://doi.org/10.1016/j.soilbio.2013.01.012
Spaepen S, Vanderleyden J (2011) Auxin and plant-microbe interactions. Cold Spring Harbor Perspect Biol 3(4). Retrieved February 7, 2015 from https://doi.org/10.1101/cshperspect.a001438
Spain A, Alm E (2003) Implication of microbial heavy metal tolerance in the enivornment. Rev Undergraduate Res 2:1–6
Srivastava S, Verma PC, Chaudhary V, Singh N, Abhilash PC, Kumar KV (2013) Inoculation of arsenic-resistant Staphylococcus arlettae on growth and arsenic uptake in Brassica juncea (L.) Czern. Var. R-46. J Hazard Mater 262:1039–1047. https://doi.org/10.1016/j.jhazmat.2012.08.019
Taghavi S, Garafola C, Monchy S, Newman L, Hoffman A, Weyens N, van der Lelie D (2009) Genome survey and characterization of endophytic bacteria exhibiting a beneficial effect on growth and development of poplar trees. Appl Environ Microbiol 75(3):748–757
Talos K, Pager C, Tonk S, Majdik C, Kocsis B, Kilar F, Pernyeszi T (2009) Cadmium biosorption on native Saccharomyces cerevisiae cells in aqueous suspension. Acta Univ Sapientiae Agric Environ 1:20–30
Tang L, Zeng GM, Shen GL, Li YP, Zhang Y, Huang DL (2008) Rapid detection of picloram in agricultural field samples using a disposable immunomembrane-based electrochemical sensor. Environ Sci Technol 42(4):1207–1212
Tangahu BV, Abdullah S, Rozaimah S, Basri H, Idris M, Anuar N, Mukhlisin M (2011) A review on heavy metals (As, Pb, and Hg) uptake by plants through phytoremediation. Int J Chem Eng. https://doi.org/10.1155/2011/939161
Tchounwou PB, Yedjou CG, Patlolla AK, Sutton DJ (2012) Heavy metal toxicity and the environment. Molecular, clinical and environmental toxicology. Springer, Basel, pp 133–164
Tetaz TJ, Luke RK (1983) Plasmid-controlled resistance to copper in Escherichia coli. J Bacteriol 154(3):1263–1268
Tiwari S, Lata C, Chauhan PS, Nautiyal CS (2016) Pseudomonas putida attunes morphophysiological, biochemical and molecular responses in Cicerarietinum L. during drought stress and recovery. Plant Physiol Biochem 99:108–117. https://doi.org/10.1016/j.plaphy.2015.11.001
Tiwari S, Prasad V, Chauhan PS, Lata C (2017) Bacillus amyloliquefaciens confers tolerance to various abiotic stresses and modulates plant response to phytohormones through osmoprotection and gene expression regulation in rice. Front Plant Sci 8:1510. https://doi.org/10.3389/fpls.2017.01510
Tunali S, Çabuk A, Akar T (2006) Removal of lead and copper ions from aqueous solutions by bacterial strain isolated from soil. Chem Eng J 115:203–211
Vasudevan P, Padmavathy V, Tewari N, Dhingra SC (2001) Biosorption of heavy metal ions. J Sci Ind Res 2001(60):112–120
Verma PK, Verma S, Pande V, Mallick S, Tripathi RD, Dhankher OP (2016) Overexpression of rice glutaredoxin OsGrx_C7 and OsGrx_C2. 1 reduces intracellular arsenic accumulation and increases tolerance in Arabidopsis thaliana. Front Plant Sci 7:740
Vijayaraghavan K, Yun KS (2008) Bacterial Biosorbents and Biosorption. Biotechnol Adv 26(3):266–291. https://doi.org/10.1016/j.biotechadv.2008.02.002

Wang JL, Chen C (2009) Biosorbents for heavy metals removal and their future—a review. Biotechnol Adv 2009(27):195–226
Wu SC, Cheung KC, Luo YM, Wong MH (2006) Effects of inoculation of plant growth-promoting rhizobacteria on metal uptake by Brassica juncea. Environ Pollut 140(1):124–135
Wuana RA, Okieimen FE (2011) Heavy metals in contaminated soils: a review of sources, chemistry, risks and best available strategies for remediation. ISRN Ecol 2011
Zaidi S, Usmani S, Singh BR, Musarrat J (2006) Significance of Bacillus subtilis strain SJ-101 as a bioinoculant for concurrent plant growth promotion and nickel accumulation in Brassica juncea. Chemosphere 64(6):991–997
Zubair M, Shakir M, Ali Q, Rani N, Fatima N, Farooq S, Nasir IA (2016) Rhizobacteria and phytoremediation of heavy metals. Environ Technol Rev 5(1):112–119

# Microbiome Analysis of Pesticide Contaminated Soils Its Impact on the Microbial Isolates

S. Sowmiya Sri, V. Suji, and G. Sophia Reena

**Abstract** The study aims to analyse the bacterial pattern, to isolate the bacteria from polluted soil site and to identify the bacteria using 16s rDNA gene sequencing. The present research is carried out to isolate and identify the bacteria using molecular study. The bacteria is isolated using plating agar. The DNA extracted and PCR carried out. Further, the gene sequence was analyzed and the sequence was submitted to NCBI and accession number was allotted as MT126476. The bacteria identified as Enterobacter cloacae, the current interest in using omics technologies for elucidating the microbial diversity from potential contaminated site for harnessing microorganism for potential bioremediation perspectives.

## 1 Introduction

Soil bacteria from micro-aggregate in the soil acts as a binder for soil particles with their secretions. Each of these micro-organisms groups has its characteristics that will define their functions in the soil. Bacteria are most significant in ecosystems in the soil including improved soil structure and soil aggregation, reprocess of soil nutrients. Soil is rich in microorganisms which are capable of producing commercially viable metabolites. The traditional approach is random screening in which bacteria are isolated, grown and their activity spectrum was assessed. Soil bacteria are very important in biogeochemical cycles and have been used for crop for decades (Zhao et al. 2018). The numbers and species of microbes in soil depend on environmental factors like nutrient availability, soil texture, presence of moisture content in soil and type of vegetation cover, and their number changes according to the type of environmental factors (Liu et al. 2012).

Contamination of soils with an array of pesticides, due to the widespread application of these pesticide in the ecosystem, their comprehensive toxicological effect on human health risk assessment and ecotoxicological implications. Although many

S. Sowmiya Sri · V. Suji · G. Sophia Reena (✉)
Department of Information Technology, PSGR Krishnammal College for Women, Coimbatore, Tamil Nadu 641004, India
e-mail: sophiareena@psgrkcw.ac.in



M. Prashanthi et al. (eds.), *Bioremediation and Green Technologies*, Environmental Science and Engineering, https://doi.org/10.1007/978-3-030-64122-1_3

pure-cultures sourced deep work has been published on the degradation of pesticide, there is a dearth of understanding on the impact; these xenobiotic-extreme environments on the indigenous microbes augmented and attenuated due these contaminations. Since culture-based approach expose only less than 1% of the total genome, metagenomic profiling is mandatory to expose the entire microbial diversity within contaminated sites, thus the need and role of biotechnological techniques. Hence work highlighting the impact of long-term pesticide exposure on the diversity of microbial communities are available only on a miniscule scale (Regar et al. 2019). For example, an extended heavy metal contaminated site in China was scrutinized to understand the role and function of microbial communities. Many taxa including Proteobacteria, Bacteroidetes, and Firmicutes illuminated via this study had the resistance against assorted heavy metals (Chen et al. 2018). Other studies conducted at Tibetan (steppe) exposed that the soil parameters and minerals had an impact on the microbial diversity and its attributes (Zhou et al. 2019). Procurement of the entire microbiome especially in polluted soil has gained traction in recent years and this study is one such attempt to glen potential and prospective microorganisms from pesticide contaminated soil.

## 2 Materials and Methods

### *2.1 Collection of Samples*

A soil sample was collected from the polluted site. The collected sample was kept in a sterile container, placed in a sterile bag, transferred to the laboratory and stored at 4 °C until further use. The sample was taken as 1 g and ground using mortar and pestle into small pieces and further proceeded to serial dilution. After serial dilution technique, the sample was introduced into the nutrient broth for growth of the culture (Choi et al. 2015).

### *2.2 Isolation of Bacteria*

Isolating bacteria from the soil is an important step, once they are isolated, bacteria can be further analysed to determine things, such as their species and their function in the soil environment (Sayyadifar et al. 2012). The bacteria are isolated using specific media like starch medium and Skim milk agar medium (Table 1).

**Table 1** Results from isolation media

| S. No. | Media used | Result |
|---|---|---|
| 1 | Skim milk | *Enterobacter cloacae* isolated |
| 2 | Starch | No microorganism |

# 3 Result and Discussions

## 3.1 DNA Extraction and PCR Sequencing of Bacterial Sample

The isolated foodborne pathogens were processed to identify the bacteria using molecular biology techniques. The steps to be followed for the extraction of DNA from the bacterial sample in the laboratories are shown in the Fig. 1. Bacterial Genomic DNA Spin-50 kit was used. The solution was prepared with the necessary components to make the RNase solution. About 100 mL of bacterial culture was taken and added with the required components. The DNA concentration was determined by both UV spectrophotometer and quantitative analysis on Agarose gel. Finally, sequencing was obtained with the aid of the ABI sequencing machine (Babalola 2003).

## 3.2 Enterobacter Cloacae

A species of gram-negative, facultatively anaerobic, rod-shaped bacteria that occurs in water, sewage, soil, meat, hospital environments, and on the skin and the intestinal tract of man and animals as a commensal. Enterobacter cloacae is a facultative anaerobic bacterium and are considered to be an important cause of nosocomial infection. However, the isolated *E.* cloacae showed higher furfural-tolerant cellular growth, compared to industrial relevant strains such as *Escherichia coli* and *Corynebacterium glutamicum.* Here, we report the genome sequence of *E.* cloacae isolated from polluted soil. The genomic DNA sequence of E. cloacae will provide valuable genetic resources for engineering of industrially relevant strains (Shankar et al. 2012).

## 3.3 Phylogenetic Tree

Phylogenetic study was performed to map the molecular level similarity of DNA sequence among the various species. Gene sequence of Enterobacter Cloacae was taken for the study. This study has been performed to enlarge the molecular level gene similarly identification to enhance the various Enterobacter species (Gong et al. 2015).

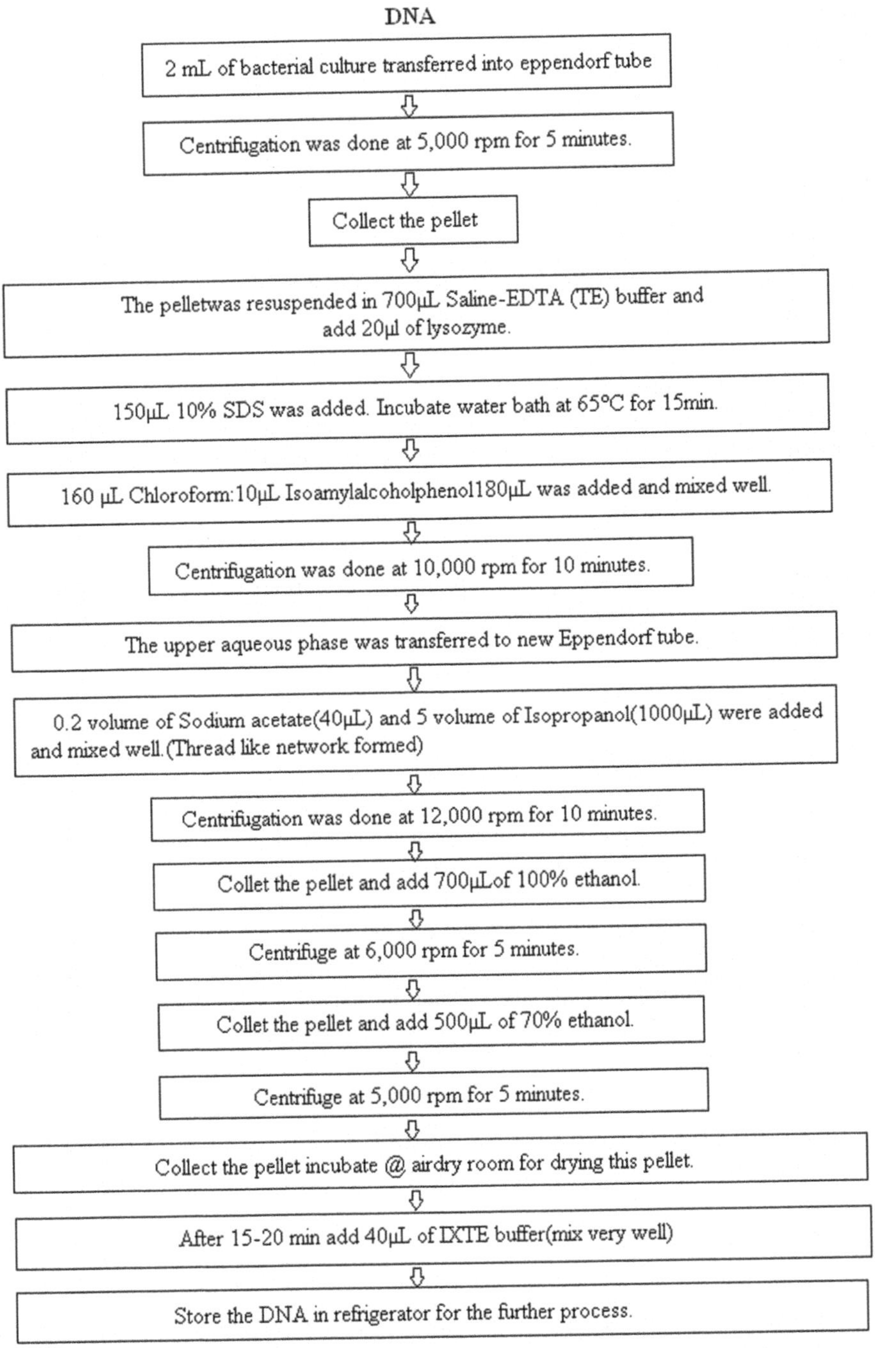


**Fig. 1** Flow of DNA extraction

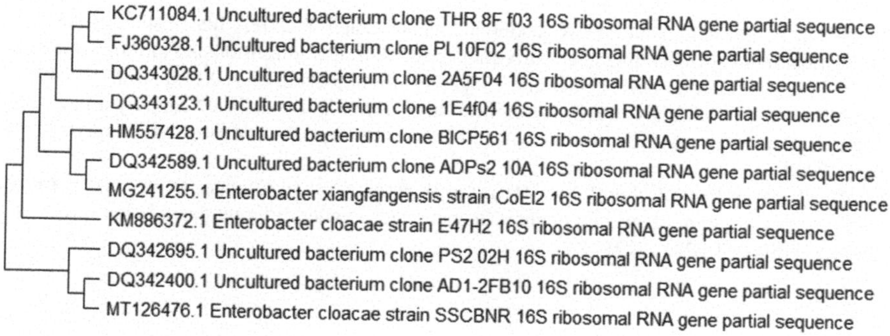


**Fig. 2** Phylogenetic tree

# 4 Conclusion

The current study examined the microbiome composition in pesticides contaminated soil. The contaminated sites were found to possess least microbial diversity, the soil bacteria were isolated from and identified as *Enterobacter cloacae* using molecular and gene sequencing study. The gene sequence was submitted to NCBI and the accession number was granted as MT126476. Our study elucidates the impact of pesticide contamination on the presence of indigenous bacteria and further, introducing this isolate to any bioremediation will enhance the outcome and hence can be used as a better strategy for remediation of a polluted site.

# References


Babalola OO (2003) Molecular techniques: An overview of methods for the detection of bacteria. Afr J Biotechnol 2(12):710–713

Chen Y, Jiang Y, Huang H, Mou L, Ru J, Zhao J, Xiao S (2018) Long-term and high-concentration heavy-metal contamination strongly influences the microbiome and functional genes in Yellow River sediments. Sci. Total Environ. 637:1400–1412

Choi SY, Gong G, Park HS, Um Y, Sim SJ, Woo HM (2015) Extreme furfural tolerance of a soil bacterium Enterobacter cloacae GGT036. J Biotechnol 193:11–13. https://doi.org/10.1016/j.jbiotec.2014.11.001

Gong G, Um Y, Park TH, Woo HM (2015) Complete genome sequence of Enterobacter cloacae GGT036: a furfural tolerant soil bacterium. J Biotechnol 193:43–44. https://doi.org/10.1016/j.jbiotec.2014.11.012

Liu W, Wang X, Wu L, Chen M, Tu C, Luo Y et al (2012) Isolation, identification and characterization of *Bacillus amyloliquefaciens* BZ-6, a bacterial isolate for enhancing oil recovery from oily sludge. Chemosphere 87(10):1105–1110

Regar RJ, Gaur VK, Bajaj A, Tambat S, Manickam N (2019) Comparative microbiome analysis of two different long-term pesticide contaminated soils revealed the anthropogenic influence on functional potential of microbial communities. Sci Total Environ 681:413–423

Sayyadifar F, Morteza H, Munawar T, Jayasimha R (2012) Molecular identification of bacteria isolated from meat sample. Int J App Biol Pharm Technol 3(4):34–39

Shankar M, Ponraj P, Illakkiam D, Rajendhran J, Gunasekaran P (2012) Inactivation of the transcriptional regulator-encoding gene sdiA enhances rice root colonization and biofilm formation in Enterobacter cloacae GS1. J Bacteriol 195(1):39–45. https://doi.org/10.1128/JB.01236-12

Zhao YN, Wang HF, Liu JX (2018) Detoxification of *Jatropha curcas* seed cake by a new soil-borne Enterobacter cloacae strain. Lett Appl Microbiol 67(2):197–204. https://doi.org/10.1111/lam.13012

Zhou H, Zhang D, Jiang Z, Sun P, Xiao H, Yuxin W, Chen J (2019) Changes in the soil microbial communities of alpine steppe at Qinghai-Tibetan plateau under different degradation levels. Sci Total Environ 651:2281–2291

# Impact of Sugar Mill Effluent on the Quality of Groundwater in Cuddalore District, Tamil Nadu, India

A. Arivoli, X. Agnello J. Naveen, and Sathiamoorthi Thangavelu

**Abstract** The sugar mill is the backbone of rural, agricultural, and socio-economic development in India. Many industries are directly or indirectly dependent on the sugar industry which in turn is responsible for the overall development of the state. The discharge of this effluent into water bodies or on the soil causes serious problem of water pollution resulting in severe damage to the flora and fauna and further environmental degradation. The study focuses on the impact of sugar mill effluent on groundwater used for domestic purpose. The quality of groundwater was studied for the period January–December 2019. In this analysis, the parameters such as pH, electrical conductivity, temperature, alkalinity, hardness, chloride, total solids, total dissolved solids, total suspended solids, dissolved oxygen and chemical oxygen demand were studied in the closest proximity of 1–1.5 km from the sugar mill as per Environmental Impact Assessment guidelines (Central pollution control board). The results reveal a significant variation in samples of the same parameters. The study shows that water quality parameters nearby the sugar industry exceeds the permissible limits.

**Keywords** Groundwater quality · Physico-chemical · Sugar industry effluents

## 1 Introduction

The sugar industry plays an important role in the economic development of the Indian subcontinent, however the released effluents produce a high degree of organic pollution in both the aquatic and terrestrial environments (Solomon 2005a; Roy

A. Arivoli (✉)
Department of Environmental Science, Government Arts College, Ariyalur, Tamil Nadu, India
e-mail: arivoli82@gmail.com

X. A. J. Naveen
Department of Environmental Science, Thanthai Roever Institute of Agriculture and Rural Development (TRIARD), Perambalur, India

S. Thangavelu
Department of Microbiology, Alagappa University, Karaikudi, Tamil Nadu, India

 
M. Prashanthi et al. (eds.), *Bioremediation and Green Technologies*, Environmental Science and Engineering, https://doi.org/10.1007/978-3-030-64122-1_4

et al. 2007; Ayyasamy et al. 2008). The sugar mill effluent are large discharges from processes such as crushing, refining, mill house, boiling house and condensate water (Khan et al. 2003; Saranraj and Stella 2012). The effluents discharged from the sugar cane factory into the land and different surface water bodies not only affect the water quality and soil but also pollute the groundwater due to the percolation of some water-soluble pollutants (Siddiqui and Waseem 2012; Poddar and Sahu 2017). The environmental challenges for sugar mills are associated with liquid waste, gaseous emission, solid waste, and noise pollution (Solomon 2005b). This situation leads to serious public health problems. Unmanaged organic waste generated from industries, municipalities and the agricultural sector decomposes, thereafter, resulting in the large scale contamination of land, water, and air in the environment (Chapman and Kimdtsch 1996; Hampannavar and Shivayogimath 2010; Sahu 2018). Irrigation with such effluent increases nitrogen, phosphorus, potassium, organic carbon and heavy metal content in the soil and crops (Hati et al. 2007; Chopra et al. 2012; Kumar and Chopra 2013; Kumar and Chopra 2014). In Cuddalore district, industrialization and urbanization have major impact on groundwater environment. Both surface and subsurface water sources are continuously being polluted due to such developmental activities (Newcomb and Rimstidt 2002). The contamination of both aquatic and terrestrial ecosystems caused by the excessive quantity of waste released in the outline of effluent produces is positively alarming. These pollutants not only change the physico chemical characteristics of the beneficiary aquatic bodies but also affect the aquatic flora and fauna (Deshmukh 2014). However, there survives insignificant data and information possible about the outcome of sugar mill effluent on the quality of groundwater of this area. Considering these serious issues, the present study was carried out to investigate the possible effects of such effluents on the quality of groundwater in the proximity of a sugar factory.

## 2 Study Area

The Cuddalore district is bounded by the Latitudes 11°11″ and 12°5″ N Longitudes 78°38″ and 80°00″ E covering an area of 3678 km$^2$. Cuddalore District consists of three Revenue Divisions viz., Cuddalore, Chidambaram, and Vridhachalam comprising ten taluks viz., Cuddalore, Chidambaram, Kattumannarkoil, Virudhachalam, Thittakudi, Bhuvanagiri, Panruti, Veppur, Srimushnam and Kurchipadi (Fig. 1). There are 5 Municipalities viz. Cuddalore, Nellikuppam, Panruti, Chidhambaram and Virudhachalam. The headquarters of the Cuddalore district is Cuddalore (11°44′45″ N and 79°45′56″ E), A large industrial town that has experienced coastal development at a rapid rate. Cuddalore is situated about 160 km south of Chennai, the state capital. The district is drained by Gadilam and Pennaiyar rivers in the north, Vellar and Kollidam River (Coleroon) in the south (MOEF 2019).

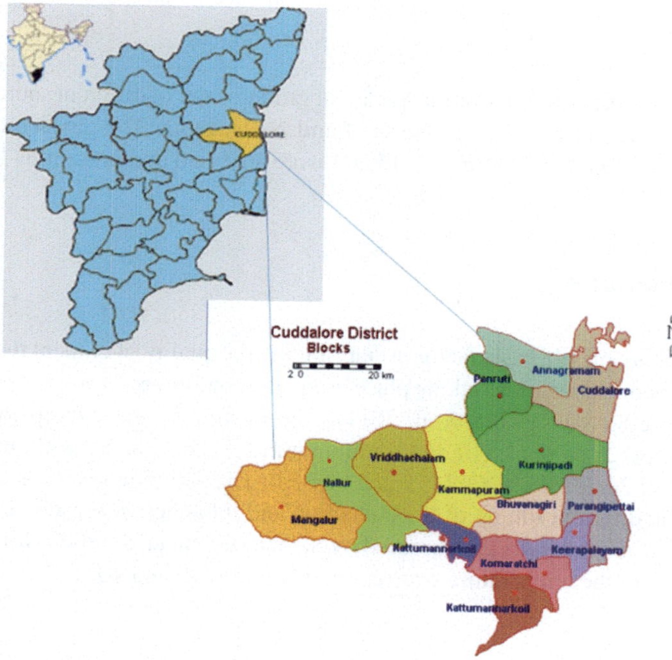


**Fig. 1** Location of study area and groundwater sampling site network in the study area

## 3 Materials and Methods

For the present investigations, three taluks of Cuddalore District (TN) were selected. There are ten sampling points (S1, S2, S3, S4, S5, S6, S7, S8, S9, S10, i.e.) with three from each taluk from January to December 2019 (Fig. 1). The groundwater samples were collected from the hand pump without any air bubbles. The sampling sites were located near the stream and a few meters away from the effluents stream (Aquifers). In addition, water samples were collected from the wells close to the stream and progressively away from it. Accordingly, the closest proximity of 1–1.5 km from the sugar mill was prescribed by the Environmental Impact Assessment (Central Pollution Control Board) and adopted for our study (Valappil et al. 1994; Banham and Brew 1996; Paliwal 2006).

The samples were collected in one-liter polyethylene bottles of one-liter capacity with care to air bubbles. The pH, electrical conductivity (EC), and temperature were measured in the field. The physico-chemical parameters such as pH, EC, TS, TDS, TSS, Alkalinity, Total Hardness, DO, BOD, and COD were analyzed for further studies. The analysis was carried out in the laboratory using the procedures given by APHA (1998).

# 4 Results

This study was taken up to assess the quality of groundwater of different sources in the surrounding areas of Cuddalore district, Tamil Nadu, India. The physico-chemical parameters of treated and untreated effluent were analysed and present in Table 1.

## 4.1 *Temperature*

The temperature plays a basic factor in determining the quality of effluent its effect on certain biochemical reactions taking place in aquatic conditions for aquatic organisms has harmful effects (Chandra et al. 2011). The maximum water temperature was observed to be 25.5 °C at S10 and a minimum of 24.5 °C at S5 with an average range value of 24.7–25.5 °C (Table 2 and Fig. 2a). The variation in temperature may be due to the different timing of the collection and influence of season (Jayaraman et al. 2003). Similar results are observed by Salequzzaman et al. (2008) for different locations where the temperature was reported between 30 and 40 °C.

## 4.2 *pH*

The pH values of all groundwater samples are found to be in the range of 7.25–7.85 (Table 2). The highest value of 7.85 is observed at station S9 whereas the lowest value

**Table 1** The physico-chemical parameters of treated and untreated sugar mill effluent

| S. No. | Parameter | Untreated | Treated | Bureau of Indian Standards (BIS 2012) |
|---|---|---|---|---|
| 1 | Color | Dark brown | Light brown | – |
| 2 | Temperature (°C) | 48 °C | 30 °C | – |
| 3 | pH | 4.8 | 6.9 | 6.5–9.0 |
| 4 | EC (μS/cm) | 2300 | 1520 | |
| 5 | TS (mg/l) | 4500 | 3000 | 2700 |
| 6 | TDS (mg/l) | 3000 | 2500 | 2100 |
| 7 | TSS (mg/l) | 1500 | 500 | 600 |
| 8 | Alkalinity (mg/l) | 300 | 220 | – |
| 9 | Chlorides (mg/l) | 178 | 140 | 600 |
| 10 | Total hardness (mg/l) | 680 | 430 | 300 |
| 11 | Dissolved Oxygen (mg/l) | 1.10 | 3.6 | 4–6 |
| 12 | COD (mg/l) | 385 | 190 | 250 |
| 13 | BOD (mg/l) | 98 | 76 | 50 |

**Table 2** Comparison of physico-chemical parameter of the collected groundwater samples with Bureau of Indian Standards (BIS)

| Parameters | S1 | S2 | S3 | S4 | S5 | S6 | S7 | S8 | S9 | S10 | BIS |
|---|---|---|---|---|---|---|---|---|---|---|---|
| Temperature °C | 24.9 | 25.1 | 24.9 | 24.8 | 24.7 | 24.8 | 25.3 | 24.7 | 25.1 | 25.5 | 7–8.5 |
| pH | 7.45 | 7.62 | 7.48 | 7.68 | 7.25 | 7.72 | 7.38 | 7.47 | 7.85 | 7.57 | – |
| EC (μS/cm) | 0.3 | 0.83 | 0.57 | 0.73 | 0.31 | 0.42 | 0.66 | 0.41 | 0.35 | 1.09 | – |
| TS (mg/l) | 2500 | 1500 | 1500 | 2000 | 1500 | 2000 | 1000 | 1000 | 1500 | 1000 | 500 |
| TDS (mg/l) | 1500 | 500 | 500 | 1500 | 500 | 1500 | 500 | 500 | 500 | 500 | – |
| TSS (mg/l) | 1000 | 1000 | 1000 | 500 | 1000 | 500 | 500 | 500 | 1000 | 500 | – |
| Total Alkalinity (mg/l) | 150 | 125 | 210 | 175 | 150 | 50 | 100 | 150 | 150 | 150 | 200 |
| Chloride (mg/l) | 26.58 | 85.08 | 87.22 | 40.76 | 19.49 | 26.58 | 51.4 | 28.36 | 31.9 | 125.84 | 250 |
| Hardness (mg/l) | 156 | 378 | 214 | 356 | 162 | 196 | 244 | 192 | 198 | 496 | 300 |
| DO (mg/l) | 5.85 | 6.82 | 5.8 | 6.45 | 6.02 | 6.68 | 6.85 | 6.65 | 8.32 | 8.02 | 6 |
| COD (mg/l) | 145 | 165.5 | 150.6 | 170 | 180.6 | 190.2 | 170.5 | 165.2 | 230.5 | 265.4 | 250 |

**Note:** S-Samples, BIS-Bureau of Indian Standards

of 7.25 is observed at station S5 in Fig. 2b. The permissible limit of pH for drinking water is 7.0–8.5 according to World Health organization (WHO). The groundwater samples in the study area are within the acceptable limit of WHO. There is no abnormal change of pH in the groundwater samples (El-Gohary et al. 1998).

## 4.3 *Electrical Conductivity (EC)*

The electrical conductivity values for all the groundwater samples are recorded within the range of 0.30–1.09 μS/cm. The highest value of 1.09 is observed at station S10 whereas the lowest value of 0.30 μS/cm is observed at station S1 (Fig. 2c). Electrical conductance is a good indication of total dissolved solids, which is a measure of salinity that affects the taste of potable water. Several factors like temperature, ionic mobility, and ionic valences also influence the conductivity. The electrical conductivity value for all the groundwater samples is above the permissible limit.

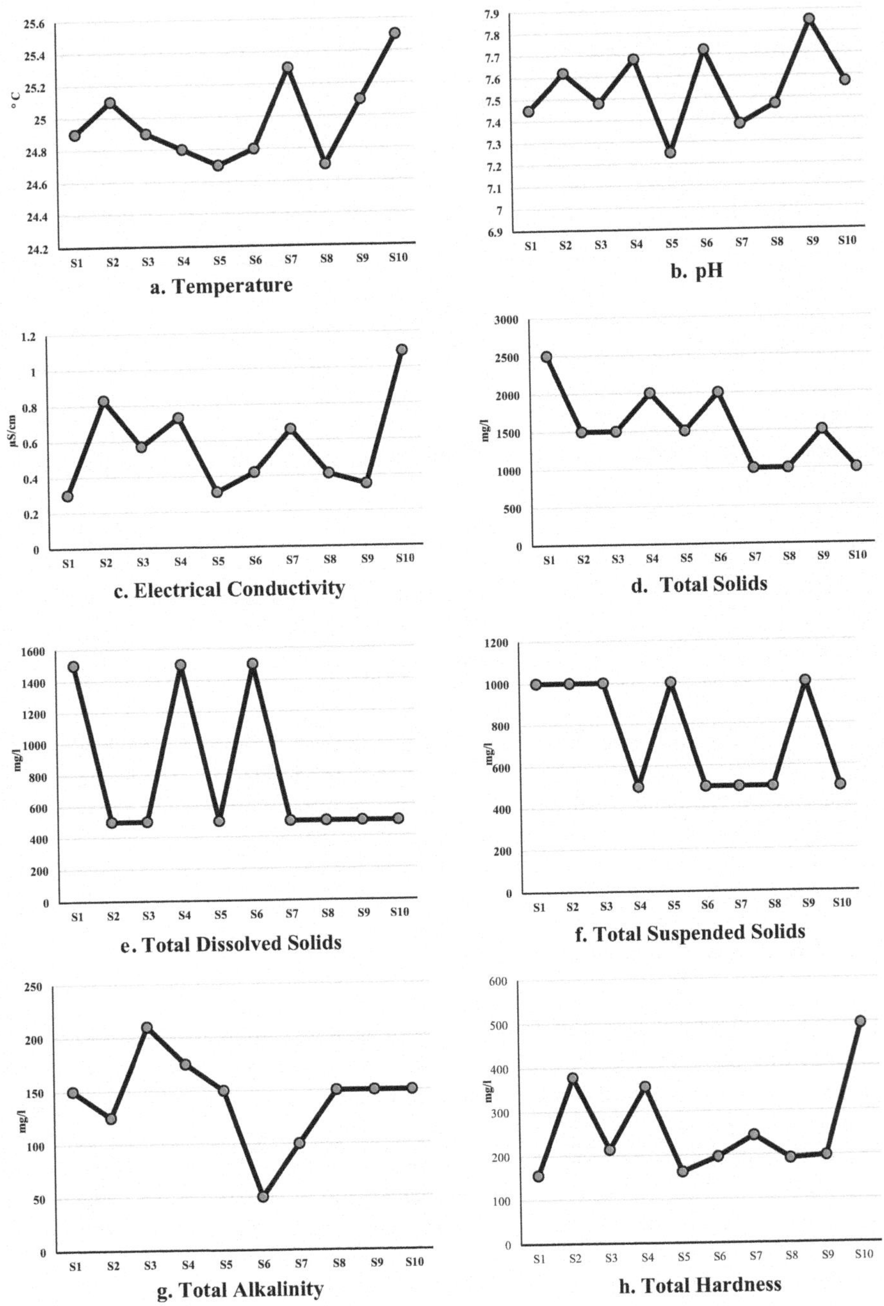


**Fig. 2 a–k** Graphs of physico chemical parameters of the collected groundwater samples

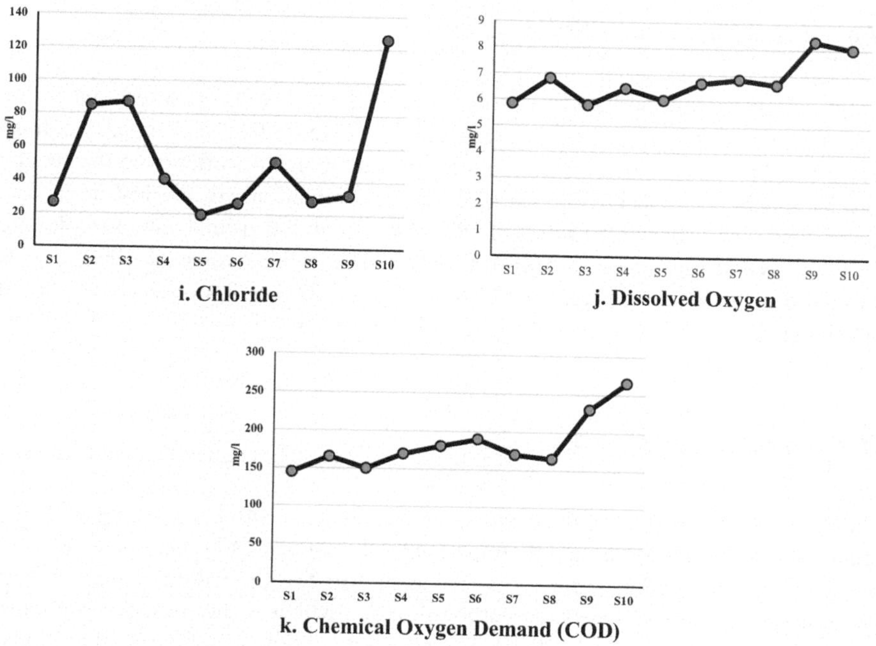


**Fig. 2** (continued)

## 4.4 *Total Solids (TS)*

Total Solids (TS) values for all the groundwater samples are recorded within the range of 1000–2500 mg/l. The highest value of 2500 mg/l is observed at station S1 whereas the lowest value of 1000 mg/l is observed at station S7 in Fig. 2d.

## 4.5 *Total Dissolved Solids (TDS)*

The total dissolved solids in water are due to the presence of sodium, potassium, calcium, magnesium, manganese, carbonates, bicarbonates, chlorides, phosphate, organic matter, and other particles (Chapmanan and Kimstach 1996). The values of the total dissolved solids for all the groundwater samples vary between 500 and 1500 mg/l. The maximum allowable limit of total dissolved solids in groundwater for domestic purposes is 1500 mg/l (WHO). The maximum value (1500 mg/l) is recorded at station S6 and the minimum value (500 mg/l) is recorded at station S7 as shown in Fig. 2e. According to the classification of drinking water based on TDS values. In this study, the TDS value for all the groundwater samples is well within the permissible limit of 1500 mg/l except the sampling station S6 (1500 mg/l).

## 4.6 Total Suspended Solids (TDS)

Total suspended solids (TSS) and total dissolved solids (TDS) correspond to the non-filterable and filterable residue, respectively. "Fixed solids" and "volatile solids" correspond to the remainder after oven-drying, and to the loss after oven-drying at a given temperature, respectively. The latter two determinations are now less frequently carried out. Total suspended solids (TSS) values for all the groundwater samples are recorded within the range of 500–1000 mg/l. The highest value of 1000 mg/l is observed at station S1 whereas the lowest value of 500 mg/l is observed at station S4 in Fig. 2f.

## 4.7 Total Alkalinity (TA)

Alkalinity is the measure of the capacity of the water to neutralize a strong acid. The alkalinity in the water is generally imparted by the salts of carbonates, silicates, etc. together with the hydroxyl ions in a free state. Most of the natural waters contain substantial amounts of dissolved carbon dioxide, which is the principal source of alkalinity (Anwar and Vanita 2014). The alkalinity varies from 50 to 210 mg/l was found above the permissible limit (200 mg/l) prescribed by BIS. The highest value of 210 mg/l is observed at station S3 whereas the lowest value of 50 mg/l is observed at station S6 in Fig. 2g.

## 4.8 Total Hardness (TH)

Total hardness is a measure of the capacity of water to the concentration of calcium and magnesium in water and is usually expressed as the equivalent of $CaCO_3$ concentration (Anwar and Vanita 2014). In the study, the total hardness of the water samples ranges between 156 to 496 mg/l of samples were found above the standard limit (200 mg/l) prescribed by BIS. The highest value of 496 mg/l is observed at station S10 whereas the lowest value of 156 mg/l is observed at station S1 in Fig. 2h. Hard water is useful in the growth of children, if within the permissible limit. However, hard water is a nuisance because of mineral buildup on fixtures and poor soap/detergent performance. The high degree of hardness in the study area can be attributed to the disposal of untreated, improperly treated sewage, and industrial wastes.

## 4.9 Chloride

Chloride occurs in all-natural waters in widely varying concentrations. The chloride contents normally increase as the mineral content increases (Dubey 2003). Chloride concentrations ranged from 22.0 to 421.0 mg/l with an average value of 156.33 mg/l (Fig. 2i). The chloride values exceeded the maximum limit (250 mg/l) prescribed by BIS. At concentrations above 250 mg/l, the water acquires objectionable salty taste. However, no adverse health effects on humans have been reported from the intake of water containing the highest content of chloride (Pawar et al. 1998; Agale et al. 2013). If the water with high chloride concentration is used for construction purposes, this may corrode the concrete.

## 4.10 Dissolved Oxygen (DO)

Dissolved Oxygen (DO) is the ability of the surface and groundwater to purify itself through the biochemical process. However, temperature, pressure, and chemical constituents may affect Dissolved Oxygen (DO) in water. Dissolved oxygen (DO) in potable water has a crucial feature, since; it greatly influences the solubility of metals, which are essential for biological life. DO values, in the study samples, are found to be in the range of 5.80–8.29 mg/l, almost all samples within the permissible limit (4 ppm). The highest DO value of 8.29 mg/l is observed at station S9 whereas the lowest value of 5.80 mg/l is observed at station S3 in (Fig. 2j). (Avasan and Rao 2001) observed that the DO of sugar mill ranged between and 2.0. (Reddy et al. 2015) recorded dissolved oxygen range as 4.61–6.68 mg/l in the winter season.

## 4.11 Chemical Oxygen Demand (COD)

The Chemical Oxygen Demand (COD) is a measure of the oxygen equivalent of the organic matter in a water sample that is susceptible to oxidation by a strong chemical oxidant, such as dichromate. The COD is widely used as a measure of the susceptibility to oxidation of the organic and inorganic materials present in water bodies and the effluents from sewage and industrial plants. The test for COD is non-specific, in that it does not identify the oxidizable material or differentiate between the organic and inorganic material present (Chapmanan and Kimstach 1996). The Chemical Oxygen Demand concentration in the groundwater of the area varies from 145 to 265 mg/l. The higher concentration of COD is observed to be 265 mg/l at station S10 whereas the lowest value of 145 mg/l is observed at station S1 in Table 2 and Fig. 2k.

## 5 Discussion

From the study, it is observed that the chloride value is higher in groundwater due to the salt content which is naturally higher in soil (Abrol et al. 1988). Parameters differ in the sample area and there is a variation for every 1.5 km. The groundwater pollution level is higher in the proximity zone of 0–3 km. Advanced technology can be used in the effluent treatment process (ETP) to achieve zero liquid discharge (Yaqub and Lee 2019). Recommended by the Central Pollution Control Board (CPCB) by the industry like Ultra-filtration (Ghosh and Balakrishnan 2003; Hamachi, Gupta and Aim 2003), Reverse osmosis (RO) (Hinkova et al. 2002), Multigrade filters (Umashanker and Bishnoi 2018). Green Chain management—Green belt can be introduced on a large scale to protect and absorbs groundwater stability and has a phytoremediation concept (Chinsinga 2017). A Closed-loop system for water saving is suggested with conventional treatment technology (Gupta and Ali 2001). A public awareness program under Corporate Social Responsibility (CSR) can be arranged and a need for periodic testing of groundwater in the locality of the sugar factory can be focused.

## 6 Conclusion

The present study investigated the impacts of the sugar industry on groundwater quality of the area around the Cuddalore district. Groundwater sampling is done at ten different locations around the sugar mill. The sugar mill effluent or effluent from industries affect the quality of groundwater as well as soil where the parameters such as pH, TDS, hardness, and chlorides are found beyond the permissible limit. The groundwater sample is compared with the prescribed standard of drinking water by BIS and are not suitable for drinking purpose. The study suggested measures to improve the groundwater quality and imposes the necessity for proper disposal of industrial effluent.

## References

Abrol IP, Yadav JSP, Massoud FI (1988). Salt-affected soils and their management (No. 39). Food and Agriculture Org

Agale MC, Patel NG, Patil AG (2013) Impact of sugar industry effluents on the quality of groundwater from Dahiwad Village, Dist-Dhule (MS). Arch Appl Sci Res 5(2):58–60

Anwar KM, Vanita A (2014) Analysis of groundwater quality using statistical techniques: a case study of Aligarh city (India). Int J Tech Res Appl 2(5):100–106

APHA, AWWA, WPCF (1998) Standard methods for the examination of water and waste waters 14th edn. America Public Health Association

Avasan M, Rao SR (2001) Effect of sugar mill effluent on organic resources of fish. Pollut Res 20(2):167–171

Ayyasamy PM, Yasodha R, Rajakumar S, Lakshmanaperumalsamy P, Rahman PK, Lee S (2008) Impact of sugar factory effluent on the growth and biochemical characteristics of terrestrial and aquatic plants. Bull Environ Contam Toxicol 81:449–454

Banham W, Brew D (1996) A review of the development of environmental impact assessment in India. Proj Appraisal 11(3):195–202

BIS (2012) Bureau of Indian standards drinking water specifications. BIS, 10500, 2012

Chandra R, Gupta M, Pandey A (2011) Monitoring of river Ram Ganga: physicochemical characteristic at Bareilly. Recent Res Sci Technol 3(6):16–18

Chapman D, Kimstach V (1996) Water quality assessments—a guide to use of biota, sediments and water in environmental monitoring, pp 1–60

Chinsinga B (2017) The green belt initiative, politics and sugar production in Malawi. J South Afr Stud 43(3):501–515

Chopra AK, Srivastava S, Kumar V, Pathak C (2012) Agro-potentiality of distillery effluent on soil and agronomical characteristics of Abelmoschus esculentus L. (okra). Environ Monit Assess 185(8):6635–6644

Deshmukh KK (2014) Environmental impact of sugar mill effluent on the quality of groundwater from Sangamner, Ahmednagar, Maharashtra, India. Res J Recent Sci 3:385–392

Dubey, N (2003) A comparative status of quality of drinking water of Bhopal city filtration plants and ground water with special reference to heavy metals and organo chemical. Ph.D. Thesis, Barkatullah University, Bhopal

El-Gohary AF, Fayza Nasr A, Hawaary S (1998) Performance assessment of a wastewater treatment plant producing effluent for irrigation in Egypt. Environmentalist 18:87–93

Ghosh AM, Balakrishnan M (2003) Pilot demonstration of sugarcane juice ultrafiltration in an Indian sugar factory. J Food Eng 58(2):143–150

Gupta VK, Ali I (2001) Removal of DDD and DDE from wastewater using bagasse fly ash, a sugar industry waste. Water Res 35(1):33–40

Hamachi M, Gupta BB, Aim RB (2003) Ultrafiltration: a means for decolorization of cane sugar solution. Sep Purif Technol 30(3):229–239

Hampannavar US, Shivayogimath CB (2010) An aerobic treatment of sugar industry wastewater by upflow anaerobic sludge blanket reactor at ambient temperature. Int J Environ Sci 1:631–639

Hati KM, Biswas AK, Bandyopadhyay KK, Misra AK (2007) Soil properties and crop yields on a vertisol in India with application of distillery effluent. Soil Tillage Res 92(1–2):60–68

Hinkova A, Bubnık Z, Kadlec P, Pridal J (2002) Potentials of separation membranes in the sugar industry. Sep Purif Technol 26(1):101–110

Jayaraman PR, Ganga Devi T, Vasudena Nayar T (2003) Water quality studies on Karamana River, Thiruvananthapuram District South Kerala, India. Pollut Res 22(1):89–100

Khan M, Kalsoom U, Mahmood T, Riaz M, Khan AR (2003) Characterization and treatment of industrial effluent from the sugar industry. J Chem Soc Pak 25(3):242–247

Kumar V, Chopra AK (2013) Accumulation and translocation of metals in soil and different parts of French bean (*Phaseolusvulgaris* L) amended with sewage sludge. Bull Environ Contam Toxicol 92(1):103–108

Kumar V, Chopra AK (2014) Pearl millet (*Pennisetum Glaucum* L.) response after ferti-Irrigation with sugar mill effluent in two seasons. Int J Recycl Org Waste Agr 3(3):1–13

Ministry of Environment, Forest and Climate Change (2019) Impact assessment division. Government of India, New Delhi-110003

Newcomb WD, Rimstidt JD (2002) Trace element distribution assessment using public domain data. Appl Geochem 17:49–57

Paliwal R (2006) EIA practice in India and its evaluation using SWOT analysis. Environ Impact Assess Rev 26(5):492–510

Pawar NJ, Pondhe GM, Patil SF (1998) Groundwater pollution due to sugar-mill effluent, at Sonai, Maharashtra, India. Environ Geol 34(2–3):151–158

Poddar PK, Sahu O (2017) Quality and management of wastewater in sugar industry. Appl Water Sci 7(1):461–468

Reddy SSG, Raju AJ, Kumar BM (2015) Phytoremediation of sugar industrial water effluent using various hydrophytes. Int J Environ Sci 5(6):1147–1158

Roy PR, Prasad J, Joshi AP (2007) Effect of sugar factory effluent on some physico-chemical properties of soils- A case study. J Environ Sci Eng 49(4):277–282

Sahu O (2018) Assessment of sugarcane industry: Suitability for production, consumption, and utilization. Ann Agrarian Sci 16(4):389–395

Salequzzaman M, Tariqul SM, Tasnuva Islam A, Kashem MA, al Masud MM (2008) Environmental impact of sugar industry—a case study on Kushtia sugar mills in Bangladesh. J Innov Dev Strat 2(3):31–35

Saranraj P, Stella D (2012) Bioremediation of sugar mill effluent by immobilized bacterial consortium. Int J Res Pure Appl Microbiol 2(4):43–48

Siddiqui WA, Waseem M (2012) A comparative study of sugar mill treated and untreated effluent—a case study. Orient J Chem 28(4):1899–1904

Solomon SK (2005a) Environmental pollution and sugar industry in India its management in: an appraisal. Sugar Tech 7(1):77–81

Solomon SK (2005b) Environmental pollution and its management in sugar industry in India: an appraisal. Sugar Tech 7(1):77–81

Umashanker S, Bishnoi NR (2018) Assessment of removal efficiency of total suspended solids (TSS) from secondary treated effluent using a multigrade multimedia filtration technique. Ann Agri Bio Res 23(1):74–77

Valappil M, Devuyst D, Hens L (1994) Evaluation of the environmental impact assessment procedure in India. Impact Assess 12(1):75–88

Yaqub M, Lee W (2019) Zero-liquid discharge (ZLD) technology for resource recovery from wastewater: a review. Sci Total Environ 681(1):551–563

# Bioremediation Studies of Contaminated Sites

# Biosorption of Ni(II) Using Seeds of *Mirabilis Jalapa*

**V. Anitha, A. Balakumaran, V. V. Ramprakash, A. V. Vinothini, P. Muthukumaran, and Aravind Jeyaseelan**

**Abstract** The release of effluents from automobile industries contains various metal ions which has to be removed effectively. This can be done eco-friendly using biosorption. Ni (II) ions were adsorbed using seeds of *Mirabilis jalapa.* The seeds showed maximum adsorption activity without activation. Optimization studies were carried out to achieve maximum removal efficiency. The parameters included in the study were pH, contact time, metal ion concentration, RPM, adsorbent dosage concentration and temperature. The equilibrium data showed that the adsorption fits Harkins–Jura isotherm model. Kinetics studies revealed that the adsorption follows pseudo-first-order kinetics. Further, optimization studies were carried out using Box-Behnken design, and the obtained data reveals that the model is significant. Modification in the functional groups were studied using FTIR analysis, and the results were interpreted.

**Keywords** Biosorption · *Mirabilis jalapa* seeds · Nickel ion · Optimization

## 1 Introduction

The presence of a huge number of industries around water sources has led to water pollution on a large scale. Only 1% of the total water on our planet is suitable for drinking and yet industries pump in tonnes of toxic heavy metal waste every day, which makes it extremely hazardous. Nickel, the metal chosen for study, constitutes a major portion of the effluent released from finishing industries such as electro galvanic industries. Tap water faucets, nickel-steel alloy cookware and coins are some main sources of nickel toxicity in water (Cempel and Nikel 2006; Aikpokpodion et al. 2010). Electroplating industries use nickel in large quantities and recycling it can

V. Anitha · A. Balakumaran · V. V. Ramprakash · A. V. Vinothini · P. Muthukumaran · A. Jeyaseelan (✉)
Kumaraguru College of Technology, Coimbatore, Tamil Nadu, India
e-mail: aravindj@dgct.ac.in; dr.j.aravind@gmail.com

A. Jeyaseelan
Dhirajlal Gandhi College of Technology, Salem, Tamil Nadu, India



M. Prashanthi et al. (eds.), *Bioremediation and Green Technologies*, Environmental Science and Engineering, https://doi.org/10.1007/978-3-030-64122-1_5

be an expensive process. Conventional methods of recycling nickel such as reverse osmosis, electro-dialysis, precipitation, redox reaction and ion-exchange have certain drawbacks and are not economically viable (Al-Rub et al. 2004). Nickel exposure poses a considerable threat to the environment and human health. Acute exposure to nickel can lead to nausea and vomiting. Long term frequent acute exposures can lead to more serious conditions like kidney failure and respiratory problems. Nickel is a carcinogen and can cause lung and prostate cancer (Parmer and Thakur 2013; Cayllahua and Torem 2011). In this paper, we propose a solution to remove nickel from effluents by the process of Bio-sorption. Bio-sorption is achieved by complex mechanisms like covalent and non-covalent forces existing between the adsorbent and the adsorbate. The advantage of bio-sorption is its eco-friendly nature when compared to other conventional methods of adsorption (Aravind et al. 2015).

In the current work, powdered seeds of the plant *Mirabilis jalapa* was explored as an adsorbent, *Mirabilis jalapa*, commonly known as four o'clock flower of the Nyctaginaceae family is found in warm temperate and tropical zones. It is a perennial tuberous plant that is primarily grown as an ornamental plant. The seeds are invasive and require no special modes of transmission for germinating (Yu and Zhou 2009). Since these seeds are produced in large numbers in a short duration of time, it was taken for biosorption studies (Fu and Wang 2011).

## 2 Materials and Methods

### 2.1 Collection and Preparation of Adsorbent

The seeds of *Mirabilis jalapa* shrub were collected and were subjected to drying to remove the total moisture content from the seeds. The dried seeds were powdered finely using a blender machine. The cotyledon part of the seed was removed since it leaches out with water wash. Only the pod part of the seeds was retained and washed thoroughly with distilled water. After the wash, the seeds were kept dry in a hot air oven at 50 °C for 2 h (Yu and Zhou 2009).

### 2.2 Characterization of Adsorbent

#### 2.2.1 Fourier Transformation Infrared Spectroscopy (FTIR)

FTIR is an analytical technique used to identify molecular components and structures using infrared radiation. When a sample to be analyzed is radiated with IR radiation, the molecules get excited and the radiation that are adsorbed characteristic of its molecular structure. The signal transmitted is analyzed with the detector to produce interferogram which are later analyzed to obtain a spectrum which is presented as graphs. FTIR analysis was performed for samples obtained before and after nickel adsorption. The spectral range taken for our sample is 4000–1000 $cm^{-1}$ (Aravind et al. 2015).

## 2.3 *Metal Analysis*

### 2.3.1 Preparation of Metal Solution

A stock solution of Ni(II) ions were prepared for 1000 ppm using $NiCl_2$, the stock was diluted to 100 ppm which is used for further studies. The pH of the solution was adjusted using 0.1 N HCl or NaOH solutions.

### 2.3.2 Nickel Analysis

DMG (dimethylglyoxime) method was used for Nickel(II) spectrophotometric analysis. A blank was prepared for Ni(II) solution. To different standard concentrations of Ni(II), 2 mL of 20% (w/v) potassium sodium tartrate solution, 10 mL of 4% (w/v) ammonium per sulfate, 2.5 mL of 5 M sodium hydroxide solutions and 0.5 mL of 1% DMG (w/v) solutions were sequentially added. After an incubation period 45 mins, spectrometric measurement was performed at 450 nm. The amount of nickel present in the sample was determined using the calibration curve (Aravind et al. 2015).

## 2.4 *Removal Efficiency*

The nickel removal percentage can be obtained using the formulae:

$$\text{Removal efficiency }(\%) = (\text{Ci} - \text{Co}/\text{Ci}) \times 100 \quad (1)$$

where Ci represents the initial metal ion concentration and Co represents the final metal ion concentration.

## 2.5 *Batch Adsorption Studies*

Adsorption studies were performed initially by taking the appropriate amount of stock solution dissolved in water to obtain desired dilutions in a conical flask. They were subjected to initial studies by maintaining a certain pH, rpm for a certain period. Then the samples were filtered, and the solution was assayed using DMG method and analyzed spectrophotometrically at 450 nm.

### 2.5.1 Effect of pH

A set of experiments were performed at varying level of pH ranging from acidic to alkaline conditions by taking 100 ppm of Nickel chloride solution. The varying range of pH was obtained by adding 0.1 N HCl or NaOH solutions to which 0.1 g of adsorbent was added. Agitation was given using an orbital shaker at 120 rpm for 45mins and the sample was analyzed spectrophotometrically at 450 nm (Vijayaragavan et al. 2008).

### 2.5.2 Effect of Contact Time

Adjusting the pH to 6.5 by using 0.1 N HCl or NaOH solutions, effect of contact time on Ni adsorption was studied using 100 ppm Ni solution to which 0.1 g of adsorbent was added, and adsorbance was obtained with samples taken at different intervals of time (15, 30, 45, 60, 75, 90 mins).

### 2.5.3 Effect of rpm

The effect of rpm studies was conducted at different rpm levels (100, 120, 140,160, 180 rpm) by optimizing the pH at 6.5 for the time period of 60 min. The samples are filtered to remove the adsorbent and OD readings are obtained at 450 nm.

### 2.5.4 Effect of Adsorbent Dosage Concentration

Varying concentrations of adsorbent (0.1, 0.2, 0.3, 0.4, 0.5 g) was added to 50 ml of 100 ppm Ni solutions. The pH of the sample was adjusted to 6.5 and kept in a shaker at 160 rpm for 60 min. The samples were filtered and spectrophotometrically read at 450 nm.

### 2.5.5 Effect of Initial Metal Ion Concentration

Varying the concentration of Ni(II) ions (50, 100, 150, 200, 250 ppm) was prepared to which 0.1 g of adsorbent was added and the experiments were performed under optimized conditions. The samples are filtered to remove the adsorbent and read spectrometrically at 450 nm.

**Table 1** Design of variables

| S. No. | Parameter | Assigned Variable | +1 value | -1 value |
|---|---|---|---|---|
| 1 | pH | A | 6 | 8 |
| 2 | Contact time (Min) | B | 30 | 60 |
| 3 | Ni concentration (ppm) | C | 50 | 150 |

#### 2.5.6 Effect of Temperature

The temperature optimization studies were performed with different ranges of temperature (30, 35, 40, 45 and 50 °C) at above optimized conditions and the samples were filtered to remove adsorbent and are analyzed spectrophotometrically at 450 nm.

### *2.6 Response Surface Methodology*

The purpose of exploring response surface methodology is to study the relationship between the variables used and to optimize these variables in order to reach desired adsorption levels. RSM was performed by taking 3 main factors of our experiment such as pH, contact time, metal ion concentration with design variables (Table 1) and a set of 17 experiments were performed. The design of the experiment is shown in Table 2. The experiment was designed using the quadratic equation given below:

$$Y = 457.51250 - 35.825A - 9.50583B - 0.59125C + 1.015AB + 0.1545AC - 0.010333BC - 2.85A^2 + 0.051222B^2 - 1.5510^{-3}C^2 \quad (2)$$

where Y = % Ni Removed, A = pH, B = Contact time, C = Ni dose.

## 3 Results and Discussion

### *3.1 Batch Adsorption Studies*

Adsorption process is highly influenced by the pH of the solution. The adsorption efficiency of Ni carried out at different pH, maximum adsorption of 84.28% was obtained for a pH of 7, below which there is a competence between Ni and $H^+$ ions in the solution and at higher pH, there is a chance of formation of soluble hydroxide which may attribute to the reduction of adsorption. Adsorption experiments were carried out at different contact time (30–105 min) at pH of 7. It was observed that equilibrium was reached at 60 min beyond which there is no appreciable change was

**Table 2** Response surface methodology design and outcome

| Run | Ph | Contact time (mins) | Metal ion (ppm) | Response (%) | Predicted value |
|---|---|---|---|---|---|
| 1 | 8 | 45 | 50 | 26.4 | 34.99 |
| 2 | 7 | 60 | 150 | 51.5 | 53.1 |
| 3 | 7 | 45 | 100 | 49.8 | 49.8 |
| 4 | 7 | 45 | 100 | 49.8 | 49.8 |
| 5 | 7 | 60 | 50 | 94.7 | 97.08 |
| 6 | 8 | 60 | 100 | 87.7 | 76.74 |
| 7 | 7 | 45 | 100 | 49.8 | 49.8 |
| 8 | 6 | 60 | 100 | 68.5 | 75.49 |
| 9 | 6 | 45 | 150 | 44.3 | 35.71 |
| 10 | 7 | 30 | 150 | 35.7 | 33.32 |
| 11 | 7 | 45 | 100 | 49.8 | 49.8 |
| 12 | 8 | 30 | 100 | 18 | 11.01 |
| 13 | 7 | 45 | 100 | 49.8 | 49.8 |
| 14 | 7 | 30 | 50 | 47.9 | 46.3 |
| 15 | 6 | 45 | 50 | 89 | 79.64 |
| 16 | 8 | 45 | 150 | 12.6 | 21.96 |
| 17 | 6 | 30 | 100 | 59.7 | 70.66 |

observed. This may be due to saturation of the adsorbent surface resulting in lack of available sites for Ni in the solution for binding. In batch adsorption studies, effective mixing is to be required for effective contact of the adsorbent with Ni ions in the aqueous streams. Adsorption of Nickel at various RPM (100–180 rpm) gave an initial increase in adsorption. Maximum adsorption was achieved at a rpm of 160 beyond which a decline in adsorption was observed. This decrease in adsorption may be attributed to the shear force that occurs at higher rpm which may disrupt the binding of Ni onto the adsorbent. The adsorption capacity of *M jalapa* seeds were tested with different concentrations of Ni ranging from 50 to 250 ppm. The results showed an increase in adsorption capacity. Consequently, a decrease in removal efficiency was also observed as this loss in removal efficiency may be due to the saturation of the binding sites present on *M jalapa* seeds or vice versa (Gohulavani and Andal 2013).

Biosorption studies were carried out using wide ranges of temperature. The results revealed that adsorption is maximum at 308 K. The adsorbing capacity of seeds tend to decrease with increase in temperature since increase in temperature favours desorption (Fertu and Gavrilescu 2012; Flores-Garnica et al. 2013).

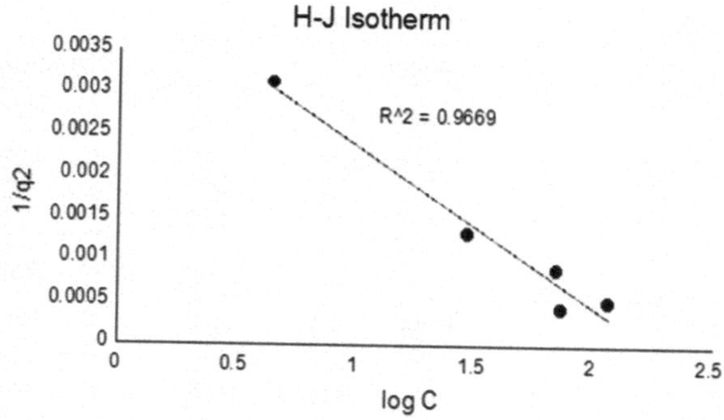


**Fig. 1** H–J isotherm

## 3.2 Equilibrium Adsorption Isotherm Studies

Evaluations of adsorption isotherms are important for adsorption process design. The equilibrium data obtained were fitted in different isotherm models to establish an appropriate representation of the dynamic adsorptive separation process occurring within the system. The isotherms chosen for the study would include Freundlinch, Lamgmuir, Temkin, Dubinin–Radushkevich (D–R) and H–J isotherms. H–J isotherm model fits well with the equilibrium data. This implies the formation of mono-layer films on the heterogeneous pore surface of the adsorbent (Fig. 1) (Malkoc and Nuhogl 2005; Murugesan et al. 2009; Kalavathy et al. 2009; Kalavathy and Rose 2010).

## 3.3 Kinetics Studies

Adsorption kinetics study was performed to identify the type of reaction process occurring in the system. Pseudo first-order kinetics and pseudo second-order kinetics models were studied by fitting the data obtained from effect of contact time (Fig. 2) (Shakirullah et al. 2006; Nuhgolu and Malkoc 2009; Shrestha et al. 2014).

As depicted in Fig. 2, pseudo-second-order kinetics has an $R^2$ value of 0.9433 which imply that adsorption of Ni onto seeds follows pseudo-first-order kinetics.

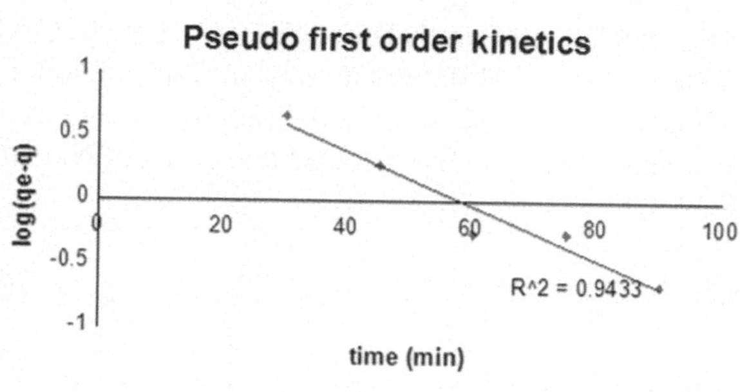


**Fig. 2** Pseudo first order kinetics

**Table 3** Kinetic and isotherm parameters

| S. No. | Model | Parameters | $R^2$ Value |
|---|---|---|---|
| 1 | *Isotherms* | | |
| 1.1 | Freundlinch | K = 11.8986<br>n = 3.5249 | 0.8883 |
| 1.2 | Langmuir | b = 5.5817<br>Q = 39.8406 | 0.8721 |
| 1.3 | Temkin | A = 1.646101663<br>B = 8.1754 | 0.7923 |
| 1.4 | Dubinin-Radushkevich | K = -0.0264<br>qm = 37.1922 | 0.7065 |
| 1.5 | Harkins-Jura | A = 526.3158<br>B = 2.2632 | 0.9669 |
| 2 | *Kinetics* | | |
| 2.1 | Pseudo first order | qe = 16.573<br>k1 = 0.0486 | 0.9433 |
| 2.2 | Pseudo second order | $q_e = 38.7597$<br>$k_2 = 0.23 \times 10^{-2}$ | 0.9061 |

A similar result was obtained by Thamilarasu et al. (2011). Adsorption process following pseudo first order imply that physisorption is predominant in the system.

Increase in Ni concentration from Ni concentration from 50 to 150 ppm resulted in a decrease of Ni removal, results also indicated a better removal capacity of Ni at lower pH. When contact time increased, adsorption increases but adsorption decreases with an increase in Ni concentration. Hence contact time and metal ion concentration had agnostic effects on Ni adsorption (Aravind et al. 2015) (Tables 3 and 4).

## 3.4 FTIR Analysis

FTIR analysis gave information on the vibrational frequencies of functional groups Fig. 3a and b shows the IR spectra of the adsorbent before and after adsorption in the range of 4000–400 $cm^{-1}$. The peaks of C–H, C=N, C–O, C–S are found as stretching vibrations and C–H deformation were responsible for adsorption. It was also seen that there were masking of certain groups like O–H, P–H, C–N, N=N, C-H were observed which may be attributed to binding of Ni to the surface of *M. jalapa* seeds (Table 5).

**Table 4** ANOVA for response surface quadratic model

| Source | Sum of squares | df | Mean square | *F* value | *p*-value Prob > *F* |
|---|---|---|---|---|---|
| Model | 7851.01 | 9 | 872.33 | 9.02 | 0.0042 |
| A-pH | 1705.28 | 1 | 1705.28 | 17.63 | 0.004 |
| B-Contact Time | 2488.65 | 1 | 2488.65 | 25.72 | 0.0014 |
| C-Metal ion concentration | 1621.65 | 1 | 1621.65 | 16.76 | 0.0046 |
| AB | 927.2 | 1 | 927.2 | 9.58 | 0.0174 |
| AC | 238.7 | 1 | 238.7 | 2.47 | 0.1602 |
| BC | 240.25 | 1 | 240.25 | 2.48 | 0.1591 |
| A2 | 34.2 | 1 | 34.2 | 0.35 | 0.5708 |
| B2 | 559.27 | 1 | 559.27 | 5.78 | 0.0472 |
| C2 | 63.22 | 1 | 63.22 | 0.65 | 0.4455 |
| Residual | 677.21 | 7 | 96.74 | - | - |
| Lack of fit | 677.21 | 3 | 225.74 | - | - |
| Pure error | 0 | 4 | 0 | - | - |
| Cor total | 8528.22 | 16 | - | - | - |

## 4 Conclusion

Biosorption studies were carried out using seeds of *Mirabilis jalapa* on Ni(II) ions. Nickel is second most heavy metal ion that disturbs the equilibrium of the aquatic system. This can be removed effectively using the substrate under study. To achieve substrate's maximal removal efficiency, optimization studies were carried out using different parameters. These studies revealed that seeds exhibit highest adsorption capacity at pH 7 with contact time of 60 mins at 160 rpm. Under above conditions, the substrate showed maximal removal of 85% of Ni(II) ions present. RSM studies were carried out using 3 varying factors. The obtained responses were near the predicted responses, and the inter-relationship between these factors were studied. Isotherm studies were carried to understand the type of adsorption and the obtained equilibrium data fits well with H-J model. This confers that substrate exhibit multi-layered adsorption occuring within heterogeneous pores. Further, kinetics studies revealed that the data follows pseudo-first-order kinetics. Changes in the functional groups using FTIR analysis revealed that certain groups were masked and certain groups were modified which are indicated by shifted peaks. This confirms the adsorption of Ni(II) ions.

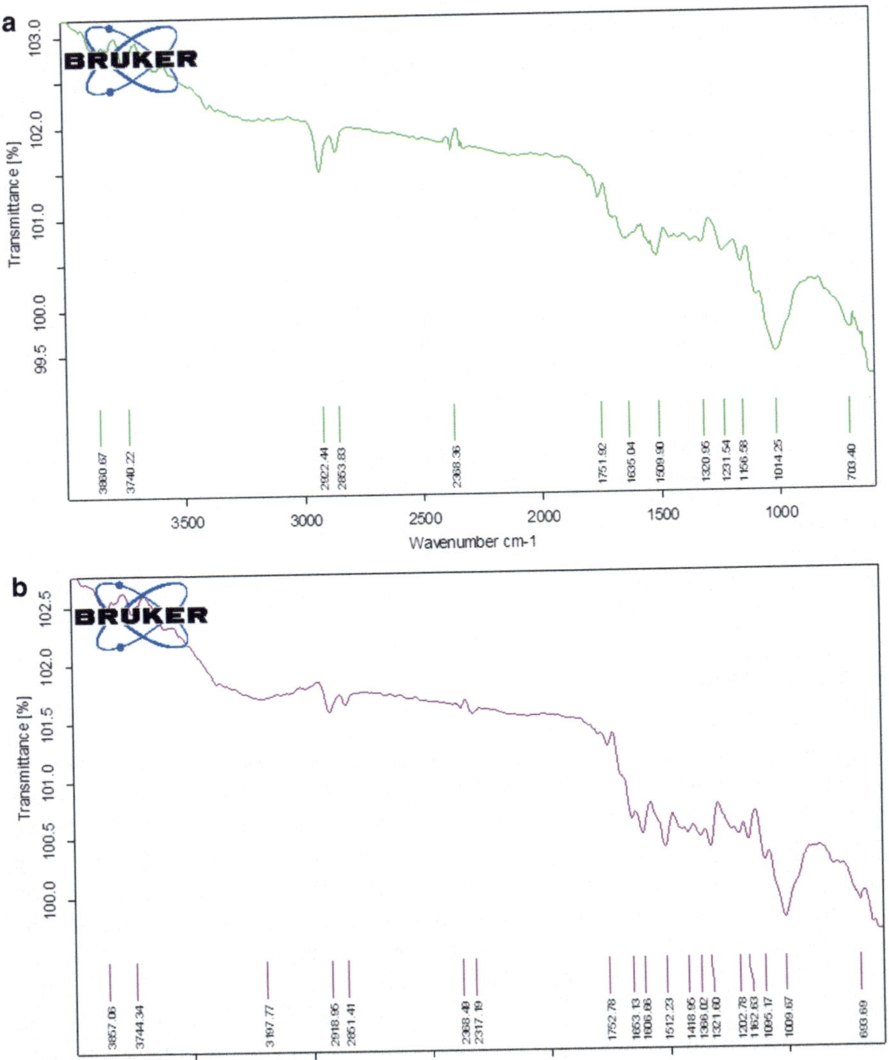


**Fig. 3** **a** FTIR spectrum before adsorption. **b** FTIR spectrum after adsorption

**Table 5** FTIR Spectral Analysis of *M. jalapa* seeds before and after analysis

| Peak | Before adsorption | After adsorption | Difference | Inference |
|---|---|---|---|---|
| 1 | 3197.77 | – | | Stretched O–H bond. Masked after adsorption |
| 2 | 2918.95 | 2922.44 | −3.49 | C-H sym. stretching |
| 3 | 2851.41 | 2853.83 | −2.42 | C-H stretching |
| 4 | 2317.19 | – | | Stretched P–H bond for group masked after adsorption |
| 5 | 1653.13 | 1635.04 | +18.09 | C = N stretching |
| 6 | 1606.66 | – | | C-N group masked |
| 7 | 1418.95 | – | | N = N group masked |
| 8 | 1366.02 | – | | Weak, deformed C-H group masked |
| 9 | 1202.78 | 1231.54 | −28.76 | C–H deformation |
| 10 | 1162.63 | 1156.58 | 6.05 | C–H deformation |
| 11 | 1095.17 | – | | Group containing C–H deformation |
| 12 | 1009.67 | 1014.25 | −4.58 | C–O stretching |
| 13 | 693.69 | 703.40 | −9.71 | C–S stretching vibration |

## References


Aikpokpodion PE, Ipinmoroti RR, Omotoso SM (2010) Biosorption of Ni(II) from aqueous solutions using waste tea (*Camella cinencis*) materials. Am Eurasian J Toxicol Sci 2:72–82

Al-Rub FA, El-Naas MH, Benyahia F, Ashour I (2004) Biosorption of Nickel on blank alginate beads, free and immobilized algal cells. Process Biochem 39:1767–1773

Aravind J, Lenin C, Nancyflavia C et al (2015) Response surface methodology optimization of Nickel (II) removal using pigeon pea pod biosorbent. Int J Environ Sci Technol 12:105–114

Cayllahua JEB, Torem ML (2011) Biosorptive flotation of nickel and aluminum ions from aqueous solution. Desalination 279:195–200

Cempel M, Nikel G (2006) Nickel: a review of its sources and environmental toxicology. Pol J Environ Stud 15(3):375–382

Fu F, Wang Q (2011) Removal of heavy metals from waste water: a review. J Environ Manage 92:407–418

Fertu DIT, Gavrilescu M (2012) Application of natural zeolites as sorbent in the cleanup of aqueous streams. Environ Eng Manage J 11(4):867–878

Flores-Garnica JG, Morales-Barrera L, Pineda-Camacho G, Cristiani-Urbina E (2013) Biosorption of Ni(II) from aqueous solutions by *Litchi chinensis* seeds. Bioresour Technol 136:635–643

Gohulavani G, Andal NM (2013) Removal of Nickel from aqueous solution using *Azaridachta indica* seed shell powder as adsorbent. Res Rev J Chem 2(1):78–84

Kalavathy HM, Regupathi I, Pillai MG, Miranda LR (2009) Modelling, analysis and optimization of adsorption parameters for $H_3PO_4$ activated rubber wood sawdust using response surface methodology (RSM). Colloids Surf B 70(1):35–45

Kalavathy HM, Rose ML (2010) *Moringa oleifera*—a solid phase extractant for the removal of Copper, Nickel and Zinc from aqueous solutions. Chem Eng J 158:188–199

Malkoc E, Nuhoglu Y (2005) Investigations of Nickel(II) removal from aqueous solutions using tea factory waste. J Haz Mat 127:120–128
Murugesan S, Rajiv S, Thanapalan M (2009) Optimization of process variables for a biosorption of nickel (II) using response surface method. Korean J Chem Eng 26(2):364–370
Nuhgolu Y, Malkoc E (2009) Thermodynamic and kinetic studies for environmentally friendly Ni(II) biosorption using pomace of olive oil factory. Bioresour Technol 100:2375–2380
Parmer M, Thakur LS (2013) Heavy Metal Cu, Ni and Zn: toxicity, health hazard and their removal techniques by low cost adsorbents: a short overview. Int J Plant Anim Environ Sci 3:143–157
Shakirullah M, Ahmad I, Shah S (2006) Sorption studies of nickel ions onto sawdust of *Dalbergia sissoo*. J Chin Chem Soc 53(5):1045–1052
Shrestha RM, Varga M, Varga I, Yadav AP, Pokharel BP, Pradhananga RR (2014) Removal of Ni(II) from aqueous solutions by adsorption onto activated carbon prepared from Lapsi (*Choerospondias axillaris)* seed stone. J Inst Eng 9(1):166–174
Thamilarasu P, Sivakumar P, Karunakaran K (2011) Removal of Ni(II) from aqueous solutions by adsorption onto *Cajanus cajan* L Milsp seed shell activated carbon. Indian J Chem Technol 18:414–420
Vijayaragavan K, Lee WM, Yun YS (2008) Evaluvation of fermentation waste (*Corneybacterium glutamicum*) as a biosorbent for treatment of Ni bearing solutions. Biochem Eng J 41:228–233
Yu Z, Zhou Q (2009) Growth responses and cadmium accumulation of *Mirabilis jalapa* under interaction between cadmium and phosphorus. J Haz Mat 167(1–3):38–43

# Bioremediation of Heavy Metals and Toxic Chemicals from Muttukadu Lake, Chennai by Biosurfactant and Biomass Treatment Strategies

**C. Elizabeth Rani, V. Balaji Ayyadurai, and K. K. Kavitha**

**Abstract** Bioremediation is the promising strategy of bringing back the environment to the natural stage by means of biological treatment. Biosurfactants are microbial products which possess many applications over traditional methods. Hence biosurfactants can be widely used for bioremediation process. In this study, a novel biosurfactant- rhamnolopid was applied for bioremediation of heavy metals, environmental toxic pollution removal and stain decolorizer. For this, water sample was collected from Muttukadu lake to check the amount of heavy metals present in it by Atomic absorption spectroscopy (AAS) and also to screen the toxic functional groups in the water sample by FTIR analysis. In additional to this stain removal application of rhamnolipid type of biosurfactant was also revealed. *Pseudomonas aeroginosa* was isolated from the metal contaminated automobile industry at Kandigai. Selected strain was subjected to phenotypic characterization and following that biosurfactant was produced by acid precipitation method. The produced biosurfactant was subjected to preliminary assays and FTIR for structural confirmation of rhamnolopid type of biosurfactnt. Before the treatment of heavy metals and after the treatment of heavy metals and toxic functional groups were examined by Atomic absorption spectroscopy and FTIR respectively. After the treatment of biosurfactant the amount of Chromium and Lead was decreased. Toxic compounds like Chlorofluro alkanes and alkanes were completely removed. To this treatment, a comparitive study was carried out with respect to biosurfactant and biomass treatment strategy. Stain removal application was evaluated using commercially available Azo dye. Stain removal was noticed by physical examination after the treatment of biosurfactant to the stained cloth pieces.

**Keywords** Bioremediation · Biosurfactant · Heavy metals · Chlorofluro alkane · Atomic absorption spectroscopy · FTIR · Azo dye

---

C. Elizabeth Rani (✉) · V. Balaji Ayyadurai
Department of Biotechnology, Hindustan College of Arts and Science, Chennai 603103, Tamil Nadu, India
e-mail: elizabeth.juneius7@gmail.com

K. K. Kavitha
Department of Environmental and Herbal Science, Tamil University, Thanjavur 613010, India



M. Prashanthi et al. (eds.), *Bioremediation and Green Technologies*, Environmental Science and Engineering, https://doi.org/10.1007/978-3-030-64122-1_6

## 1 Introduction

Workers of welding industries, tanneries, factories manufacturing chromate containing paints are exposed to hexavalent chromium that increases the risk of developing serious adverse health effects was reported by Karanth et al. (1999). Heavy metals are considered as major type of environmental pollutant. Similarly, Dhanya et al. (2005) reported that Thirupur district of Tamil Nadu was contaminated with heavy metals and the experimental studies revealed that contamination was due to the release of effluent from textile industries.

Some industrial effluent may also contains toxic chemical compounds which will affect the entire aquatic system. Those compounds may aggregate in waste water, river, lakes, etc. Synthetic surfactants have applications in the processing of organic compounds, enhanced oil recovery, etc. Surfactant or surfact-active compounds are amphiphilic in nature which decreases the surface and surface interfacial tension. Chemically derived surfactant posses wide range of environmental pollutions was clearly explained by (Swetal and Vaidehi 2018).

Many types of biosurfactants can be produced by a variety of microbes. The biosurfactant produced by bacteria are the major classes which includes lipopeptides, lipoproteins, glycolipids, surfactins, neutral lipids, phospholipids, etc., (Rosenberg and Ron 1999). The microorganisms which produce biosurfactant are considered as a nature gift because of the following properties; performance under extreme conditions, environmental friendly, large scale production, etc., (Banat et al. 2000). Mostly bacteria and yeasts are well known to produce a wide range of surface active agents (Ebrahimi and Tashi 2012).*Pseudomonas aeruginosa* is a Gram-negative, rod shaped bacterium. It is commonly found in flora of skin, water, air, soil, etc. Recent studies showing the applications of this strain in various bioremediation processes. Yassin et al. (2016), reported the importance of glycerol as the sole carbon source for the production of rhamnolipid type of biosurfactant by *Pseudomonas aeruginosa*. Zolgharnein (2010) experimental results showing that *Pseudomonas spp.*, are potent candidate for the absorption of heavy metals. In addition, the maximum absorbance of heavy metal was reported for zinc, chromium, lead and cadmium.

The objective of this study was the production of rhamnolipid type of biosurfactant and to apply in various fields as the bioremedial strategy and also to compare the treatment process with biomass of P.aeroginosa.

# 2 Methods

## 2.1 Sample Collection

Water sample was collected from Muttukadu lake, Chennai, for the estimation of heavy metals and its concentrations and also to detect the presence of toxic functional groups. The water sample was collected in a large sterile screw cap bottle. Estimation of heavy metals in untreated sample was carried out by using Atomic absorption spectroscopy and the chemical groups were determined by FTIR analysis respectively.

## 2.2 Isolation of Indigenous Bacteria from Metal/Iron Contaminated Water Sample

For the isolation of indigenous microbe, the water sample was collected from metal contaminated effluent at Kandigai, Chennai. Nutrient broth was prepared in sterile conical flask and sterilized by autoclaving at 121 °C for 20 min. Meanwhile the water sample were serially diluted by tenfold dilution. Sterile petriplates were taken and marked as $10^{-2}$, $10^{-3}$, $10^{-4}$, $10^{-5}$. 1 ml of the diluted samples were added to the sterile petriplates respectively and molten nutrient agar was poured then allowed for solidification. Then the plates were incubated at 37 °C for 24–48 h. Morphologically differing various colonies were observed and further pure cultured.

## 2.3 Identification of Isolated Bacteria

Based on Bergey's manual of systematic Bacteriology the unknown candidate was identified. The unknown bacteria was identified by morphological characteristics, Gram's staining followed by biochemical characterization.

## 2.4 Production of Biosurfactant by Selected Strain

The selected strain was used to produce biosurfactant by submerged fermentation in a stirred tank reactor and purified by acid precipitation method. 100 ml of isolation medium for *Pseudomonas spp.*, (cetrimide broth) was prepared, additionally 2% glycerol was supplied as the carbon source and sterilized by autoclaving at 121 °C for 20 min, 15 lbs. Loop full of selected culture was inoculated and incubated for 3–4 days at 37 °C in a shaking incubator and 130 rpm was maintained.

### 2.5 Separation of Biosurfactant by Solvent Extraction Method

The entire broth was centrifuged at 10,000 rpm for 10 min to remove the cells. The supernatant was transferred into a sterile screw cap bottle to which acid precipitation method was employed by reducing the pH to 2.0 by slowly adding the conc. Hcl. After the acidification of the cell free extract, it was allowed to incubate overnight in fridge. After the incubation, the content was centrifuged at 10,000 rpm for 10 min. The surfactant in the broth was extracted in chloroform: methanol (2:1) solvent system. The cell free broth was transferred into a sterile separating funnel following that equal volume of above mentioned solvent mixture was added. The contents were vigorously agitated for proper mixing. Then it was kept undisturbed for overnight. Next day all the different layers were collected in different petriplates and allowed for complete evaporation.

## 3 Screening of Biosurfactant

### 3.1 Haemolysis Test

The first screening method for the biosurfactant synthesizing bacteria was the haemolysis test (Carrillo et al 1996). The selected strain was streaked on to blood agar medium and the plates were incubated for 24–48 h at room temperature. Physical examination was done to check the clear zone formation around the streaked areas.

### 3.2 Oil Spreading Test

Oil spreading assay was conducted by the method described by (Morikawa et al. 1993). In brief, 20 ml of distilled water was poured onto a clean petri plate followed by the addition of 20 μl of crude oil to the surface of water. To this, 10 μl of cell free culture broth was overlayed to the crude oil surface. If biosurfactant is present in the cell free extract, the oil will be displaced and a clear zone and diameter of this indicates the surfactant activity. The diameter of the clearing zone of the oil surface indicating the activity of biosurfactant. A negative control was maintained by adding a drop of distilled water to the oil surface (Morikawa et al. 1993).

### 3.3 Hydrocarbon Overlay Agar Method

The hydrocarbon overlay agar test was performed to determine the ability of bacteria to utilize hydrocarbon as the carbon source. Nutrient medium was poured onto sterile petriplates and allowed for solidification. 50 μl of Crude oil was spread on the nutrient agar medium. Spread plate technique was used to overlay the selected organism to the top of crude oil layer. The plates were incubated at 37 °C for 24–72 h. A colony surrounded by emulsified halo was considered as positive for the biosurfactant production.

### 3.4 Emulsification Index (E24)

Emulsification activity of the isolates were determined by emulsification index (E24) for crude oil. 1.5 ml of crude oil was added to a clean test tube, to this equal volume of cell free broth was added and vortexed at high speed for about 2–3 min. The content was allowed to stand for 24 h to examine the emulsification activity. The percentage of emulsification activity was calculated by the following equation: (Sarubbo et al. 2007): E24 = Height of emulsion formed × 100/total height of the solution.

### 3.5 Drop Collapse Test

To determine the biosurfactant activity, 1 drop of crude oil was placed onto the clean cavity slide followed by 1 drop of cell free extract. After 1 min, the shape of drop becomes flat after adding supernatant indicating ability of biosurfactant production. A negative control was maintained by adding 1 drop of water to the crude oil sample.

### 3.6 Thin Layer Chromatography (TLC)

Screening of biosurfactant activity was evaluated by Thin layer chromatography technique. Silica gel plates were prepared, samples were spotted by using Pasteur pipette. It was allowed to run in the solvent system consisting of chloroform: methanol: water in (90:10:0.05) ratio respectively. Anthrone reagent was sprayed to detect the presence of glycolipid type of biosurfactant as yellow spot.

## 4 Characterization of Biosurfactant by FTIR

The chemical nature of the extracted biosurfactant was determined by Fourier transform infrared spectroscopy (FTIR). A pinch of crystalline biosurfactant was mixed with sodium phosphate buffer (neutral pH). This sample was then analyzed by spectrometry. The spectrometric analysis was carried out by LCQ quadrupole iontrap mass spectrometer utilizing electrospray ionization. The sample was into the mass spectrometer at a flow rate of 10 μl/minute and the auxiliary gas flow, nitrogen were maintained at 5 and 50 ml/minute respectively. The spray voltage was set at 5kv and heated capillary temperature was 250 °C. Built- in plotter were used to obtain the IR absorption spectra. IR absorption spectra were collected over the of 550–4500 $cm^{-1}$ with a resolution of 4 $cm^{-1}$. All spectra were obtained from 180 scans. The spectrum was studied to reveal the chemical nature of the biosurfactant.

## 5 Bioremediation of Heavy Metals by Biosurfactant in Comparison with Biomass as Treatment Strategies

10 ml of the collected water sample was taken to determine the different types of heavy metals present in it and also to determine the concentrations of the heavy metals.

**Sample-1** raw water sample (untreated).

**Sample-2** (treated with biosurfactant). 100 micro litre of biosurfactant was added to 10 ml of the water sample.

**Sample-3** (biomass treated). 100 micro litre of pure culture of *Pseudomonas aeroginosa* was added to 10 ml of the water sample.

After the addition of biosurfactant and biomass individually, the samples were mixed thoroughly for 2–3 min by vortex. Then it was incubated at room temperature for 30 min. After 30 min all the 3 sample were centrifuged at 10,000 rpm for 5 min and the supernatant was separated for further investigations. The estimation of heavy metals was done by Atomic absorption spectroscopy (AAS). Before and after the treatment of heavy metals were studied. Biosurfactant treatment strategy was also compared with the treatment with biomass.

## 6 Determination of Toxic Chemical Compounds and Treatment by Biosurfactant and Biomass

In this study FTIR was used to determine the toxic functional groups present in the water sample. Before the treatment and after the treatment with biosurfactant and biomass was studied. Previously used untreated, treated with biosurfactant and treated with biomass samples were taken to FTIR analysis. No changes were made in the previously treated strategy and the same samples were taken for FTIR to determine the presence of functional groups in untreated and the absence of functional groups in the treated samples with biosurfactant and biomass respectively.

## 7 Biosurfactant as Stain Decolorizer Against Azo Dye (Red)

To check the ability of the biosurfactant as a vital ingredient in detergent, it was studied by artificially stained the white cloth pieces with commercially available Azo dye (red). For this study, initially 0.05% of 3 drops of synthetic azo dye was used to stain on white cloth pieces to reveal the biosurfactant potential on stain removal or decolorizer. This was done by adding Equal volume of produced biosurfactant in the stained cloth and control was maintained. The stained cloths were kept undisturbed hardly for about 30 min and changes were observed visibly.

## 8 Results

### *8.1 Isolation of Indigenous Bacteria*

Pour plate technique of the collected water sample gave various colonies with distinct colony morphologies. The morphology was studied and based on its appearance the selected organism was streaked on nutrient agar plates and allowed to incubate for overnight at 37 °C in and hence pure culture of selected strain was achieved.

### *8.2 Identification of Isolated Bacteria*

**Phenotypic characterization:**

The Gram stained cells were observed under the light microscope at 100 × using oil immersion. The colour and shape of the cells were studied. Gram negative rod shaped cells were observed as pink colour (Table 1). Other phenotypic methods as per Bergy's manuel was carried out and the results were tabulated.

From the above observation and results the unknown organism was identified as ***Pseudomonas aeruginosa***.

**Table 1** Biochemical characterization of *Pseudomonas sp.*: Phenotypic characterization of unknown bacteria by biochemical assays

| S. No. | Name of the test | Observation | Result |
|---|---|---|---|
| 1 | Gram's staining | Pink rods | Positive |
| 2 | Capsule | No capsules | Negative |
| 3 | Motility test | Tumbling movement | Positive |
| 4 | Indole test | No red ring formation | Negative |
| 5 | Methyl red test | Yellow colour was observed | Negative |
| 7 | Voges proskauer test | No colour changes | Negative |
| 8 | Citrate utilization test | Persian blue colour developed | Positive |
| 9 | Catalase test | Bubble formation was observed | Positive |
| 9 | Oxidase test | Disc turned to purple colour | Positive |
| 10 | Urease test | Pink colour was developed | Positive |

## 8.3 Production and Separation of Biosurfactant

Biosurfactant production from the selected strain was achieved and the extraction of crude biosurfactant was done by using solvent extraction method. The solvent mixture of chloroform: methanol (2:1) ratio had served as the appropriate proportion for the extraction of biosurfactant. (Banat et al. 2010) reported that, chloroform and methanol solvent mixture is a preferable procedure for the extraction of biosurfactant. The resultant crude product was observed in the form of crystalline honey like viscous material in the petri plates. Similarly (Dubey and Juwarkar 2001), reported that the *Pseudomonas aeruginosa* produced biosurfactant was in the form of crystals. Figure 1 Showing the presence of biosurfactant in crystalline form.

## 8.4 Screening of Biosurfactant

### 8.4.1 Haemolysis Test

The haemolytic activity of the selected strain was observed. Clear zones were thoroughly examined around the streaked areas on blood agar medium.

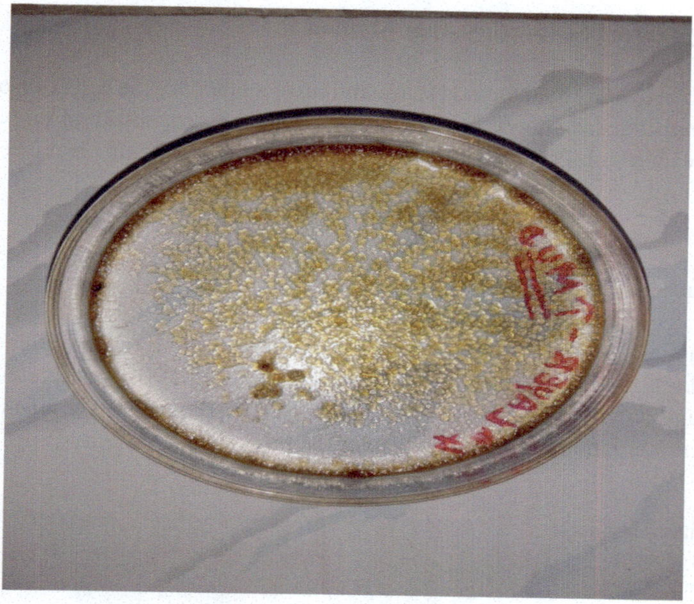

**Fig. 1** Biosurfactant produced in the form of crystalline

### 8.4.2 Oil Spreading Test

The oil spreading assay was successfully conducted for the crude biosurfactant obtained from solvent extraction to reveal the biosurfactant activity. The diameter of the oil displacement was noted as 4 cm diameter. The zone of oil displacement correlates with the biosurfactant activity (Swetal and Vaidehi 2018) (Fig. 2).

**Fig. 2** Demonstration of oil spreading assay

### 8.4.3 Hydrocarbon Overlay Agar Method

Hydrocarbon overlay agar method is a quantitative method for the detection of hydrocarbon emulsifiers. It shows the hydrocarbon degrading capability of on organism. Clear zones were observed and the test organism showed 0.8 mm clearance zone of diameter.

### 8.4.4 Emulsification Index (E24)

For this assay, crude oil was used as hydrophobic substrate to study the emulsification activity of the crude sample. Emulsification index >30% considered as high bisurfactant activity was reported by (Nayarisseri et al. 2018). The results of the extracted crude sample showing 66.6% for the biosurfactant activity (Fig. 3).

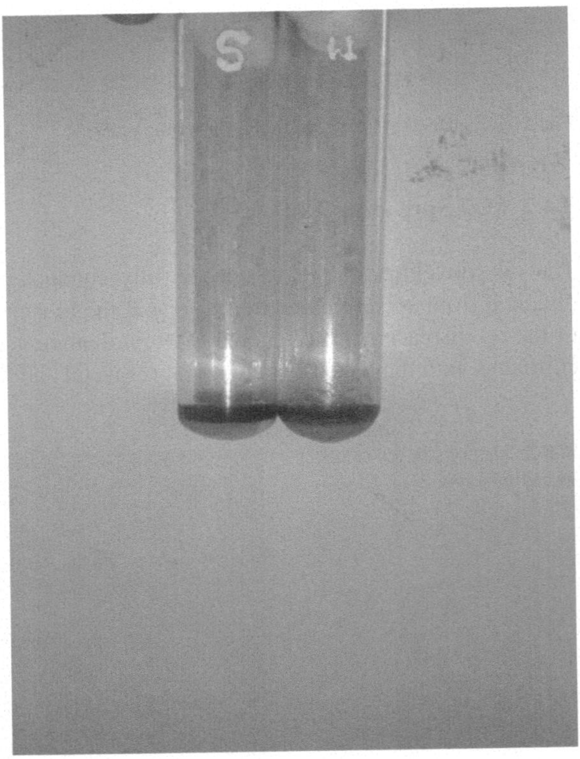


**Fig. 3** Emulsification test for the screening of biosurfactant

### 8.4.5 Drop Collapse Test

The changes in the shape of crude oil sample was noted when added with extracted crude biosurfactant in a cavity slide. This drop collapse may due to the force between hydrophobic substrate and the liquid. The shape changes from spherical to flattened indicates the presence of biosurfacatnt in the crude extract.

### 8.4.6 Thin Layer Chromatography (TLC)

Yellow colour spot was appeared when sprayed with anthrone reagent. It indicating the interaction between the anthrone reagent with the rhamnose sugar present in the crude sample.

## *8.5 Characterization of Biosurfactant by FTIR*

The crude biosurfactant sample showing a strong and broad peak at 3279 $cm^{-1}$ indicating the O–H stretching vibrations of (free hydroxyl groups of rhamnose ring) in the crude sample.Strong bonding of C=O stretch was noted at absorption of 1638 $cm^{-1}$ representing ester bonds and carboxylic acid function in the sample was reported by (Rath et al. 2016). Another maximum adsorption was noted on peak value of 2115 $cm^{-1}$ showing C–C stretching of alkyne compounds. Medium intensity of peak around 1540 $cm^{-1}$ determines C=O group in the sample was reported by (Mishra and Trivedi 2019).The frequency around 1452 and 1397 $cm^{-1}$ indicating the presence of aliphatic bonds $CH_3$, $CH_2$, C–H stretching was observed. Similarly the peaks around 1115 and 1016 $cm^{-1}$ representing C–O–C stretching in the rhamnose was reported by (Rahman et al. 2002). These characteristic adsorption bands together demonstrate the chemical nature of the sample was rhamnolipid as glycolipid type of biosurfactant (Fig. 4).

## 9 Bioremediation of Heavy Metals by Biosurfactant in Comparison with Biomass as Treatment Strategies

The untreated waste water sample was subjected to detect selectively four types of heavy metals like Chromium, Lead, Zinc and Copper. Among these only the chromium and lead was detected by Atomic absorption spectroscopy. Hence the Muttukadu lake Chennai, water sample does not contain zinc and copper or they may be present in trace amount which cannot be detected by Atomic absorption spectroscopy. All the three following samples were tested for the presence of above mentioned heavy metals.

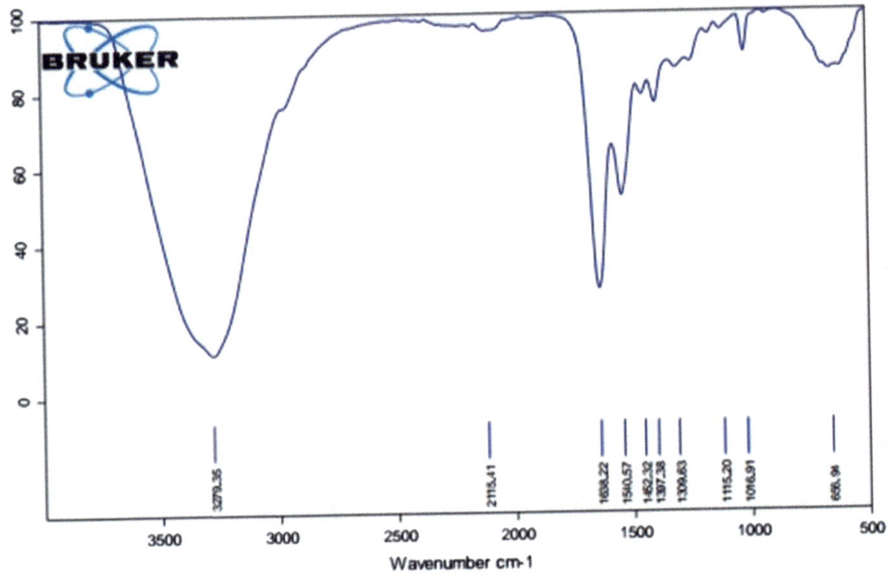


**Fig. 4** FTIR results of produced biosurfactant

**Sample-1**: (The untreated waste water) shows the presence of Chromium and Lead in the concentrations of 4.71 ppm and 4.55 ppm respectively. As the result both the heavy metals are almost detected at same concentrations.

**Sample-2**: (100 µl of biosurfactant treated waste water) after the treatment the concentration of chromium was reduced from 4.91 ppm to 3.55 ppm. Also the concentration of lead after the treatment with biosurfactant was reduced from 4.55 ppm to 2.28 ppm respectively (Fig. 5).

**Fig. 5** Diagram showing the results of biosurfactant as bioremediation of heavy metals

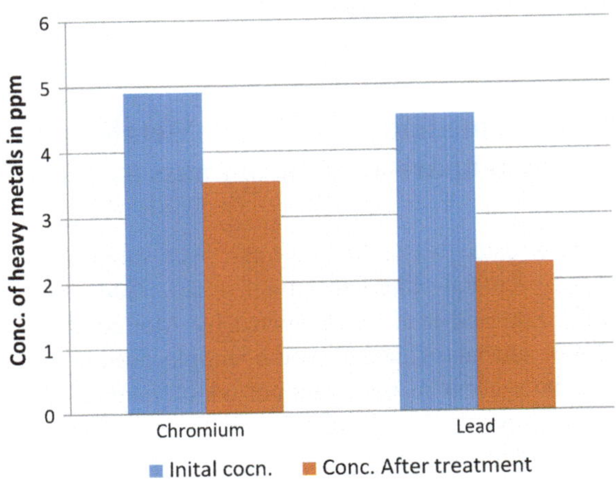

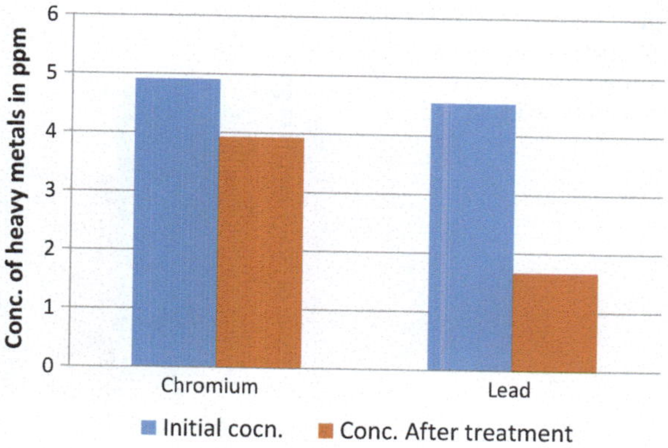


**Fig. 6** Bar diagram shows the bioremediation of removing heavy metals by biomass as treatment strategy

**Sample-3**: (100 μl of biomass treated waste water) before and after the treatment of biomass with respect to the heavy metals was noted. Chromium was reduced from 4.71 ppm to 3.93 ppm and lead was reduced from 4.55 ppm to 1.68 ppm respectively (Fig. 6).

## 10 Comparison of Biosurfactant and Biomass Treatment Strategies

In this study, both the treatment methods showed significant results and also has an application in the removal of heavy metals was proved. Moreover Biosurfactant treatment method was reliable with respect to Chromium type of heavy metal. But maximum removal of Lead type of heavy metal was noticed by biomass treatment when compared with biosurfactant as treatment strategy.

Hence, a specific heavy type of metal may require a specific treatment methodology for the maximum removal of heavy metals was established by this present study (Fig. 7).

### *10.1 Statistical Data of Heavy Metal (Cr) Removal in Percentage*

The initial concentration of Cr was detected as 4.91 ppm but, after with biosurfactant treatment (0.1 ml) shows 27.6% removal of chromium. So it means 0.368 ml is enough to remove the total concentration (4.91 ppm) of Cr. Similarly 0.1 ml of biomass achieved 20% removal of Cr. It shows that exactly 0.5 ml is required for the complete removal (4.91 ppm) of chromium. This sufficient data will be useful for the removal of Cr in different concentrations by treating the exact volume of the treatment

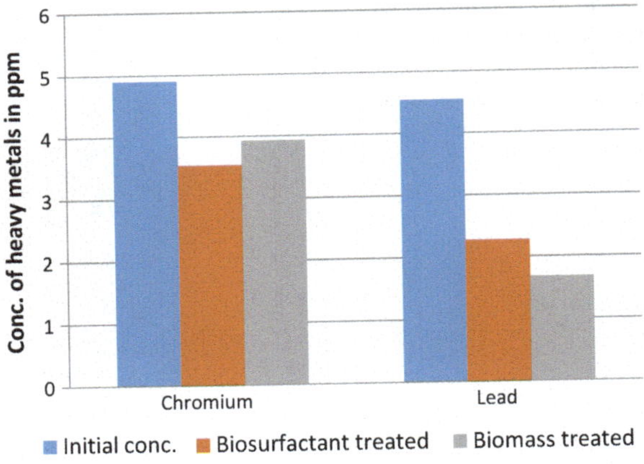


**Fig. 7** Diagram showing the comparison of both the treatment strategies of chromium and lead

strategy. The below mentioned tabular column exhibits the complete information with respect to chromium heavy metal. All the above mentioned before and after the treatment of heavy metal was determined by Atomic absorption spectroscopy (Table 2).

The raw (untreated water sample) showed the presence of lead type of heavy metal and the concentration was detected as 4.55 ppm. After the treatment with respect to biosurfactant the concentration was decreased to 2.28 and 1.68 ppm when treatment with biomass as another comparative treatment methodology. When the biosurfactant was applied it showed 49.8% of lead removal. Maximum of 64% of Pb removal was achieved by biomass as treatment strategy. These values were obtained by Atomic absorption spectroscopy (Table 3).

The above data shows that 0.2 ml of biosurfactant is required for the removal of total Pb (4.55 ppm). Also that 0.162 ml is enough for compete removal of lead heavy metal when biomass is applied as treatment strategy. This experimental outcome reveals that exact quantity of biosurfactant required for certain concentration of heavy metals. So that future remediation of heavy metals can be planned accordingly without any additional usage of biosurfactant.

**Table 2** Tabular column showing the results of heavy metal removal in percentage

| S. No. | Treatment method | Quantity (ml) | Initial conc. of Cr (ppm) | Conc. of Cr (after treatment) (ppm) | Removal rate (%) | Exact quantity of treatment required for complete removal (ml) |
|---|---|---|---|---|---|---|
| 1 | Biosurfactant (BS) | 0.1 | 4.91 | 3.55 | 27.6 | 0.368 |
| 2 | Biomass(BM) | 0.1 | 4.91 | 3.93 | 20 | 0.5 |

**Table 3** Lead type of heavy metal removal data in percentage for both the biosurfactant and bioass treatment methods

| S. No. | Treatment method | Quantity (ml) | Initial conc. of Pb (ppm) | Conc. of Pb (after treatment) (ppm) | Removal rate (%) | Exact quantity of treatment required for complete removal (ml) |
|---|---|---|---|---|---|---|
| 1 | Biosurfactant (BS) | 0.1 | 4.55 | 2.28 | 49.8 | 0.2 |
| 2 | Biomass (BM) | 0.1 | 4.55 | 1.68 | 64 | 0.162 |

**Table 4** Table showing the presence of functional groups detected by FTIR

| S. No | Wave number $cm^{-1}$ | Functional groups | Possibility of compound class |
|---|---|---|---|
| 1 | 3327 | O–H stretching | Alcohol |
| 2 | 2214 | C–N stretching | Nitriles |
| 3 | 2117 | C–C stretching | alkyne |
| 4 | 1637 | C–C stretching | aromatic vibration |
| 5 | 1370 | $CH_3$–CH bend | Alkanes |
| 6 | 1283 | C–O–C stretch | Ethers |
| 7 | 1045 | C–O stretch | Alcohol |

## *10.2 Determination of Toxic Functional Groups in the Raw Water Sample by FTIR Analysis*

The raw sample was analyzed for the presence of any toxic compounds by FTIR was successfully done and the following are the detected compounds.

The absorption at 3327 $cm^{-1}$ corresponding to O–H stretch indicating the presence of alcoholic compounds in the waste water sample. C–N stretch of Nitrile groups were detected at 2214 $cm^{-1}$ wave number. Peak value at 2117 $cm^{-1}$ resembles C–C stretch of alkynes and 1637 $cm^{-1}$ indicates C–C stretch which represents aromatic skeleton vibration was reported by (Ladwani et al. 2012). The stretch 1370 $cm^{-1}$ attributes to $CH_3$–CH bend. Ethers were observed at 1238 $cm^{-1}$ by C–O–C stretch and peak at 1045 cm $^{-1}$ indicating C–O stretch of alcohol.

In this present study the FTIR results showing the presence of alcohols, alkynes, nitriles, etc. are in the form of PAH was found in Muttukadu lake, Chennai. Due to the presence of alcoholic and alkyne functional groups may be responsible for the color of the waste water as the organic compounds were detected in multi-complex in nature. Because of this it becomes more toxic to the aquatic system and humans. Hence a specific treatment strategy is required for the removal of above mentioned contaminants as it was reported by Ladwani et al. (2012) (Fig. 8).

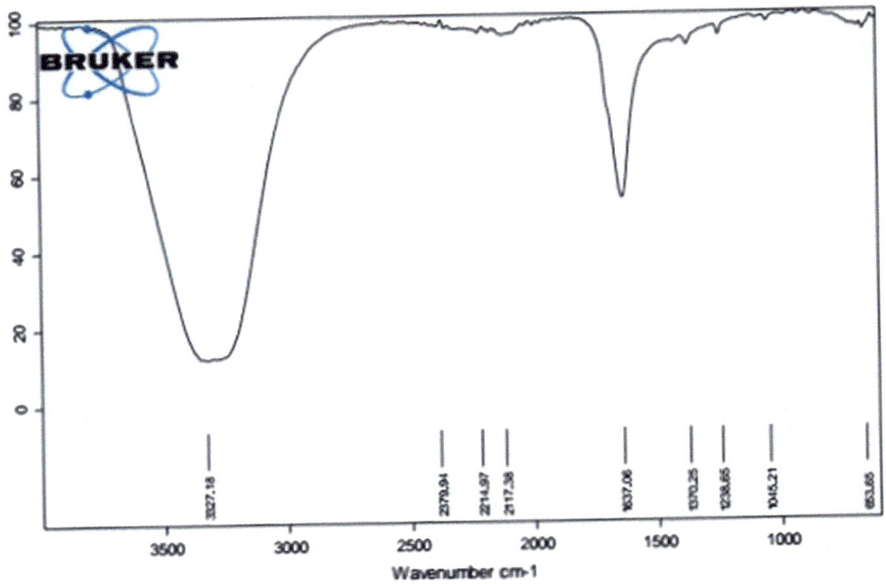


**Fig. 8** FTIR results showing the presence of toxic functional groups in untreated water sample

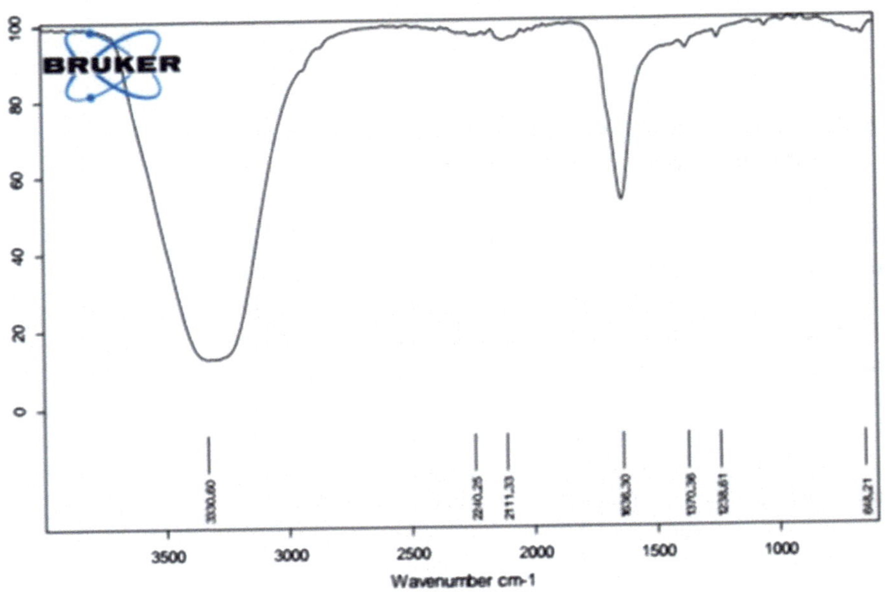


**Fig. 9** Diagram represents the FTIR spectrum of biosurfactant treated water sample

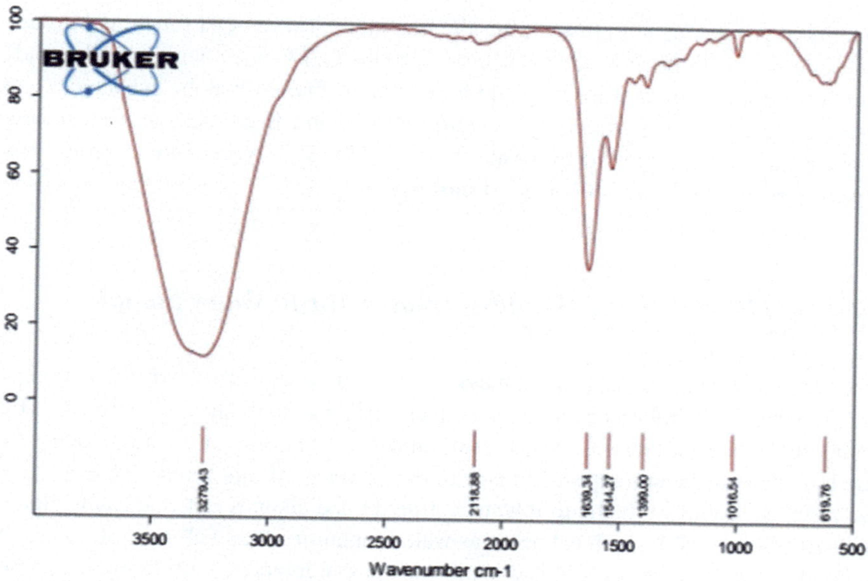


**Fig. 10** FTIR results of biomass treated waste water sample

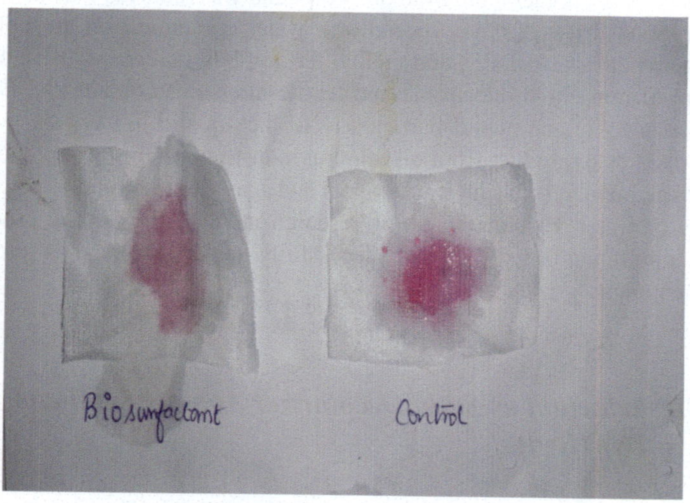


**Fig. 11** Picture showing the efficacy of biosurfacatnt as stain removal of azo dye

### *10.3 FTIR Results of Biosurfactant Treated Waste Water Sample*

The absorption at 1045 $cm^{-1}$ showed the presence of alcoholic functional group in untreated sample which was completely removed by biosurfactant treatment as this functional group was not present after the treatment. These –OH functional groups may interacts with other aromatic benzene groups and forms phenolic compounds which resulting in greater toxicity and long lasting. These phenolic substances can also be a carcinogen particularly affecting RBC's and liver damage even at low concentrations was reported by (Anku et al. 2017). This type of toxic group was removed by the treatment of 0.1 ml of biosurfactant.

## 10.4 FTIR Results of Biomass Treated Waste Water Sample

The treatment with respect to biomass showed the removal of two toxic groups are Aknenes, Alcohol and Ethers respectively. Alkenes may attaches with catalytic compounds like palladium, nickel, platinum, etc., resulting in water toxicity by catalytic dehydrogenation process. Due to the presence of this compound can alters the taste and odour of the natural water. Triple boned alkenes can also easily attach with alkyenes (C=C) which is known as water contaminant and also becomes tough to breakdown. In addition to this alkenes may easily interacts with freely available halogen compounds and proceed water contamination. This type of unstable, strong triple bond compound was eliminated by biomass (0.1 ml) as treatment strategy.

Simultaneously another class of toxic compound was removed is ethers. Usually ethers are found in two forms especially in water contamination are polybrominated diphenyl ethers (PBDES) and methyl tertiary butyl ether (MTBE). Typically the effluent of household, electronic and textile industries contain PBDES which is known as a toxic compound particularly to the aquatic life system. MTBE is widely derived as byproduct from oil refining industries. In this group, the volatile compound methanol reacts with water vapors and again settles in water which affects the quality of water. Existing scientific research indicates this cyclic process and termed as a pollutant or toxic group. Successfully this compound was removed by biomass treatment.

## 10.5 Biosurfactant as Stain Decolorizer Against Synthetic Azo Dye (Red)

To the clean white cloth pieces the azo dye stained + biosurfactant added possess positive results. When visibly observed the biosurfactant stained cloth piece showed almost 50% stain removal ability. It was concluded by comparing with control which was contained only the azo dye. Hence, another application of the produced biosurfactant was revealed by this study.

## 11 Conclusion

From this study, the types of heavy metals and toxic functional groups present in the Muttukadu Lake, Chennai was determined. It reveals that bioremediation using biosurfactant can be a suitable treatment strategy for Cr type of heavy metal meanwhile more Pb was eliminated when treated with biomass. The present study acts as an evidence to show that each heavy metal requires unique treatment strategy. On the other hand, biomass possessed good results by eliminating three toxic functional groups as mentioned above. In addition, rhamnolipid kind of biosurfactant produced by *P. aeruginoa* has wider applications including removal of the synthetic azo dye.

## References

Banat IM, Makkar RS, Cameotra SS (2000) Potential commercial applications of microbial surfactants. Appl Microbiol Biotechnol 53:495–508
Banat IM, Franzetti A, Gandolfi I, Bestetti G, Martinotti MG et al (2010) Microbial biosurfactants production, applications and future potential. Applied Microbiol Biotechnol 87:427
Carrillo PG, Mardaraz C, Pitta-Alvarez SI, Giulietti AM (1996) Isolation and selection of biosrfactant producing bacteria. World J Microbiol. Biotechnol 12:82–84
Dhanya G, Swetha S, Madhavan NK, Sukumaran RK, Pandey A (2008) Response surface methodology for the optimization of alpha amylase production by Bacillus amyloliquefaciens. Bioresour Technol 99:4597–4602
Dubey K, Juwarkar A (2001) Distillery and curd whey wastes as viable alternative sources for biosurfactant production. World J Microb Biot 17(1):61-69
Ebrahimi A, Tashi N (2012) Isolation of biosurfactant producing bacteria from poultry breast skin. Nat Pharm Prod 7(3):93–96
Karanth N, Deo P, Veenanadig N (1999) Microbial production of biosurfactants and their importance. Curr Sci 77(1):116–126
Ladwani KD, Ladwani KD, Manik VS, Ramteke DS (2012) Impact of domestic wastewater irrigation on soil properties and crop yield. Int J Sci Res Pub 2:1–7
Mishra A, Trivedi RK (2019) Synthesis and characterization of biosurfactant using waste from oil processing industry as substrate by Pseudomonas aeruginosa (MTCC 424). RJC 1011–1021
Morikawa M, Daido H, Takao T, Murata S, Shimonishi Y, Imanaka T (1993) A new lipopeptide biosurfactant produced by Arthrobacter sp. strain MIS38. J Bacteriol 175:6459–6466
Rahman KS, Banat IM, Thahira J, Thayumanavan T, Lakshmanaperumalsamy P (2002) Bioremediation of gasoline contaminated soil by a bacterial consortium amended with poultry litter, coir pith and rhamnolipid biosurfactant. Bioresour Technol Rep 81:25–32
Rath K, Singh AB, Chandan S, Vatsala RS (2016) Isolation and characterization of a biosurfactant producing strain pseudomonas aeruginosa SMVIT 1 from oil contaminated soil. J Sci Ind Res 681–686
Rosenberg E, Ron EZ (1999) High- and low-molecular-mass microbial surfactants. Applied Microbiol Biotechnol 52:154–162
Sarubbo LA, Farias CBB, Campos-Takaki GM (2007) Co-Utilization of canola oil and glucose on the production of a surfactant by Candida lipolytica. Curr Microbiol 54:68–73
Swetal M, Vaidehi C (2018) Production of Biosurfactant (Rhamnolipid) and its efficacy to remove Oil and Ink stains. Int J Sci Res Rev 7(3):631–637
Yassin AS, Eraqi WA, Ali AE, Amin MA (2016) Utilization of crude glycerol as a substrate for the production of Rhamnolipid by Pseudomonas aeruginosa. Biotech Res Int 1–10

# Degradation Efficacy of *Pinus radiata* Don Needle Leaf, Twig and Bark by Wood Degrading Fungi in Forest Ecosystem

**M. N. Abubacker and M. Prince**

**Abstract** The degradation of lignin, holocellulose and hot water soluble content of *Pinus radiata* needle leaf, twig and bark by wood degrading fungi was studied in virgin forest ecosystem of Doddabetta, Nilgiris for a period of 180 days. The study revealed that the maximum percentage of lignin degradation of needle leaf material with *Heterobasidium annosum* was 8.8, twig with the same fungus was 18.3, and the bark with 19.5. For holocellulose needle leaf degradation with a maximum of 10.3 for *Thelopora terrestris* fungus for twig with the same fungi shows 12.1 and for bark *Polyporus squamosus* with 16.9 of degradation. Hot water soluble content was maximum of 18.7% with *Colricia perennis* fungus for the needle leaf, for twig 10.5% for the same fungi and for bark with 9.0% for *Polyporus squamosus* for a period of 180 days. The elemental status of mixed samples of needle leaf, twig and bark constitute the forest litter of *P. radiata* were inoculated with *H. annosum*, *T. terrestris*, *P. squamosus* and *C. perennis* mycelia. The SEM–EDX analysis revealed the maximum availability of elements at 180 days of degradation when compared with initial stage.

**Keywords** Degradation · Holocellulose · Lignin · Hot water soluble content · SEM–EDX

## 1 Introduction

Forests represent approximately 27% of the world's land area and wood is the predominant, commercial product from forests. Wood is used for the manufacture of wood-based products. Logging operations in forests usually generate abundant amounts of waste such as residual wood, branches, twigs, leaves and bark. The

M. N. Abubacker
PG and Research Department of Biotechnology, National College, Tiruchirappalli 620001, Tamil Nadu, India

M. Prince (✉)
Department of Biology, Sarala Birla Academy, Bengaluru 560083, Karnataka, India
e-mail: princem@saralabirlaacademy.org



M. Prashanthi et al. (eds.), *Bioremediation and Green Technologies*, Environmental Science and Engineering, https://doi.org/10.1007/978-3-030-64122-1_7

waste amounts for more than 60% of the total biomass (Kuhad et al. 1997). Forest is the suitable-substratum where the micro and macro fungi not only live but also to degrade the leaf litter fallen from the trees and supply most valuable nutrients back to the forest soil. A large amount of plant waste (leaf litter) is being continuously accumulated on the forest soil. A part of the waste gets into the soil and an intensive degradation process starts. Intensity of this decay process depends on different environmental factors such as: species of plant, sort of soil, moisture, temperature and microbial associations. In order to increase the efficiency of these metabolic processes, the scientific society is making strong efforts in this area. Moreover, the aim of the latest researches are oriented to intensify plant remnants decay and modify metabolites enriching the soil by useful biologically active substances and finding materials able to enrich forage and food by important biologically valuable additions (Kelley 1992; Varnaite 2001; Abubacker and Prince 2013, 2015).

## 1.1 Forest Biomass

The composition of lignocellulose materials forms the wood wall structure. The major component of lignocellulose material is cellulose, along with lignin and hemicellulose. Cellulose and hemicellulose are macromolecules formed from different sugars; whereas lignin is an aromatic polymer synthesized from phenylpropanoid precursors. The composition and percentages of these polymers vary from one plant species to another. Moreover, the composition within a single plant varies with age, stage of growth, and other conditions (Jeffries 1994; Krishna and Mohan 2017).

Hemicellulose is a complex carbohydrate polymer and makes up 25–30% of total dry weight of wood. It is a polysaccharide with a lower molecular weight than cellulose. It consists of D-xylose, D-mannose, D-galactose, D-glucose, L-arabinose, 4-O-methyl-glucuronic, D-galacturonic and D-glucuronic acids. Sugars are linked together by β-1,4- and occasionally β-1,3-glycosidic bonds. The principal component of hardwood hemicellulose is glucuronoxylan whereas glucomannan is predominant in softwood. Structures of hemicelluloses are described in several reviews the main difference with cellulose is that hemicellulose has branches with short lateral chains consisting of different sugars. In contrast to cellulose, they are easily hydrolyzable polymers. They do not form aggregates, even when they are co-crystallized with cellulose chains (Jeffries 1994).

Lignin (along with cellulose) is the most abundant polymer in nature. It is present in the cellular cell wall, conferring structural support, impermeability, and resistance against microbial attack and oxidative stress. Structurally, lignin is an amorphous heteropolymer, non-water soluble and optically inactive; it consists of phenylpropane units joined together by different types of linkages (Jeffries 1994).

### 1.2 Decomposition

Litter decomposition is a fundamental process to ecosystem, responsible of carbon and nutrient cycling. The chemical composition and the structural characteristics of litter determines the "decomposability" resulting in fast or slow degradation and the fraction of recalcitrant residue that determine the buildup of soil organic matter (Berg and McClaugherty 2008; Baldrin and Lindahi 2017). Decomposition process of leaf litter is generally divided into early and late stages with different dominant organic chemical components limiting the decomposition rate. In the early stages of decomposition soluble components decay or are leached away very rapidly; cellulose and hemicellulose decompose faster in nutrient rich versus nutrient poor litter; thus the length of the early stage may range from a few months to more than a year. As litter decomposes, concentrations of lignin and nitrogen (N) increase (Berg et al. 1997). Litters with low initial nitrogen concentration may accumulate more nitrogen than those with a high level (Hobbie and Vitousek 2000; Pirjo Koivusarri Mysore et al. 2019). Site factors also contribute to nitrogen immobilization into decomposing litter and nitrogen accumulation in litter may be enhanced by soil N levels (Virzo De Santo et al. 1998; Sun et al. 2020). Fungi are the best-known microorganisms capable of degrading these three polymers. The substrates being insoluble and fungal degradation have to occur exocellularly, either in association with the outer cell envelope layer or extracellularly.

The present work is focused on the degradation efficacy of *Pinus radiata* Don, needle leaf, twig and bark by wood degrading fungi in forest ecosystem. The name of the Fungi are *Colricia perennis, Heterobasidium annosum, Polyporus squamosus, Thelophora terrestris.*

## 2 Materials and Methods

### 2.1 Study Area

The area of the study for this research work was identified based on the undisturbed virgin forest ecosystem of Doddabetta, Nilgiris, Tamil Nadu, India, which is 86 km from Coimbatore situated at 11°24′08.7° N and 76°44′12.2° E with a elevation of 8652 ft (2637 m) and 9 km from Ooty town (Satellite Images-Fig. 1). The area surrounding Doddabetta is mostly forest, Sholas cover the hollows of its slopes, which receives both South-west and North-east Monsoon rains. The average South-west Monsoon rain fall 900–1050 mm, the average North-east monsoon rain fall 350–450 mm. The present research work was carried out with the Gymnosperm forest *Pinus radiata* Don. of Doddabetta, Nilgiris, India (Fig. 1).

**Fig. 1** **a** Study area. **b** Pine forest *with Pinus radiata* Don

## 2.2 Decomposition Study by Litterbag Method

The leaf, twig and bark used for decomposition study was picked from the respective forest floor and placed in sterile polythene bags. The fungal fruit body was also collected in sterile polythene bags and brought to the laboratory and inoculated in the malt-extract agar medium and the growth mycelia was used for further experimental work. They were cultured in medium modified by Miura and Kudo (1970) contains glucose 0.1%, $KH_2PO_4$ 0.1%, $MgSO_4$, $7H_2O$ 0.02%, KCl 0.02%, $NaNO_3$ 0.2%, Yeast extract 0.02%. The decomposition study pertaining to holocellulose, lignin and hot water soluble content was studied with a litter bag methods (Crossley and Hoglund 1962) as well as Erlenmeyer flask method.

## 2.3 Determination of Lignin Content

TAPPI Standard method, T 222 OS-74 (2009) procedure was followed for the determination of lignin content. One gram of air-dried sample was weighed out accurately in a weighing bottle and transferred to a 50 ml beaker, then 10 ml of 72% sulphuric acid was added carefully with a pipette and the mixture was stirred with a glass rod.

The mixture was moved quantitatively with a wash bottle of a 500 ml round-bottle flask and diluted with water until the final volume was 300 ml. The solution was then refluxed for 3 h, filtered in a glass filter and dried in an oven at 105 °C for 12 h. The crucible was cooled in a desiccator for 15 min and then weighed accurately. The glass filter containing the lignin was reported as a percentage by weight of the dried sample. Lignin content was calculated using

$$\text{Lignin content}(\%) = \frac{\text{Oven dried weight of lignin}}{\text{Oven dried weight of initial sample}} \times 100 \quad (1)$$

## 2.4 Determination of Holocellulose Content

Holocellulose content was determined with reference to ASTMD 1104-56, (1978) method, in which 1 g of sample was placed in an Erlenmeyer flask (250 ml) and 150 ml of distilled water was added. While slowly shaking, 1 g of $NaClO_2$ and 0.2 ml of acetic acid were added and the flask was covered with glass and boiled at 70–80 °C for 60 min. Again, 1 g of $NaClO_2$ and 0.2 ml of acetic acid were added and boiled. After cooling, the sample was filtered using a filter flask and washed with hot water until free of acid. Afterward, the insoluble portion was dried in an oven at 105°c for 4 h, cooled in a desiccator and weighed repeatedly until obtaining a constant weight. Holocellulose content was calculated as follows.

$$\text{Holocellulose content}(\%) = \frac{\text{Oven dried weight of holocellulose}}{\text{Oven dried weight of initial sample}} \times 100 \quad (2)$$

The soluble of treated leaf, twig and bark samples were examined with reference to ASTMD1110-87 (2007). Two-gram sample was oven-dried and placed into a 250 ml Erlenmeyer flask containing 200 ml of distilled water. A reflux condenser was attached to the flask and the apparatus was placed in a gently boiling water bath for three hours with constant shaking. Special attention was given to insure that the level of the solution in the flask remained below that of the boiling water. Samples were then removed from the water bath and filtered by vacuum suction into a glass filter of known weight. The residue was washed with hot water before the glass-filter was oven-dried at 103 ± 2 °C. The glass-filter was then cooled in a desiccator and weighed until a constant weight was obtained. The following formula was used to obtain the hot water soluble of the sample

$$\text{Hotwatersoluble}(\%) = \frac{W_1}{W_1 - W_2} \times 100 \quad (3)$$

where

$W_1$ weight of oven – dry test sample (g).
$W_2$ Weight of oven – dry sample after extraction with hot water (g).

### 2.5 SEM–EDX Elemental Analysis

Scanning Electron Microscopy equipped with Energy Dispersive X-ray (SEM–EDX) analysis was performed to determine the cellular and sub-cellular structure of degrading biomass and the elemental levels at different degradation stages (Vinod and Sashidhar 2010). The leaf, twig and bark mixed degraded samples were dried and ground into fine powder and then placed in the steel stab with carbon tape and sputter coated with gold particle for 50 s in high vacuum conditions. Elemental analysis and Scanning Electron Microscopic images coupled with energy dispersive X-ray consisting 3.5 and 2.5 nm resolution for tungsten filament and LaB6 and EDX detector resolution 133 eV of the degraded samples was assessed.

## 3 Results and Discussion

Degradation of lignin, holocellulose, hot water soluble content of *Pinus radiata* Don needle leaf, twig and bark by wood degrading fungi. The degradation abilities of lignin, holocellulose, hot water soluble content of *Pinus radiata* needle leaf, twig and bark by specific wood degrading fungi of *P. radiata* forest ecosystem was presented in (Table 1).

The lignin content of needle leaf as control 23.8% at 90 days, the treated samples with fungi showed the decrease in the lignin content 17.1–19.5% and after 180 days it was 13.5–15.3%, depending on the fungal genera (Table 1).

The most efficient lignin degrading fungi of needle leaf was found to be *Colricia perennis* with 6.7% at 90 days and 8.8% at 180 days of degradation.

For twig, the initial lignin content of control was 52.0% which is decreased to 38.0–44.2% at 90 days treated sample with fungi and at 180 days it was 50.8% for control, was decreased from 32.5 to 38.0% depending on the fungal genera (Table 1). The most efficient lignin degrading fungi for twig was *Heterobasidion annosum* with 14.0% at 90 days and 18.7% at 180 days of degradation.

In the case of bark, the control sample have shown 90.2% at 90 days, the treated samples with fungi showed the decrease in the lignin control, i.e., 68.0–82.0% and at 180 days it was 66.5 to 78.3% as against the control 88.0%. The most efficient lignin degrading fungi of bark was found to be 15.6% for *Heterobasidion annosum* at 90 days, but at 180 days it was found to be 11.5% for *Colricia perennis* (Figs. 2, 3, 4, 5 and 6).

**Table 1** Degradation of lignin, holocellulose and hot water soluble content of *Pinus radiata* Don needle leaf, twig and bark by wood degrading fungi

| Name of the fungi | | Leaf | | | | Twig | | | | Bark | | | |
|---|---|---|---|---|---|---|---|---|---|---|---|---|---|
| | | 90 days | % | 180 days | % | 90 days | % | 180 days | % | 90 days | % | 180 days | % |
| *Colricia perennis* | LC | 17.1 ± 0.44 | 6.7 | 14.9 ± 0.40 | 7.4 | 39.4 ± 0.34 | 12.6 | 37.8 ± 0.36 | 13.0 | 80.4 ± 0.65 | 10.2 | 66.5 ± 0.50 | 21.5 |
| | HC | 16.1 ± 0.44 | 5.1 | 15.0 ± 0.15 | 5.5 | 41.3 ± 0.08 | 3.6 | 35.3 ± 0.39 | 9.3 | 68.0 ± 0.20 | 4.8 | 59.5 ± 0.45 | 12.5 |
| | HWSC | 36.0 ± 0.50 | 5.7 | 72.5 ± 0.06 | 18.7 | 64.2 ± 0.32 | 9.9 | 84.0 ± 0.25 | 10.5 | 68.5 ± 0.55 | 4.2 | 81.5 ± 0.40 | 6.5 |
| *Heterobasidion annosum* | LC | 18.2 ± 0.14 | 5.6 | 13.5 ± 1.00 | 8.8 | 38.0 ± 0.60 | 14.0 | 32.5 ± 0.10 | 18.3 | 74.6 ± 0.40 | 15.6 | 68.5 ± 0.45 | 19.5 |
| | HC | 18.0 ± 0.30 | 3.2 | 12.5 ± 0.51 | 8.0 | 40.0 ± 0.40 | 4.9 | 36.2 ± 0.38 | 8.4 | 60.0 ± 0.50 | 12.8 | 58.0 ± 0.65 | 14.0 |
| | HWSC | 31.5 ± 0.50 | 1.2 | 62.3 ± 0.10 | 8.5 | 68.2 ± 0.41 | 13.9 | 82.0 ± 0.30 | 8.5 | 67.1 ± 0.40 | 2.8 | 80.1 ± 0.53 | 5.1 |
| *Polyporus squamosus* | LC | 18.5 ± 0.20 | 5.3 | 15.0 ± 0.21 | 7.3 | 40.2 ± 0.18 | 11.8 | 36.0 ± 0.10 | 14.8 | 81.8 ± 0.08 | 8.4 | 70.0 ± 0.18 | 18.0 |
| | HC | 17.6 ± 0.10 | 3.6 | 14.5 ± 0.40 | 6.0 | 42.5 ± 0.20 | 2.4 | 39.0 ± 0.12 | 5.6 | 61.0 ± 0.20 | 11.8 | 55.1 ± 0.02 | 16.9 |
| | HWSC | 34.0 ± 0.38 | 3.7 | 62.5 ± 0.40 | 8.7 | 68.0 ± 0.48 | 13.7 | 77.0 ± 0.15 | 3.5 | 67.8 ± 0.40 | 3.5 | 84.0 ± 0.41 | 9.0 |
| *Thelophora terrestris* | LC | 19.5 ± 0.30 | 4.3 | 15.3 ± 0.40 | 7.0 | 44.2 ± 0.40 | 7.8 | 38.0 ± 0.10 | 12.8 | 82.0 ± 0.40 | 8.2 | 78.3 ± 0.17 | 9.7 |
| | HC | 17.5 ± 0.45 | 3.7 | 10.2 ± 0.40 | 10.3 | 39.0 ± 0.10 | 5.9 | 32.5 ± 0.30 | 12.1 | 67.3 ± 0.20 | 5.5 | 60.0 ± 0.38 | 12.0 |
| | HWSC | 32.5 ± 0.40 | 2.2 | 71.5 ± 0.33 | 17.7 | 62.1 ± 0.42 | 7.8 | 78.9 ± 0.65 | 5.4 | 68.0 ± 0.65 | 3.7 | 78.5 ± 0.15 | 3.5 |
| Control | LC | 23.8 ± 0.38 | – | 22.3 ± 1.10 | – | 52.0 ± 0.10 | – | 50.8 ± 0.58 | – | 90.2 ± 0.18 | – | 88.0 ± 0.08 | – |
| | HC | 21.2 ± 0.10 | – | 20.5 ± 1.15 | – | 44.9 ± 0.39 | – | 44.6 ± 0.50 | – | 72.8 ± 0.40 | – | 72.0 ± 1.00 | – |
| | HWSC | 30.3 ± 0.47 | – | 53.8 ± 0.45 | – | 54.3 ± 0.38 | – | 73.5 ± 0.10 | – | 64.3 ± 0.18 | – | 75.0 ± 0.36 | – |

*LC* Lignin content, *HC* Holo Cellulose, *HWSC* Hot water soluble content

**Fig. 2** Bio degradation samples of Pinus radiata Don

## *3.1 Degradation of Lignin*

Lignocellulose is the predominant component of woody plant and dead plant materials, and the most abundant biomass on earth. Lignin and holocellulose in the biomass structure are the major energy sources available to decomposer organisms constituting 70–80% of fresh organic material (Swift et al. 1979). Lignin is a recalcitrant plant polymer and its mineralization by white rot basidiomycetes plays a major role in carbon recycling (Martinez et al. 2005). White rot fungi are wood degrading organisms capable of decomposing all wood polymers, lignin, cellulose and hemicelluloses (Hakala 2007). Holocellulose, a polysaccharide containing cellulose and hemicellulose (Petterson 1984) is a major component of wood suitable-for fungal growth. Polysaccharide content generally ranges between 60 and 80% (w/w) in hardwood (Willfor et al. 2005). However, decomposition rate of cellulose is higher than that of lignin (Fioretto et al. 2005). White-rot fungi belong to the basidiomycetes and their activity is usually related to the moisture content of wood (Blanchette 1995). The decaying fungi belong to saprophyte fungal organisms, since they live on dead or residual vegetation, decomposing them into simpler molecular compounds (Dubeux et al. 2006; Ohkuma et al. 2001).

**Fig. 3** Wood Degrading fungi source—*Pinus radiata* Don

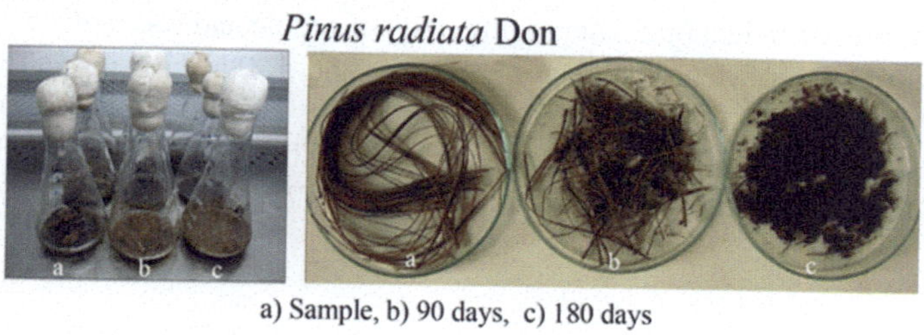


**Fig. 4** Biodegradation of leaf samples

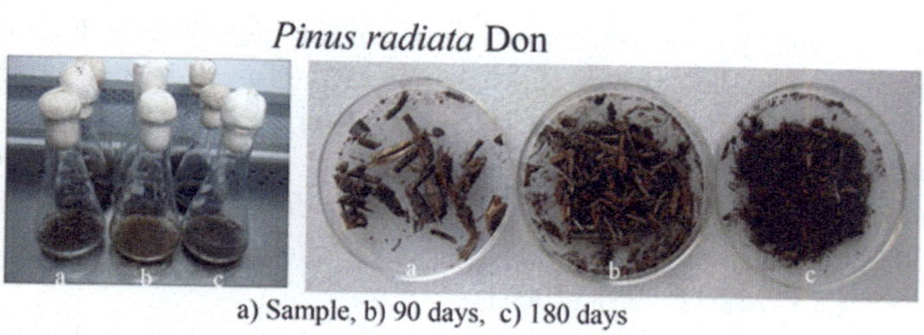


**Fig. 5** Biodegradation of twig samples

*Pinus radiata* Don

a) Sample, b) 90 days, c) 180 days

**Fig. 6** Biodegradation of bark samples

Fungi require a carbon source, macronutrients such as nitrogen, phosphorous and potassium and certain trace elements for their growth. Carbon serves primarily as an energy source for the microorganisms, while a small fraction of the carbon in incorporated into their cells (Tuomela et al. 2000). Biomass including leaf, twigs, bark and other residual materials of forest ecosystem, naturally undergo degradation due to fungal enzymatic action. This causes increase in $CO_2$ in the environment. Therefore, it would be better for the woody materials to be recycled by biological degradation or removal of lignin (Watanabe et al. 2003). It was found that lignin content of either leaf, twigs or bark degraded effectively by naturally occurring lignin degrading fungi of forest ecosystem. The present work revealed that in *Pinus radiata* forest ecosystem consists of specific lignin degrading fungi are *Colricia perennis, Heterobasidion annosum, Polyporus squamosus* and *Thelophora terrestris* in which the most efficient lignin degrading fungi of needle leaf was *Colricia perennis* with 6.7 and 8.8% of degradation for 90th day and 180th days respectively. For twig *Heterobasidion annosum* with 14.0% at 90 days and 18.3% at 180 days of degradation. In the case of bark *H. annosum* with 15.6% and *C. perennis* 11.5% of degradation (TAPPI Standard method, T 222 OS-74 (2009)).

## 3.2 Holocellulose Degradation

The degradation ability of holocellulose of control sample of needle leaf of 90 days was 21.2% and at 180 days it was 20.5%, whereas the fungal treated samples have shown 16.1–18.0% for 90 days and at 180 days it was 10.2–15.0% degrading in the fungal genera involved in the degradation (Table 1). The most efficient holocellulose degrading fungi was *Colricia perennis* with the degradation percentage 5.1 for 90 days of degradation and 10.3% degradation by *Thelophora terrestris* for 180 days of degradation (ASTMD1110-87 (2007)).

The degradation of holocellulose for twig of *Pinus radiata*, control samples have shown 44.9% and 44.6% for 90 days and 180 days respectively, whereas the fungal treated samples have shown 39.0 to 41.3% for 90 days of degradation and 32.5–39.0% for 180 days samples with respect to the fungal genera involved in degradation (Table 1). The most efficient fungi involved in the holocellulose degradation of twig was found to be *Thelophora terrestris* with 5.9 and 12.1% for both 90 and 180 days of degradation. In the case of degradation of bark the control samples have shown 72.8 and 72.0% for 90 and 180 days. The fungal treated ones have shown 60.0 and 68.0% of degradation for 90 and 55.1 and 60.6% for 180 days, with respect to the fungal genera involved in the holocellulose degradation (Table 1). The most efficient fungi involved in the degradation of bark was found to be *Heterobasidion annosum* with 12.8% for 90 days, whereas *Polyporus squamosus* with 16.9% for 180 days of degradation (Figs. 2, 3, 4, 5 and 6).

The *Pinus radiata* forest ecosystem consists of specific holocellulose degrading fungi are *Colricia perennis, Heterobasidion annosum, Polyporus squamosus* and *Thelophora terrestris* in which the most efficient holocellulose degrading fungi of needle leaf was *Colricia perennis* with 5.1% at 90 days and *Thelophora terrestris* with 10.3% at 180 days. For twig *Thelophora terrestris* with 5.1% at 90 days and 12.1% at 180 days of degradation. In the case of bark *Heterobasidion annosum* with 12.8% and *Polyporus squamosus* with 16.9% of degradation (ASTMD 1104-56 (1978)).

### 3.3 Hot Water Soluble Content

The volume of the soluble content of *Pinus radiata* needle leaf, twig and bark in hot water is shown in Table 1. It was found that the duration of inoculation in all the sample tested was found to be increased in solubility in hot water treatment. The value of soluble content varied depending on the fungi inoculated. For needle leaf the value of soluble content was 31.5–36.0% as against control sample 30.3% in 90 days. At 180 days, the value of soluble content was 62.3–72.5% as against the control sample 53.8%. The maximum hot water soluble content was recorded in *Colricia perennis,* 5.7% for 90th day of degradation and 18.7% on the 180th day for the same fungus. In the case of twig, the percentage was between 62.1 and 68.2% as against 54.3% for control in 90 days. In 180 days, the results are between 77.0 and 84.0% when compared with the control 73.5%. The maximum hot water soluble content was found in *Heterobasidion annosum* with 13.9% on the 90$^{th}$ day and 10.5% on the 180$^{th}$ day of degradation for *Colricia perennis.* For bark sample, the percentage of soluble content was 67.1–68.5% as against the control, it was 64.3% at 90 days. In 180 days, the results are between 78.5 and 84.0% as against the control it was 75.0%. The maximum hot water soluble content was found in *Colricia perennis* with 4.2% on the 90th day and 9.0% on the 180th day of degradation for *Polyporus squamosus.*

In this study, the hot-water solubility of treated samples increased significantly with incubation time meaning that some amount of lignocellulose content was degraded. This is presumably supported by the monosaccharides in leaf, twig and bark samples like xylose, mannose and glucose (Pinto et al. 2005; Abubacker and Kirthiga 2015) which are soluble in water, besides the degradation of cellulose containing polymers and polysaccharides into simpler components like monomers through fungal activity (Blanchette et al. 1994). In both 90 and 180 days of incubation, the hot water soluble content of all treated samples increased. Perhaps the soluble matter was consumed for energy by the fungi, since lignin and holocellulose content were less decreased.

### *3.4 SEM–EDX Elemental Analysis of Mixed Degraded Samples*

The SEM–EDX elemental analysis of degradation of mixed samples were conducted for all the four forest materials namely leaf, twig and bark mixed samples. For *Pinus radiata* the degradation sampling was conducted at initial stage, 90th day and 180th day of degradation. In *Pinus radiata* the initial stage the elements available in the mixed samples of biomass shown C 64.2%, O 34.39%, Mg 0.10%, Al 0.23%, Si 0.18%, P 0.13%, S 0.17%, Cl 0.08%, K 0.40% and Fe 0.11%. On the 90th day of degradation the available elements were found to be C 7.95%, O 56.05%, Mg 0.57%, Al 11.35%, Si 12.83%, Cl 0.14%, K 0.67%, Ca 0.24%, Ti 0.44% and Fe 9.76%. On 180th day of degradation the following elements were recorded. C 32.18%, N 2.19%, O 55.26%, Na 0.15%, Mg 0.25%, Al 5.15%, Si 5.08%, P 0.55%, S 0.26%, Cl 0.23%, K 0.53%, Ti 0.10%, Mn 0.14% and Fe 2.31% (Fig. 7).

In *Pinus radiata,* the initial stage the elements available in the mixed samples of biomass shown C 64.2%, O 34.39%, Mg 0.10%, Al 0.23%, Si 0.18%, P 0.13%,S 0.17%, Cl 0.08%, K 0.40% and Fe 0.11%. On the 90th day of degradation the available elements were found to be C 7.95%, O 56.05%, Mg 0.57%, Al 11.35%, Si 12.83%, Cl 0.14%, K 0.67%, Ca 0.24%, Ti 0.44% and Fe 9.76%. On 180th day of degradation the following elements were recorded. C 32.18%, N 2.19%, O 55.26%, Na 0.15%, Mg 0.25%, Al 5.15%, Si 5.08%, P 0.55%, S 0.26%, Cl 0.23%, K 0.53%,Ti 0.10%, Mn 0.14% and Fe 2.31% (Fig. 7).

## 4 Conclusion

The rate of degradation of *P. radiata* needle leaves, twigs and barks varied depending on the fungal genera inoculated. An increase in incubation time tended reducing both lignin and holocellulose content. However, the reduction rate was not significant, therefore more time is needed to degrade lignin rather than other components in the sample. This report will help to gain the insight of lignin and holocellulose degradation in the early stage in the natural forest ecosystem**.** Fungi play a vital role in plant litter decomposition in forest ecosystem through nutrient cycling and humus formation in soil because they colonize the lignocellulosic matrix in litter that other organisms are unable to decompose (Lodge 1985; Hunt et al. 1989). The present research work with SEM–EDX elemental analysis of degraded mixed samples of

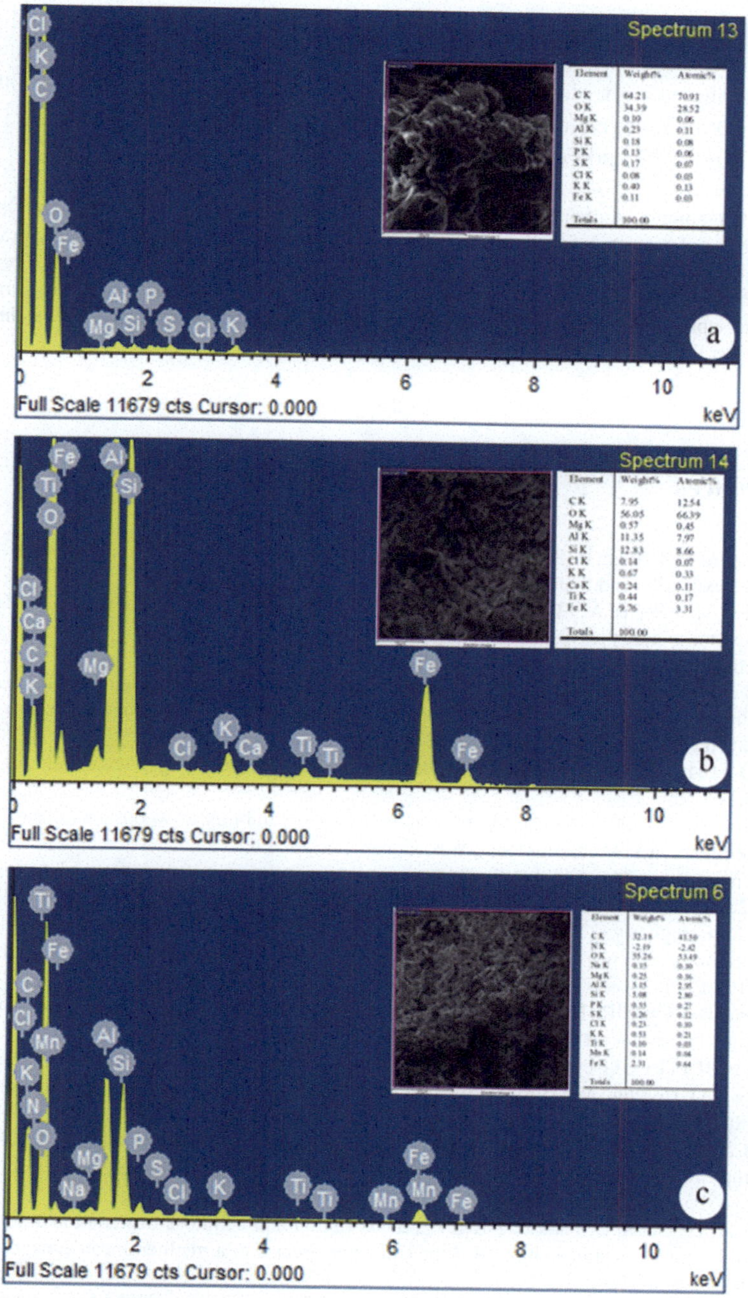


a) Initial, b) 90 days, c) 180 days

**Fig. 7** SEM–EDX: elemental analysis of degradation of mixed samples of *Pinus radiata* Don

needle leaf, twig and bark of *Pinus radiata,* showed that the specific initial stage of elements are Si 0.18% and Fe 0.11%. In the 90th day of degradation the same elements shown higher levels of 12.83% Si and 9.79% Fe, whereas in the 180th day of degradation more elements are recorded, they are Al 5.5%, P 0.55%, K 0.53%, Ti 0.10%, Mn 0.14% and Fe 2.31%.

**Acknowledgements** Author (MNA) wish to thank DST-FIST, Government of India, New Delhi for providing the infrastructure facilities to the Department of Botany, National College, Tiruchirappalli, Tamil Nadu. Authors also expresses thanks to Padmavibhushan Dr. V. Krishnamurthy, President, Sri. K. Ragunathan, Secretary and Dr. K. Anbarasu, Principal, National College, Tiruchirappalli for all the supports and encouragement given to PG and Research, Department of Biotechnology to carry over the research work.

## References

Abubacker MN, Kirthiga B (2015) Bioremediation potential of textile. *Aspergillus flavus* TEAK-07 against textile dye and their toxicity assessment. Biolife 3:54–60

Abubacker MN, Prince M (2013) Decomposition of lignin and holocellulose of *Acacia dealbata* Link (Mimosoideae) leaves, twigs and barks by fungal isolates from virgin forest ecosystem of Doddabetta belt of Nilgiris. Biosci Biotech Res Asia 10:719–726

Abubacker MN, Prince M (2015) Decomposition of lignin and holocellulose of *Pinus roxburghii* Sar (Pinaceae) needle leaves, twigs and barks by fungal isolates from virgin forest ecosystem of Doddabetta belt of Nilgiris. Biolife 3:553–560

ASTMD 1104-56 (1978). Method of test of holocellulose in wood. https://www.astm.org/standards/D1104.htm

ASTMD 1110-87 (2007) Standard test methods for water solubility of wood. American Standard Test Methods. ASTM International, West Conshohocken, PA

Berg B, McClaugherty CA (2008) Plant litter: decomposition, humus formation, carbon sequestration, 2nd edn. Springer, Heidelberg, Berlin

Berg B, McClaugherty C, Johansson M-B (1997) Chemical changes in decomposing plant litter can be systemized with respect to the litter's initial. Reports from the Departments in Forest Ecology and Forest Soil, *Swedish University of Agricultural Science. Report*, 74:85

Blanchette RA, Obst JR, Timell TE (1994) Biodegradation of compression wood and tension wood by white and brown rot fungi. Holzforschung 48:34–42

Blanchette RA (1995) Degradation of lignocellulose complex in wood. Can J Bot 73:999–1010

Crossley DAJ, Hoglund MP (1962) A litter bag method for the study of microarthropods inhabiting leaf litter. Ecology 43:571–573

Dubeux JCB Jr, Sollenberger LE, Interrante SM, Vendramini JMB, Stewart RL Jr (2006) Litter decomposition and mineralization in bahia grass pastures managed at different intensities. Crop Sci 46:1305–1310

Fioretto A, Di Nardo C, Papa S, Fuggi A (2005) Lignin and cellulose degradation and nitrogen dynamics during decomposition of three leaf litter species in a mediterranean ecosystem. Soil Biol Biochem 37:1083–1091

Hakala TK (2007) Characterization of the lignin—modifying enzymes of the selective white-rot fungus *Physisporinus rivulosus*. Helsinki University Printing House, Helsinki, Finland, p 60

Hobbie SE, Vitousek PM (July 2000) Nutrient limitation of decomposition in Hawaiian forests. Ecology 81(7):1867–1877

Hunt HW, Elliott ET, Walter DE (1989) Inferring trophic transfers from pulse-dynamics in detrital food webs. Plant Soil 115:247–259

Jeffries TW (1994) Biodegradation of lignin and hemicelluloses. In: Ratledge C (ed) Biochemistry of microbial degradation. Kluwer, Dordrecht, pp 233–277
Kelley J (1992) Upgrading of waste cereal straws. Outlook Agric 21:105–108
Krishna MP, Mohan M (2017) Litter decomposition in forest ecosystem. Rev Energy Ecol Environ 2:236–249
Kuhad RC, Singh A, Eriksson KE (1997) Microorganisms and enzymes involved in the degradation of plant fiber cell walls. Adv Biochem Eng Biotechnol 57:45–125
Lodge DJ (1985) Preliminary estimates of fungal biomass and nutrient stores in the litter and soil of a tropical rainforest. In: Agronomy abstracts, Soil Science Society of America, Chicago, pp 158–159
Martínez AT, Speranza M, Ruiz-Dueñas FJ, Ferreira P, Camarero S, Guillén F, Martínez MJ, Gutiérrez A, del Río JC (2005) Biodegradation of lignocellulosics: microbial, chemical, and enzymatic aspects of the fungal attack of lignin. Int Microbiol 8:195–204
Miura K, Kudo M (1970) An agar-medium for aquatic hypomycetes. Trans Mycol Soc Japan 11:116–118
Ohkuma M, Maeda Y, Johjima T, Kudo T (2001) Lignin degradation and roles of white rot fungi: study on an efficient symbiotic system in fungus growing termites and its application to bioremediation. Riken Rev Foe Econ Sci Res 42:39–42
Petterson RC (1984) The chemical composition of wood. In: Rowell RM (ed) The chemistry of solid wood. Americal Chemical Society, Washington, DC, pp 57–126
Pinto PC, Evtuguin DV, Neto CP (2005) Structure of hardwood gucuronoxylans: modifications and impact on pulp retention during wood kraft pulping. Carbohyd Polym 60:489–497
Baldrin P, Lindahi BD (2017) Decomposition in forest ecosystem: after decades of research still novel findings. Fungal Ecol 4:359–361
Pirzo Koivksarri Mysore V, Tejesvi MT, Markkota A, Mykra H, Pirttila AM (2019) Fungi originating from tree leaves contribute to fungal diversity of litter in streams. Front Microbiol. https://doi.org/10.3389/fMicb2019.0065
Weng SS, Di Y, Lin X, Yang G (2020) Screening of cellulose degrading fungi in forest litter and fungal effects on litter decomposition. Bio Res 15:2937–2946
Swift MJ, Heal OM, Anderson JM (1979) Decomposition in terrestrial ecosystems. J Ecol 34:45–50
TAPPI Standard Methods T 222 OS 74 (2009) Fibrous materials and pulp testing. Technical Association of Pulp and Paper Industry, Atlanta
Tuomela M, Vikman M, Hatakka A, Itavaara M (2000) Biodegradation of lignin in a compost environment: a review. Biores Technol 72:169–183
Varnaite R (2001) Bioconversion of rye straw by micromycetes. Biologija 4:34–36
Vinod VPT, Sashidar RB (2010) Surface morphology, chemical assessment of gum Kondagogu. Indian J Natural Prod Res 1(2)
Virzo De Santo A, Rutigliano FA, Berg B, Fioretto A, Fierro AR (1998) Nitrogen dynamics of decomposing needle litters in three coniferous forests of the Mediterranean area. Fresen Environ Bull 7:510–517
Watanabe T, Watanabe Y, Nakamura K (2003) Biodegradation of wood and dual cultures of selected two fungi determined by chopstick method. J Biosci Bioeng 95:623–626
Willfor SA, Sundberg A, Pranovich A, Holmbom B (2005) Polysaccharides in some industrially important hardwood species. Wood Sci Technol 39:601–617

# Biodegradation of Lead from Accumulated Municipal Solid Waste Using Bacterial Consortium for Effective Biomining

**Mihigo Felix, Manikandan Arjunan, Kavya Siddeshwar, Rajakumar Sundaram, and Prashanthi Devi Marimuthu**

**Abstract** Discharge of waste containing heavy metals through anthropogenic activities has created global environmental and health burden and remediation of contaminated sites becomes more expensive, however sustainable methods of degrading heavy metals using resistant microorganisms have become important biotechnological development. 42 lead resistant bacteria were isolated from municipal solid waste in Tiruchirappalli. Lead resistant bacteria were screened and the maximum tolerable concentration was found to be in the range of 100–260 mg/kg. Lead resistant bacteria such as *Pseudomonas spp., Klebsiella spp., Enterobacter spp., Escherichia spp.* were identified and used to make consortium and used in column test. The results show lead diminishing from 0.923 to 0.1943 ppm concerning the control sample from day 1 until day 10. The consortium used has shown capability to biodegrade lead in solid waste and may further be used for biomining and other bioremediation studies.

**Keywords** Municipal solid waste · Bioming · Bioremediation

## 1 Introduction

Environmental pollution is one of the major global threats. Anthropogenic activities are the major source of heavy metals in the environment causing environmental and public health concern through contamination by these heavy metals (Tchounwou 2012). Various industrial activities such as agriculture, mining, electroplating, mining, leather industries release heavy metals through waste and effluent discharges. Due to the higher concentration of heavy metals released in effluents, it affects aquatic life (Moten and Rehman 1998). However various bacteria and other microorganisms have acquired resistance mechanisms for adaptation in toxic

M. Felix · K. Siddeshwar · P. D. Marimuthu (✉)
Department of Environmental Science and Management, Bharathidasan University, Tiruchirappalli, Tamil Nadu, India
e-mail: prashanthidevi@gmail.com

M. Arjunan · R. Sundaram
Department of Marine Biotechnology, Bharathidasan University, Tiruchirapalli, Tamil Nadu, India

 
M. Prashanthi et al. (eds.), *Bioremediation and Green Technologies*, Environmental Science and Engineering, https://doi.org/10.1007/978-3-030-64122-1_8

and heavy metals contaminated sites hence become resistant (Rehman et al. 2008). The remediation of heavy metals contaminated environment has been proved to be expensive (Massadeh 2005) and heavy metals tolerant microorganisms have a significant role in the bioremediation of contaminated environment (Hrynkiewicz and Baum 2014). Toxic waste can be consumed by microorganisms and convert them into nontoxic waste through their metabolism and breakdown the complex compounds in waste (Thassitou and Arvanitoyannis 2001). The column test has been previously used in biogeochemical and heavy metals interaction using bacteria for transformation and biodegradation (Ayyasamy and Lee 2009).

The present work aims to biodegrade lead from accumulated solid wastes using selective lead resistant bacterial consortium in column for effective booming.

# 2 Materials and Methods

## 2.1 *Study Area*

Ariyamangalam dump yard is a Municipal Solid Waste dumpsite under Tiruchirappalli City Corporation which covers 47.70 acres of land positioned at 10°48′N and 78°43′E coordinates on the Tiruchirappalli-Thanjavur highway. The dumpyard has been under operation since 1367. Ariyamangalam dump yard is an open damping yard that receives an estimated volume of 470 tons per day of both organic and inorganic waste generated by the city. The average height of dumped waste is 4 m. The dump yard waste management is not managed and planned properly as it is freely accessible by rag pickers and different animals. The dump yard is surrounded by schools, residences, commercial complexes, and a rice mill (Fig. 1).

The solid waste at the Ariyamangalam dump yard includes residential waste, industrial as well as commercial waste from all over the city. The solid waste is composed (approximately 60%) of organic waste from the markets, residential kitchen waste, hotels, and restaurants and the remaining include recyclable waste such as plastics, paper, wood. Hospital wastes which include tissues, syringes, soiled cotton, and toxic waste such as insecticides, chemicals, bulbs, spray cans, and others.

## 2.2 *Sample Collection*

Solid waste samples were collected at the Ariyamangalam dump yard in Tiruchirappalli. Solid waste samples were systemically collected based on the amount of time they have been dumped and were categorized into three types. New solid waste, old solid waste, and older solid waste. S1 was a new solid waste that was dumped in less than 7 days. S2 was old waste which was up to 4 weeks old at the dump yard and S3 was older waste that was more than 4 weeks old disposed to the dump yard.

Each solid waste sample was approximately 500 g of mixed waste and three samples were collected from different sample points. Each solid waste sample was labeled according to its location. Samples were gently placed into hygienic and sterilized plastic zipper bags and were taken to the laboratory. 50 g of thoroughly mixed waste were taken for microbial analysis and the rest of the samples were allowed to dry at room temperature for ten days. After drying, solid waste was shredded and ground, and were further analyzed for their heavy metals content.

## 2.3 *Isolation and Identification of Bacteria*

Solid waste samples were taken to the laboratory under aseptic conditions. Samples were serially diluted and spread using sterile L-rod on pre-sterilized nutrient agar plates in triplicates (Bergey et al. 1974; Yamina et al. 2012). After incubation of inoculated plates for 24 h at 37 °C, the plates were observed for the appearance of bacterial colonies. Morphologically different 42 colonies were selected and the isolates were subcultured and maintained on nutrient agar slant and stored at 4 °C. Among 42 isolates, the heavy metals resistant bacteria were selected and were identified based on morphological characteristics, gram staining, and biochemical tests such as Indole, MR-VP, motility, and Covac's oxidase tests were performed using Bergey's manual of determinative bacteriology (Bergey et al. 1974).

## 2.4 *Heavy Metals Analysis*

Exactly 2 g solid waste was weighed using a monopan digital balance and transferred to a 500 ml round bottom flask. To this, a little amount of distilled water was added and swirled gently to make a slurry. Then to this, a mixture of each 20 ml nitric acid, 4 ml perchloric acid, and 2 ml sulphuric acid was added respectively. The sample was then digested (Guna-Soxhlet extraction mantle) for 3 h at 90 °C. The appearance of brown fumes during digestion was followed by the addition of 5 ml of 30% Hydrogen peroxide. Bumping was controlled by putting glass beads in the sample getting digested. Ensured digestion was followed by the addition of a little amount of distilled water to the unit. The digested samples were then cooled and filtered into a 50 ml standard flask. The final volume of 50 ml was made with distilled water. The samples were then transferred to pre-treated, laboratory cleaned, plastic vials, and stored at 4 °C till further analysis. The digested samples were aspirated in the Atomic Absorption Spectrophotometer (iCE 3300 AA Spectrometer) and absorbance was noted from which estimations were made for each metal concentration in mg/kg (U.S. EPA. 1996). Working standards were prepared from 1000 ppm stock solution of lead (Pb) to 2, 4, and 6 ppm.

## 2.5 Screening of Heavy Metal Resistant Bacteria

The isolates were cultured on various heavy metals concentrations and were incubated for 42 h at 37 °C. Different concentrations of Copper, Lead, and Nickel were prepared and mixed with sterilized nutrient agar, plated, and inoculated with the isolates. The medium was mixed at the ratio of 9:1. 1 ml of stock heavy metal solution of different concentrations was added to 9 ml of medium. The concentration of each element was increased every time bacterial growth was observed until when there was no growth at maximum concentration. Bacteria which grow at higher concentration were selected and there maximum tolerable concentration were recorded.

Bacteria observed to grow at higher concentrations of Copper, Lead, and Nickel among the 42 isolates screened were selected for further study in column test using bacterial consortium.

## 2.6 Bacteria Resistance to Heavy Metals

The isolates were cultured on various heavy metals concentrations and were incubated for 48 h at 37 °C. Different concentrations of Lead were prepared and mixed with sterilized nutrient agar, plated, and inoculated with the isolates. The medium was mixed at the ratio of 9:1. 1 ml of stock heavy metal solution of different concentrations was added to 9 ml of medium. Among 42 isolates screened, bacteria observed to grow at higher concentrations of Lead were selected for further study in column test using bacterial consortium.

## 2.7 Maximum Tolerable Concentration

The maximum tolerable concentration of isolated bacterial strains against their respective heavy metals was determined by gradually increasing the heavy metals by 10 ppm each time on the nutrient agar plate. The initial concentration for all heavy metals was 100 ppm. After sterilization, nutrient agar was mixed with heavy metal solution with different concentrations at the ratio of 9:1 and after solidification of media, plates were streaked with the isolated culture and incubated at 37 °C for a maximum of 48 h. The culture growth was observed. The culture growing on the initial concentration was again sub-cultured on higher concentrations by streaking on a new plate (Rojas et al. 2011). The maximum Inhibitory concentration was determined when the strains failed to grow on plates after 48 h (Yamina et al. 2012).

### 2.8 Column Test

Column tests were done for 10 days using 30 cm high and 8 cm diameter columns. Two columns were filled with 1.2 kg of ground solid waste. One column was kept as control while the other was fed with bacterial consortium (Fig. 2). Two samples were taken from the column on every alternate day for 10 days. Samples collected were digested and analyzed the presently available amount of copper, lead, and nickel after column test in comparison to control samples using Thermo Fischer Atomic Absorption Spectroscopy, and the pH of the samples were recorded (Ayyasamy and Lee 2009; Balint et al. 2013).

## 3 Results and Discussion

### 3.1 Isolation and Identification of Solid Waste Bacteria

Out of three samples of solid waste collected, 42 morphologically different bacterial colonies were isolated before screening using Bergey's manual of systematic bacteriology (David 1986). Based on the capability of heavy metal resistance, 4 isolates have shown higher resistance to lead and were selected for further study in the column. The selected isolates resistant to lead were identified as *Escherichia spp, Klebsiella spp., Pseudomonas spp.* and *Enterobacter spp.*

### 3.2 Heavy Metal Analysis in Solid Waste Samples

Preliminary analysis of heavy metals concentrations in solid waste was done for all the three samples before the identification of heavy metals resistant bacteria present in the solid waste at the dump yard. Lead concentrations in solid waste samples ranged between 12.8103–18.2907 mg/kg and were measured as 12.81 mg/kg in S1, 18.29 mg/kg in S2 and 16.08 mg/kg in S3. The highest concentration of lead was observed at sample location S2, and S1 showed lower concentration compared to other locations. The higher concentration with 18.29 mg/kg at S2 might be due to migrated leachate from lead-containing household products, lead-based paints, inclusion of chalcopyrite in batteries, and chemical for photograph processing waste dumped along the other types of municipal solid waste in the dump yard.

**Table 1** Maximum tolerable concentration (MTC)

| S. No. | Bacteria | Metal | MIC (ppm) |
|---|---|---|---|
| 1 | *Escherichia spp* | Lead | 260 |
| 2 | *Klebsiella spp.* | Lead | 250 |
| 3 | *Escherichia spp* | Lead | 260 |
| 4 | *Pseudomonas spp.* | Lead | 250 |
| 5 | *Enterobacter spp.* | Lead | 250 |

### 3.3 Determination of Maximum Tolerable Concentration

The maximum tolerable limit of municipal solid waste isolated bacterial strains against lead was determined by gradually increasing the lead by 10 ppm each time on a nutrient agar plate. The initial concentration for Lead was 100 ppm. After sterilization, nutrient agar was mixed with heavy metal solution with different concentrations at the ratio of 9:1 and after solidification of media, plates were streaked with the isolated culture and incubated at 37 °C for a maximum of 48 h. The culture growth was observed. The culture growing on the initial concentration was again subcultured on higher concentrations by streaking on a new plate (Rojas et al. 2011). The maximum tolerable concentration was determined when the strains failed to grow on plates after 48 h (Yamina et al. 2012). Table 1 shows the maximum inhibitory concentration of Lead.

### 3.4 Bacteria Consortium for Lead Removal

Several microbial cultures were combined to make a consortium based on the bacteria that had higher resistance to lead. Using bacterial consortium culture would be much more effective than a single bacterial culture. Therefore one combination of highly resistant bacteria consortia was prepared and used in column bioreactor. Table 2 shows the selected bacteria for consortium prepared.

Lead concentration from samples extracted from the column was analyzed. An interval of one day was taken for different sampling batches for 10 days. The results show a reduction of lead for the control which was not inoculated with bacteria and kept constant. Lead reduction ranged from 0.923 to 0.1943 mg/l (Fig. 3). Results show

**Table 2** Bacteria selected for preparation of consortium

| S.no | Sample code | bacteria |
|---|---|---|
| 1 | S1-3 | *Escherichia spp* |
| 2 | S1-4 | *Klebsiella spp.* |
| 3 | S1-9 | *Pseudomonas spp.* |
| 4 | S1-10 | *Enterobacter spp.* |

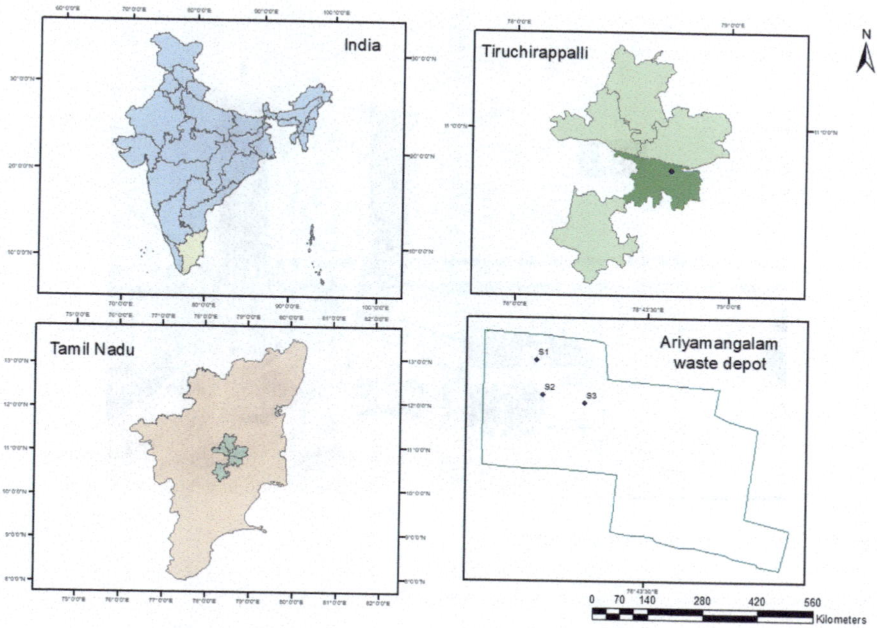


**Fig. 1** Map of study area and study site

from day 1 and 4, lead was observed with a drastic reduction and slightly reduce from day 6 to day 10. A drastic reduction might be due to enriched microbial consortium added to the waste samples in the column. Bacteria that are highly tolerant to lead could be the results of variation in their genetic makeup (Massadeh et al. 2005).

The pH of the solid waste samples has fluctuated during the time of the experiment. It was in the range of 6–6.5. Day 4 was observed to have a pH of 5. At day 10. The pH of samples was increased while the control sample was reduced to 7 (Fig. 4). The reduction of pH of the control might have been due to the continuous supply of distilled water into the column. However, the increase of the sample pH might have been due to anaerobic bacterial activities in the column whereas lower pH might be due to the increased presence of sulfate (Malin et al. 2017).

## 4 Conclusion

Tiruchirappalli dump yard daily receives 470 tonnes of mixed municipal solid waste containing heavy metals such as lead which would contaminate soil and groundwater in the area affecting the local population. The dump yard has a wide range of bacterial communities and some are resistant to lead and other heavy metals. However, the consortium of bacterial species such as *Pseudomonas spp.*, *Klebsiella spp.*, *Enterobacter spp.*, *Escherichia spp.* is considered to be effective and could be used for

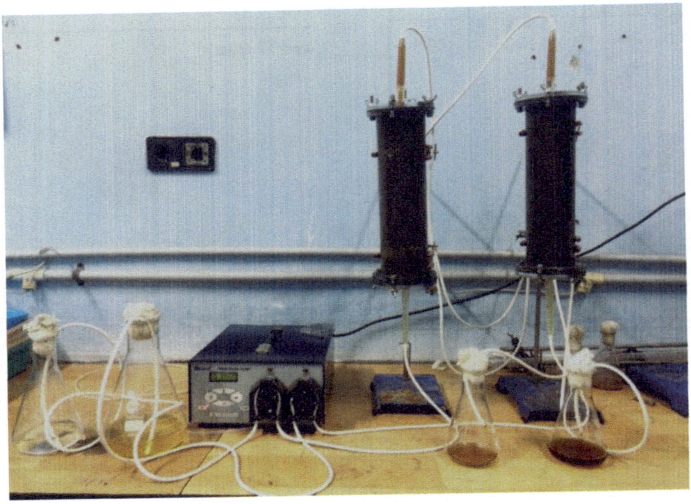

Fig. 2 Laboratory scale column setup

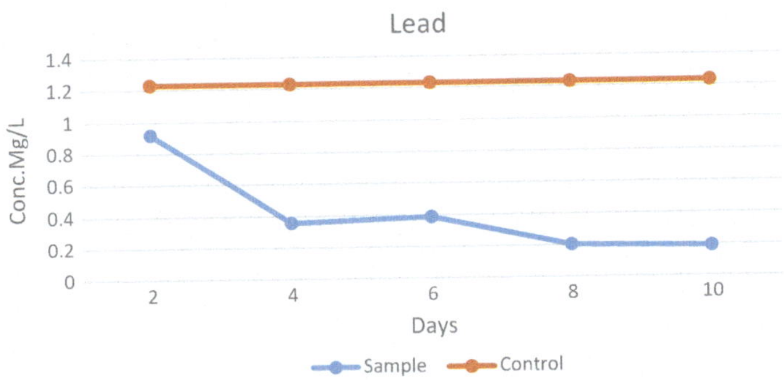


Fig. 3 Concentration of lead extracted through column test

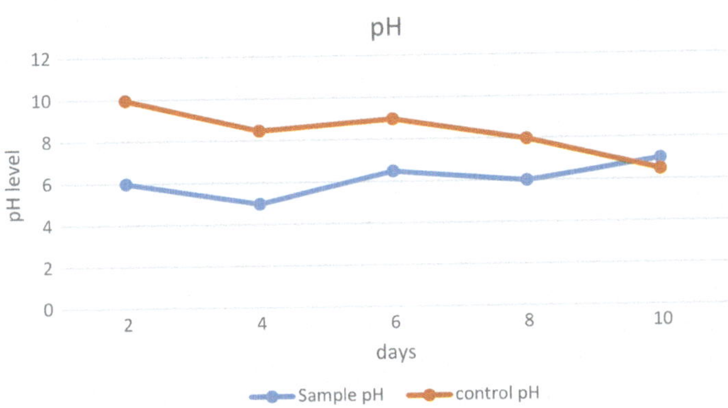


Fig. 4 pH level of the samples

bioremediation. These resistant species could also potentially be used for biomining of heavy metals.

## References

Ayyasamy PM, Lee S (2009) Redox transformation and biogeochemical interaction of heavy metals in Korean soil using different treatment columns in the presence of Shewanella sp. Chemosphere 77(4):501–509

Balint Nimirciag R, Buratto E, Ajmone-Marsan F (2013) Leaching of trace metals from soil under alternating oxic-anoxic conditions: a column study. In: E3S web of conferences, vol 1, p 19005

Bergey DH, Buchanan RE, Gibbons NE (1974) Bergey's manual of determinative bacteriology. Williams & Wilkins, Baltimore

Bergey DH (1986) Bergey's manual of systematic bacteriology, 1st edn, vol 2. The Williams & Wilkins Co., Baltimore

Hrynkiewicz K, Baum C (2014) Application of microorganisms in bioremediation of environment from heavy metals. Environ Deterioration Human Health 215–227

Malin B, Jarno M, Marja S, Mona A (2017) Microbial community structure and functions in ethanol-fed sulfate removal bioreactors for treatment of mine water. Microorganisms 5:61

Massadeh AM, Al-Momani FA, Haddad HI (2005) Removal of lead and cadmium by halophilic bacteria isolated from the dead sea shore, Jordan. Biol Trace Element Res 108(1–3):259–270

Moten AM, Rehman A (1998) Study on heavy trace metal ions in industrial waste effluents in Pakistan. Environ Exp Com Article 909

Rehman A, Zahoor A, Muneer B, Hasnain S (2008) Chromium tolerance and reduction potential of a Bacillus sp. ev3 isolated from metal contaminated wastewater. Bull Environ Contam Toxicol 81:25–29

Rojas LA, Yáñez C, González M, Lobos S, Smalla K, Seeger M (2011) Characterization of the metabolically modified heavy metal-resistant cupriavidus metallidurans strain MSR33 generated for mercury bioremediation. PLoS ONE 6(3):e17555

Tchounwou PB, Yedjou CG, Patlolla AK, Sutton DJ (2012) Heavy metal toxicity and the environment. Mol Clin Environ Toxicol 133–164

Thassitou P, Arvanitoyannis I (2001) Bioremediation: a novel approach to food waste management. Trends Food Sci Technol 12(5):185–196

U.S. EPA (1996) Method 3050B: acid digestion of sediments, sludges, and soils, Revision 2. Washington, DC

Yamina B, Tahar B, Marie Laure F (2012) Isolation and screening of heavy metal resistant bacteria from wastewater: a study of heavy metal co-resistance and antibiotics resistance. Water Sci Technol 66(10):2041–2048

# Bioremediation Approaches for Degradation of Textile Dye Effluents

**S. Lakshmana Prabu, A. Umamaheswari, and A. Puratchikody**

**Abstract** In earth, water is considered as one of the most valuable substance for living organisms for their healthy growth. Due to rapid industrialization and its advancement, in many occasions, thousands of living organisms die because of discharge of effluent/polluted water from various industries to the environment. Among the pollutants, textile industries effluents are the major causative agent damaging the ecosystem and water system in the environment significantly. Presently, almost one million tons of dyes are utilized in textile industry, whereas about 10% of dyes are released to the environment without effective effluent treatment. Among the different types of dyes, azo dyes are most predominant one in the textile industry and these azo dyes are not easily degraded by bacteria. These effluent wastewaters from azo dye textile units can deteriorate the ecosystem, quality of water, generate the pollution, eutrophication, interruption of aquatic life and can cause different disorders to the human health. In addition, metabolic intermediates of these dyes are mutagenic and carcinogenic; it can cause severe damage to different organs of the human body. Several physico-chemical techniques are utilized to treat this effluent. However, each technique has its own limitations, generating toxic and secondary waste products leading to critical situations. Recently, removal of dye from the effluent by biological method has gained much importance because of its sustainable approach, non-toxic, treatment cost effectiveness and ecofriendly in managing the pollution. In biological method, different biological elements such as microorganisms including fungi, bacteria, yeast, algae, actinobacteria and plants are utilized and it can be applicable to contaminated sites. Among the above biological elements, plants have been considered as one of potential as well as economical element in the dye effluent treatment. Consequently, successful enactment of this biological treatment can provide valuable protection this environment from this textile industry dye effluent wastewater. In this chapter, remediation techniques with respect to this textile dye effluent wastewater have been discussed deeply in order to protect and maintain this environment sustainably in near future.

---

S. Lakshmana Prabu (✉) · A. Umamaheswari · A. Puratchikody
Department of Pharmaceutical Technology, University College of Engineering (BIT Campus), Anna University, Tiruchirappalli 620024, Tamil Nadu, India
e-mail: slaxmanvel@gmail.com



M. Prashanthi et al. (eds.), *Bioremediation and Green Technologies*, Environmental Science and Engineering, https://doi.org/10.1007/978-3-030-64122-1_9

# 1 Introduction

In earth, water has been considered one of the most important substances and every living organisms like from smallest microorganism to the largest mammal must have water for their survival, no life on earth without water. In addition, water is having an important role for the healthy growth of plants, animals and in the production of many products through aerobic respiration and without water aerobic respiration could not take place. In living organisms, aerobic respiration creates the energy in the form of ATP for their life process and in earth the majority of the living organisms use aerobic respiration as an energy source. Some organisms have been made up to 95% of water, whereas almost all organisms are made of minimum of 50% water. For the normal healthy life like drinking and other purposes, the water should be clean which means it should be clear, free from any microorganisms and chemicals (https://sciencing.com/about-6384365-water-important-life-earth-.html).

Water has been collected from various sources such as Surface water, River water, Springs, Rock catchment areas and rock holes, Excavated dams, Rainwater tanks, Bores and wells, Artesian bores (https://www.health.gov.au/internet/publications/publishing.nsf/Content/ohp-enhealth-manual-atsi-cnt-l~ohp-enhealth-manual-atsi-cnt-l-ch6~ohp-enhealth-manual-atsi-cnt-l-ch6.1).

## *1.1 Origin of Synthetic Dye*

From the beginning of mankind, human being have used colorants for dyeing, painting their surrounding and for their cloths. Men have used dyes in the walls since 15,000–9,000 BC (Altanira caves in Spain). Till nineteenth century, men used natural dyes from plants, insects, lichens and mushrooms for the textile dye purpose (Shah 2018). These natural dyes are aromatic compounds and the major disadvantage of these natural dyes in the dying process is that it requires several steps.

In 1856, English chemist W.H. Perkin made an attempt to synthesis quinine, the research outcome obtained a substance having bluish in color which had a very good dyeing character. Later this bluish color substance has become a violet aniline, purple tyrant or mauveine (Aksu and Karabayir 2008). Due to diversity of these natural dye sources, application process, rapid change in the demand and advancement, this basic research made a new path for the preparation of different synthetic dyes and its utilization in near future. Nowadays in textile dying to meet the present needs nearly 100,000 synthetic dyes are produced (Robinson et al. 2001). In current scenario yearly almost 1,000,000 tons of dyes are manufactured and utilized in textile industry. The foremost reason for this increased production of dyes is to meet the present needs.

## 1.2 Classification of Dyes

Dyes are defined as molecules with electronic systems delocalized with conjugates double bonds which contain two groups such as chromophore and auxochrome. Chromophore is a group of atoms; generally it is an electron withdrawing groups and it controls the dye color. Examples for important chromophore group in dyes are –C=N–, –N=N–, –C=C–, –C=O, $-NO_2$ and –NO. Auxochrome is a group of atoms, generally it is an electron donor substituent group and it modifies the energy of the electronic system which leads to intensify the color of the chromophore. Examples for important auxochrome group in dyes are $-NR_2$, $-NH_2$, $-SO_3H$, –NHR, –COOH, –OH, $-SO_3H$, and $-OCH_3$ (Aksu and Tezer 2005; Alhassani et al. 2007).

Textile dyes were classified as azo, diazo, basic, cationic, anthraquione and metal complex based, depending on its chemical structure (Palanivelan et al. 2013). Chemical structure which is responsible for the color of the dyes was resistant to fading on exposure to water, light and many chemicals. Based on the chemical structure dyes are classified as cationic, anionic and nonionic. All these dyes have indigoid, antroquinoid and azo aromatic structures. In UV visible area, these groups allow strong π-π transitions with high extinction co-efficient. Among these three structures, most prevalent dye in the dye industry is azo aromatic. This azo dye has one or more nitrogen–nitrogen double bonds (–N==N–). Based on the azo group, azo dyes are called as mono-azo dyes, having two azo groups—di-azo dyes and tri-azo dyes with greater than three azo linkages are labeled as polyazo dyes. Cost effectiveness and ease of synthesis, these azo dyes are being increasingly utilized in the textile industries. The main disadvantages of these azo dyes that it cannot be easily degraded by aerobic, anaerobic or microaerobic reductive bacteria (Shivangi 2012). Dyes classification is shown in Fig. 1.

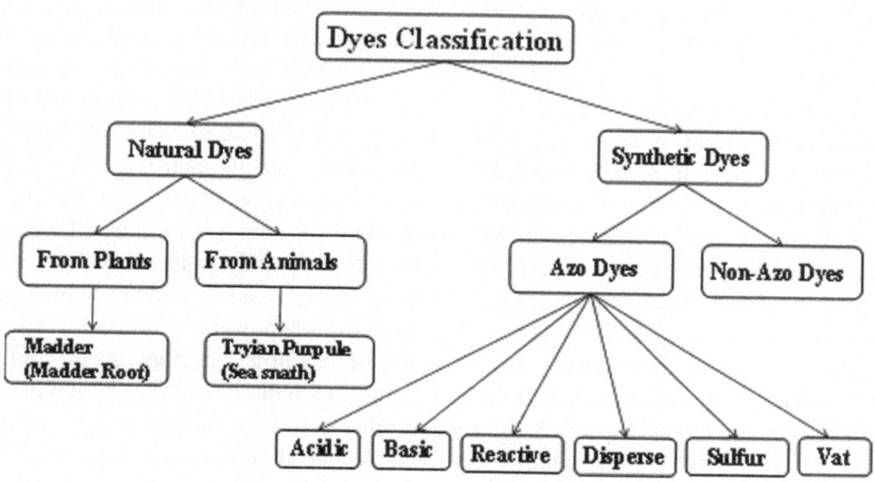


**Fig. 1** Classification of dyes

## 1.3 Importance of Textile Dye Effluent Treatment

Nowadays, water is getting contaminated due to rapid and advancement in industrialization and urbanization. In earth, in many occasions and throughout the history where thousands of living organisms have died because of polluted water discharged from various industries (https://www.health.gov.au/internet/publications/publishing.nsf/Content/ohp-enhealth-manual-atsi-cnt-l~ohp-enhealth-manual-atsi-cnt-l-ch6~ohp-enhealth-manual-atsi-cnt-l-ch6.1).

About 1,000,000 million tons of dyes are utilized in textile industries among which 10% dyes are discharged as waste into the natural resources and environment (Jadhav et al. 2010). Early study results reported that per day, about 1.6 million litres of water has been utilized for 8000 kg of fabric production in a medium sized textile industry (Khandare and Govindwar 2015). In simple 200 litres of water is essential to process about 1 kg of textile fabric.

World Bank assessed that textile industry water pollution is between 17 and 20% obtained from dyeing and finishing treatments of fabric (Holkar et al. 2016; Rani et al. 2013). The composite waste stream may have 200–600 ppm of biological oxygen demand (BOD) after treating of 1000 kg of cotton fabric, suspended solids in the range of 30–50 ppm and total solids in the range of 1000–1600 ppm in a volume of 50–160 $m^3$ (Kimmatkar et al. 2017).

In general, presence of small quantity of dye can affect the quality of water. Due to its dark color, the effluent blocks the sun light penetration leading to higher stability behaviour and toxicity of the dyes. Disposal of these colored dyes into the environment from the textile industries can deteriorate the quality of water by generating the pollution, eutrophication and disruption of aquatic life. This textile industries effluent has been considered as one of the major issue, because these azo dye are resistant to breakdown. Below 1 ppm, these colored dyes are clearly visible and significantly damages the ecosystem and water system in the environment (Gupta et al. 2003). About 72 toxic chemicals are discharged from textile dyeing industry of which, 30 chemicals cannot be removed by waste water treatment (Chen and Burns 2006). Also, most of the dyes are having complex structures, more stable to oxidation and light, resistant to breakdown. Disposal of these dyes into the environment generate hazardous and harmful metabolites into the aquatic ecosystem (Talarposhti et al. 2001). Non-biodegradable and its by-products are generated by the reaction between synthetic dyes with other chemicals in the environment. Earlier study results highlighted that textile dye effluents are highly polluted with dyes, chemicals, textile auxiliaries, surfactants, different salts, mineral oils, heavy metals and others (Hassan et al. 2009). Researchers reported that presence of different inorganic and organic compounds like vat dyes, naphthol, sulfur, enzymes, acetic acid, chromium compounds, soaps, nitrates, heavy metal includes copper, lead, mercury, nickel, arsenic, cadmium, cobalt, and some other auxiliary chemicals can cause toxic effects. Other organic compounds like chlorinated stain remover, formaldehyde as dye fixing agents, nonbiodegradable dyeing chemicals and hydrocarbon based softeners can cause carcinogenic.

In addition, these azo dyes alter the pH, reduce the dissolved oxygen level and increase the BOD, chemical oxygen demand (COD) and create a grave hazard to public health. Removal of colored dye substances from effluent is very critical and crucial one, hence effluents from textile industries has been identified and considered as a major pollutant among the all other pollutants (Bharathiraja et al. 2018).

### *1.4 Public Health Hazards/Health Impacts in Human Being*

Most of the azo dyes are carcinogenic and mutagenic (Pinherio et al. 2004). The azo dyes carcinogenic nature is due to the following mechanisms. First mechanism is based on the release of aromatic amines from the azo dye by cleavage of azo bond. These azo bonds covalently binds with DNA leads to form metabolically oxidized state. Second mechanism is based on the free aromatic group from the azo dye, without azo reduction to form metabolically oxidized form. Third mechanism is based on the direct oxidation of azo bond with electrophilic diazonium salts of the azo dyes (Bharathiraja et al., 2018).

Mutagenic nature of the azo dyes is based on the mutagenic compounds like aromatic amines which includes chloro aniline, napthylamines etc. (Shivangi 2012). In human health, presence of these azo dyes or its degraded products can cause different disorders like allergies, hemorrhage, nausea, ulceration of skin infections (skin irritation, dermatitis) and mucous membranes. Also, it can cause severe damage to reproductive system, kidney, central nervous system, brain and liver. Different organ cancers such as liver, spleen and bladder, in addition chromosomal deformities in mammalian cells were also reported for these azo dyes (Chung and Chen 2009; Ali 2010; Pooja et al. 2019).

## 2 Dye Removal (Wastewater Treatment) Techniques

Effluent treatment, disposal cost and regulations have made effluent treatment techniques play a critical role to protect the water resources as well as the environment. The utmost important objective of the wastewater (effluent) treatment is to safe guard the environment and to provide socio-economic concerns with public health (Metcalf and Eddy 1991). The main principle of the wastewater treatment process is to remove the pollutant from the effluent instead of destroying them. To design an appropriate technique, understanding the nature of effluent is a fundamental one. Textile dye effluent treatment techniques are shown in Fig. 2.

Each technique in the dye removal has its own limitations. Selection of an appropriate technique for the dye removal is based on the various factors. The factors are

- Limits of the contaminants/pollutants in the final effluent from the plant must be

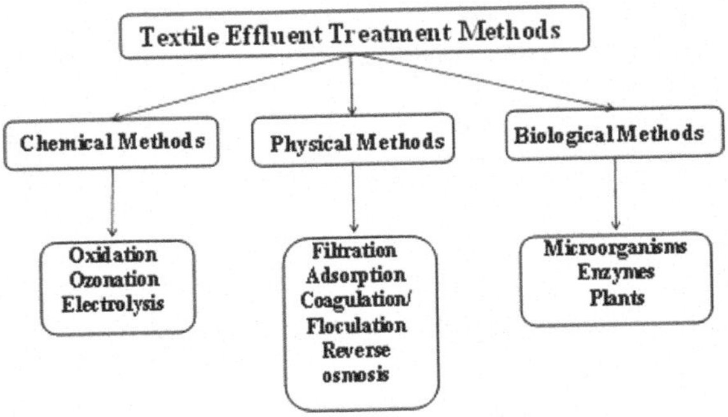


**Fig. 2** Textile dye effluent treatment techniques

- Physico-chemical properties of the pollutant in the final effluent
- Nature of the generated solid wastes from the treated water
- Economic feasibility
- Environmental fate and handling
- Running cost (Lakshmana Prabu et al. 2016).

## 2.1 Physical and Chemical Effluent Treatment Techniques

Different methods are performed for the dye-contaminated wastewater treatment, the treatment techniques are (Ahmad et al. 2015).

- Coagulation/Flocculation
- Membrane separation
- Ion exchange
- Advanced oxidation process
- Adsorption by activated carbon
- Fenton's reagent
- Ozonation
- Photochemical
- Electrochemical destruction
- Irradiation
- Electro kinetic coagulation.

## 2.2 Coagulation/Flocculation

Coagulation/Flocculation is a physico-chemical process that assists the accumulation of fine particles (colloids) contained in water to form a floc that can be easily removed through filtration water.

Coagulation—Aggregation with specific type where the aggregates are compact in nature.

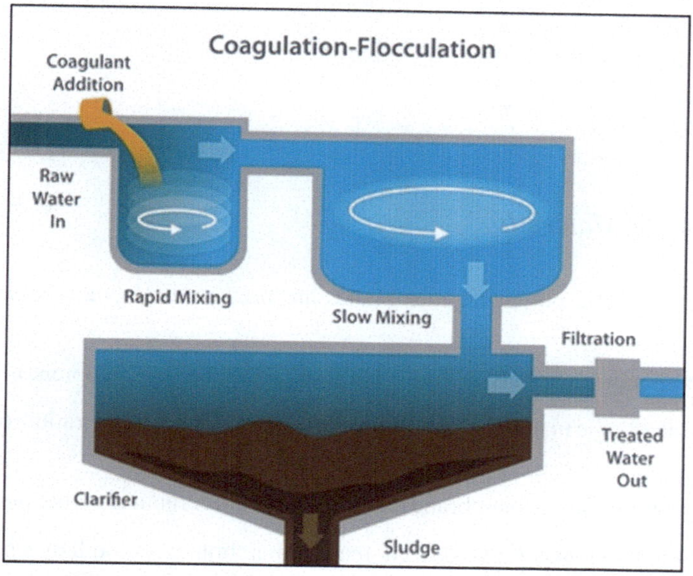


**Fig. 3** Schematic diagram for coagulation-flocculation

Flocculation—A type of aggregation where the aggregates are loose or open type.

Coagulation/Flocculation technique is used to remove the particles from the effluent also to achieve the maximum removal of COD. The mechanism involved in the Coagulation/Flocculation includes sorption (protonation of amino groups), electrostatic attraction and bridging (interrelated with polymer high molecular weight). Commonly used coagulants are ferrous sulfate, ferric chlorisulfate, ferric chloride and alum (aluminium sulfate). Though, conventional aluminium based coagulants can cause Alzheimer's disease (Hassan et al. 2009; https://www.profilt.net/en/water-treatment-principles/; https://www.columbia.edu/~ps24/PDFs/Principles%20of%20Flocculation%20Dispersion%20Selective%20Flocculation.pdf; Sabur et al. 2012). Schematic diagram for coagulation-flocculation is shown in Fig. 3.

## 2.3 *Membrane Separation*

Membrane processes is a process used to separates the specific materials based on its pore size or minute gaps which is present as a molecular arrangement in the continuous structure of the membrane. This membrane separation technique is not only used to produce purified water also used for recycling the specific contaminants from the industrial effluent waste. Based on the nature of membrane, membrane separation technique is classified as

a. Micro filtration

b. Ultra filtration
c. Nano filtration
d. Ion exchange
e. Reverse osmosis.

## 2.4 Type of Materials

Natural or synthetic polymeric membranes are used in the membrane separation process.

Natural membrane materials—Wool, cellulose and rubber (polyisoprene).

Synthetic membrane materials—polystyrene, polyamide and polytetrafluoroethylene (Teflon).

Inorganic non-polymeric membrane materials—Metal, ceramic, zeolites and carbon.

Hybrid membrane materials—Mixed matrix membranes of both inorganic and organic compounds.

Biopolar membranes—Made with different ionic charge materials (https://www.asahi-kasei.co.jp/membrane/microza/en/kiso/kiso_1.html; https://www.separationprocesses.com/Membrane/MT_Chp03.htm; Mohan et al. 2014).

## 2.5 Ion Exchange

Ion exchange—Effluent is passed through a ion exchange bed, which removes the undesirable ions and replace with less objectionable ones. Ion exchange water softener is the most common process used in water treatment. By making an appropriate design and operation, this process can remove selected ions completely. It is a very versatile and effective process to remove water soluble hazardous materials from textile dye effluent. This process can be used to remove the particular ions as well as to concentrate the pollutants.

Some common ion exchangers are

- $H^+$ (proton) and $OH^-$ (hydroxide).
- Monovalent charged ions like $Na^+$, $K^+$, and $Cl^-$.
- Divalent charged ions like $Ca^{2+}$ and $Mg^{2+}$.
- Polyatomic inorganic ions like $SO_4{}^{2-}$ and $PO_4{}^{3-}$.
- Organic bases containing amine functional groups like $-NR_2H^+$.
- Organic acids containing carboxylic group as function group $-COO^-$ (https://textilelearner.blogspot.com/2012/05/base-exchange-ion-exchange-water.html; Lin and Chen 1997a, b; Parameswaran et al. 2014; Wawrzkiewicz and Hubicki 2015).

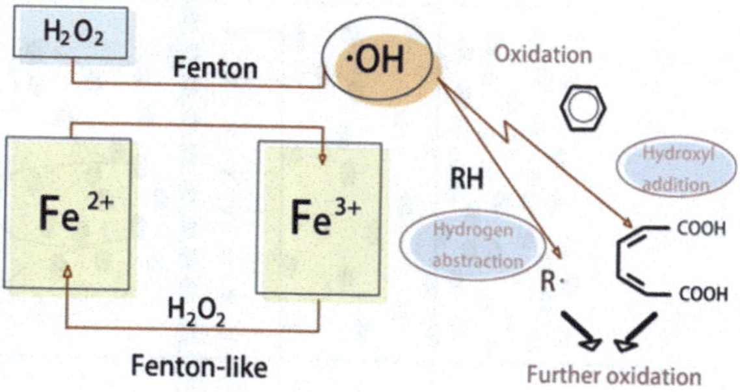


**Fig. 4** Fenton oxidation process

## 2.6 *Advanced Oxidation Process*

It is a process with a set of chemical treatment procedure to generate a powerful but relatively nonselective highly reactive oxidizing species like hydroxyl radical. This process can remove both organic and inorganic materials in the effluent by oxidation (Pesoutova et al. 2011). Fenton oxidation process (Huang et al. 2017) is shown in Fig. 4.

Types of advanced oxidation process

a. **Fenton Process**: This is a simple method for the generation of OH radicals. Hydroxyl radicals are generated by reaction between the additions of $H_2O_2$ to $Fe^{2+}$ salts (Pera-Titus et al. 2004).

$$H_2O_2 + Fe^{2+} \rightarrow OH^{\bullet} + OH^{-} + Fe^{3+} \quad (1)$$

b. **Electro-Fenton Process**: This is another method to generate hydroxyl radicals by electrochemical reduction of oxygen in presence of ferrous ions.
c. **Photo-Fenton Process**: In this method, hydroxy ions are generated by the UV radiation on the ferrous ions.

$$H_2O_2 + UV \rightarrow OH^{\bullet} + OH^{-} \quad (2)$$

## 2.7 *Adsorption by Activated Carbon*

It is a process in which effluent components are attracted to the surface of a solid adsorbed by through physical or chemical bonding. Adsorption through activated carbon is a simple, economic and very effective method to remove the pollutants. Among the different adsorbents, activated carbon has been considered as an effective adsorbent to remove both inorganic and organic pollutants from the textile effluent. Adsorption of the pollutants is due to either physical sorption (Van der Waals forces) or chemisorption (chemical bonding). Activated carbon is a crude form of graphite, amorphous in nature with highly porous in different pore size (Pera-Titus et al.

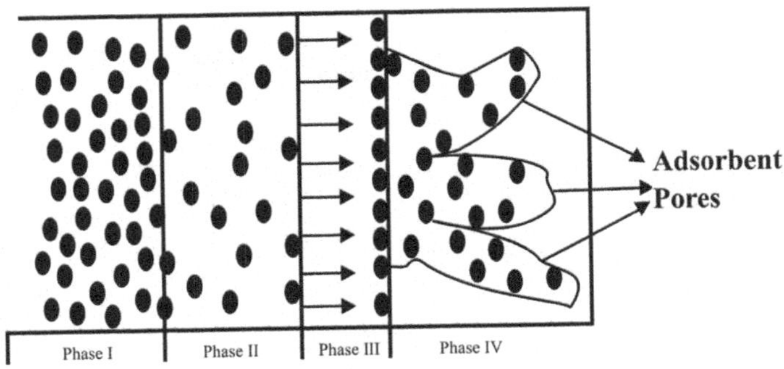


**Fig. 5** Adsorption process between the pollutants and adsorbents

2004; Foo and Hameed 2010; Popuri and Guttikonda 2015). The adsorption process between the pollutants and adsorbents (Patel 2018) are shown in Fig. 5.

## 2.8 *Ozonation*

Ozonation technique is preferably used in treating wastewater due to its oxidation property. At pH 9, it has been considered as an advanced oxidation process. Chemical reaction between the ozone and the hydroxyl radicals is the main oxidation process and preferably act on the unsaturated bonds of chromophores [conjugated double bonds] of the dye. In water, ozone decompose into ozone and radical species. In ozonation process, either directly or indirectly oxidation can take place (Shriram and Kanmani 2014).

Direct Direct reaction with the dissolved molecular ozone ($O_3$).
Indirect Indirect reaction with the radical species (HO, $HO_2$).

Factors affecting the ozonation process are

- pH,
- Ozone dose,
- UV radiation,
- concentration of dye,
- Temperature.

## 2.9 *Irradiation*

Ionizing radiation is another technique in textile effluent treatment. Gamma rays is an electromagentic radiation source with high frequency and sufficient energy. They generate ions by extricating the electrons from atoms and molecules. Further reaction of ions forms free radicals. This formed free radicals radiolysis the water and undergoes chemical reaction with the dye components (azo dyes) and convert into amide, subsequently to ammonium salts (Rahman Bhuiyan et al. 2014; Parvin et al. 2014).

### 2.9.1 Electrochemical Methods

Electrochemical methods are

1. Electrochemical oxidation
2. Electro coagulation
3. Electro floatation.

*Electrochemical oxidation*

In this method, pollutants are converted into simple forms like $CO_2$ and $H_2O$ by oxidation. Pollutants are oxidized either direct or indirect.

Direct oxidation (anodic oxidation process)—In this method, pollutants are adsorbed on the anode surface and oxidized by anodic electron transfer reaction.

Indirect oxidation—Here, electrochemically potent oxidants such as chlorine, hypochorite, hydrogen peroxide, ozone and hydroxyl ions are produced at the anode surface. Potent oxidants are formed in two stages.

First stage—Adsorbed hydroxyl radical is produced at the anode by $H^+$ in acidic solution or $OH^-$ in alkalin solution.

Second stage—Interaction between the adsorbed hydroxyl radical and metal oxide leads to produce higher oxide components (Feng et al. 2016; Radjenovic and Sedlak 2015; Martínez-Huitle and Andrade 2011).

*Electro Coagulation*

In this method, electrostatic charges on pollutants are neutralized and expedite coagulation formation, which will help to purify the water. This technique includes simultaneous formation electro coagulation followed by electro flocculation.

For electro coagulation aluminium from aluminium electrode or ion from ion electrodes are used to generate corresponding metal ions. These metal ions can form multi-charged polynuclear complex, which has an improved adsorption characteristic property leads to form coagulation of the pollutant. Instantaneously hydrogen gas is engendered from the cathode; it will help to float the coagulated particles out of water. For coagulation of chemical method, metal salts such as $Al_2(SO_4)_3$, $Fe_2(SO_4)_3$, $Fe(SO_4)$ and $FeCl_3$ are preferred. Schematic diagram of an electrocoagulation process (Cañizares et al. 2005) is shown in Fig. 6.

Steps involved in the electrocoagulation process are

1. Electrolytic reactions at the electrode
2. Coagulants formation

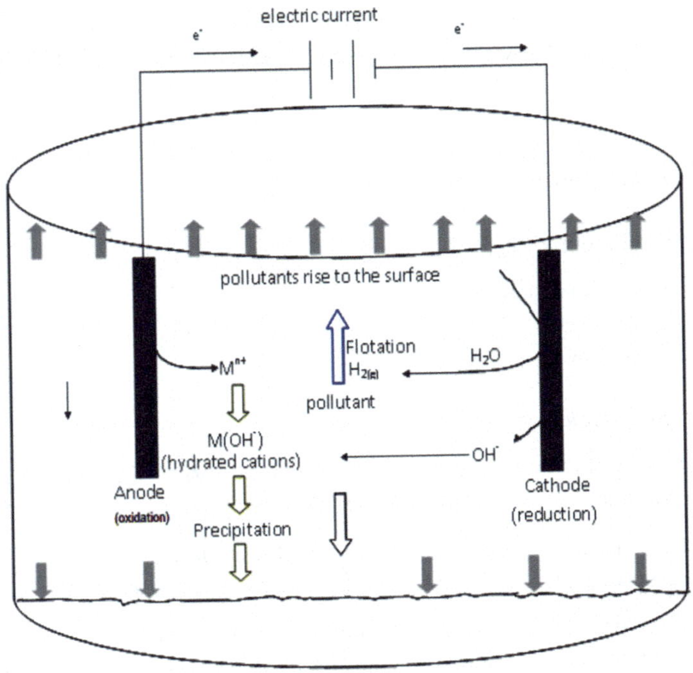


**Fig. 6** Schematic diagram of an electrocoagulation process

3. Pollutants adsorption
4. Separation of pollutants by floatation or sedimentation.

Factors affecting electro coagulation are

- Effect of pH
- Effect of Current Density
- Effect of Electrode Material
- Effect of Electrolyte (NaCl) Concentration
- Effect of electrocoagulation Time
- Electrical Energy and Electrode Consumption
- Effect of Applied Voltage
- The Effect of Inter Electrode Distance
- Effect of Operating Temperature
- Effect of Initial Dye Concentration.

*Electro Flotation*

Electro flotation is a subsequent process of electro coagulation. In electrochemical reaction, small even size bubble of gases such as hydrogen from cathode and oxygen from anode are evolved. These evolved gases will assist to floats the pollutants to the water surface.

Steps involved in the electro flotation are

1. Generation of gas bubble
2. Interaction between pollutants and gas bubble
3. Adsorption of the gas bubble on the pollutants surface
4. Rising of gas bubble along with pollutants to the surface of water.

Factors affecting the flotation rate are

1. Solution surface tension
2. Diameter of gas bubble
3. Pollutant particle size
4. Residence time of water in the electrolytic cell
5. Zeta potential of gas bubble
6. Temperature
7. pH
8. Concentration of the pollutant (Arulmurugan et al. 2007; Ukiwe et al. 2014; Abbas and Ali 2018; Butler et al. 2011; Naje et al. 2015).

#### 2.9.2 Disadvantages

All the above methods have its own limitations, but high cost and low efficiency when more number of dyes is present in the waste water stream, also not environmentally safe. Further generations of toxic and secondary waste products are the quite aggravating of these methods which needs to be tackled further.

## 3 Biological Methods

Recently, treating due effluent by biological methods has been received much consideration due to its cost, easy process, simplicity and can used to remove different categories of dye effluents. In biological methods, microorganisms like fungi, bacteria, yeast, actinobacteria, algae and plants are used. Among the different biological sources, plants (Phytoremediation) has been considered to be more potential and economical when compared to all other methods.

### *3.1 Phytoremediation*

Phytoremediation—Utilization of live plants to remove the pollutants from the environments. The term Phytoremediation is derived from Greek Phyto, plant and Latin remedium, cure. This phytoremediation method is an emerging technology, gaining much interest among the researchers due to its promising results with inexpensive. It is the in situ and eco-friendly method to remove different hazardous waste pollutants like pesticides, metal hydrocarbons and chlorinated solvents etc. (Macek et al. 2000; Susarla et al. 2002; Xia et al. 2003).

In phytoremediation method, live plants are utilized for removing the pollutants by

1. Degradation
2. Assimilate
3. Metabolize
4. Detoxification of pollutants (Vasanthy et al. 2011).

Mechanisms involved in the Phytoremediation are

1. Uptake the contaminants, aggregation and consequently metabolised in plant tissues.
2. Through leaves volatile organic hydrocarbons are transported.
3. Stimulation of microbial activity by releasing exudates and transfer biochemically nearby the roots.
4. At the root-soil surface improve the mineralization that is recognized by microbial consortia and mycorrhizal fungi (Schnoor et al. 1995).
5. Original-Enhancement of mineralization at the root–soil interface that is attributed to mycorrhizal fungi and microbial consortia associated with the root surface (Schnoor et al. 1995).

*Classification of Phytoremediation*

Based on the applicability, fundamental process and type of pollutants, phytoremediation is classified as

i. Phytodegradation
ii. Phytoextraction
iii. Phytostimulation or rhizodegradation
iv. Phytovolatilization
v. Rhizofiltration
vi. Phytostabilization (Materac et al. 2015; Ismail 2012; Sri Lakshmi et al. 2017).

*Phytodegradation*

Phytodegradation, is also named as phytotransformation. It is a metabolic process, enzymes which are secreted from the plant that catalyze and accelerate the chemical reaction to degrade the xenobiotics (Alkio et al. 2005). Phytodegradation can occur either inside the plant or outside the plant (Burns et al. 2013). In plant, microbes are colonized around the shoot and roots, these accumulated microbes can degrade the carbon substrates. Organic pollutants are degraded into small molecules and these molecules are fused into the plant tissues which will help for the plant growth (https://www.unep.or.jp/Ietc/Publications/Freshwater/FMS2/2.asp; Materac et al. 2015; Sri Lakshmi et al. 2017). Degradation of organic contaminant by phytodegradation (UNEP 2002) is shown in Fig. 7.

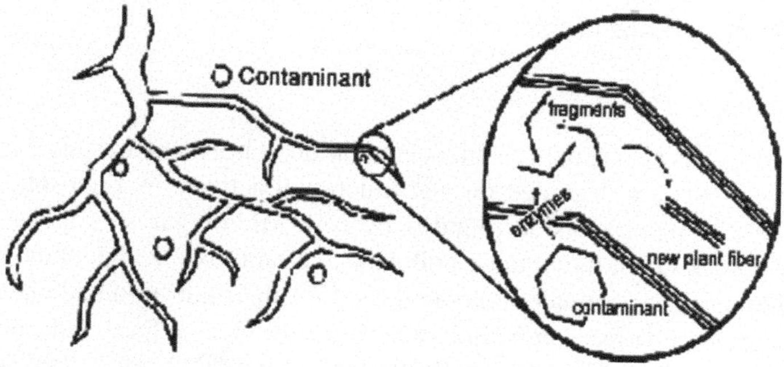


**Fig. 7** Degradation of organic contaminant by phytodegradation

*Phytoextraction*

Phytoextraction is also named as phytoaccumulation and has largest economic opportunities in phytoremediation (Raskin et al. 1997). In this phytoextraction, toxic metal pollutants from soil are uptake by plant roots, transferred and accumulated in the above ground plant body parts. When the plant attained maximum growth and metal accumulation, plants are harvested above the ground levels leads to perpetual elimination of metal from the site. By phytoextraction metals such as zinc, nickel and copper can be eliminated (Ismail 2012; https://www.unep.or.jp/Ietc/Publications/Freshwater/FMS2/2.asp; Materac et al. 2015; Sri Lakshmi et al. 2017). Extraction of metal (Nickel) by phytoextraction (https://www.unep.or.jp/Ietc/Publications/Freshwater/FMS2/2.asp) is shown in Fig. 8.

Phytoextraction is classified as

1. Chelate Assisted Phytoextraction or induced phytoextraction
2. Continuous Phytoextraction.

Phytoextraction depends on

1. Availability of metals in the soil
2. Absorbing capacity and concentrate of plants

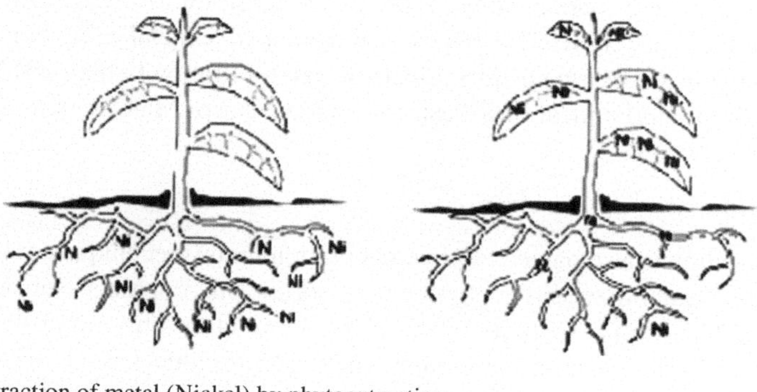

**Fig. 8** Extraction of metal (Nickel) by phytoextraction

3. Detoxification of the metals.

*Phytostimulation*

Phytostimulation is also named as rhizodegradation. It is a plant assisted degradation process; pollutants are degraded through microbial activity by rhizosphere which is present in the plant roots surrounded by soil. Microorganisms (yeast bacteria and fungai) break down the organic pollutants, certain toxic compounds, fuels and solvents into harmless simpler compounds and for their nutrition and energy. Root of the plant release some compounds which are utilized as food and nutrition by microorganism and contribute for their growth (Ismail 2012; https://www.unep.or.jp/Ietc/Publications/Freshwater/FMS2/2.asp; Materac et al. 2015; Sri Lakshmi et al. 2017).

*Phytovolatilization*

In phytovolatilization, pollutants are absorbed by the plant roots from the soil, transferred to the plant aerial parts, then finally to the leaves. In leaves, through metabolic activities theses pollutants are converted into gaseous form (volatile) and then transpired to the atmosphere.

Phytovolatilisation take place in growing trees. Metals like Se and Hg can be are removed by this phytovolatilisation. The main disadvantage of this process is elimination of metals as volatile compounds into the environment or atmosphere is safe or not (Ismail 2012; https://www.unep.or.jp/Ietc/Publications/Freshwater/FMS2/2.asp; Materac et al. 2015; Sri Lakshmi et al. 2017).

*Phytofiltration*

Phytofiltration is named based on the plant roots (rhizofilteration) or seedlings (blastofilteration) (Raskin et al. 1997). In Rhizofiltration (rhizo means "root"), plant roots absorb the pollutants (metals) from the water adjacent to the root zone. This method is used to remove the contaminants from water rather than soil. In this method, either contaminated water is collect from the waste site or growing plants hydroponically and relocating into the contaminated water area, where plant roots adsorb the contaminants from the water. Over a period of time, the roots will be saturated with the contaminants, then whole plant or roots are collected for disposal. Metal such as Pb, Cr and Zn can be removed by this method. Factors like metal being removed and plant metabolism are the two factors affects the phytofiltration (Ismail 2012; https://www.unep.or.jp/Ietc/Publications/Freshwater/FMS2/2.asp; Materac et al. 2015; Sri Lakshmi et al. 2017).

*Phytostabilization*

Phytostabilization is also named as phytorestoration. In this technique contaminants are immobilized/stabilized either physically by sorption in root surface or chemically by chemical fixation in the roots of the plant. In addition preventing the vertical migration of contaminants to the ground water, which reduces food chain process entry and dispersion through wind or water.

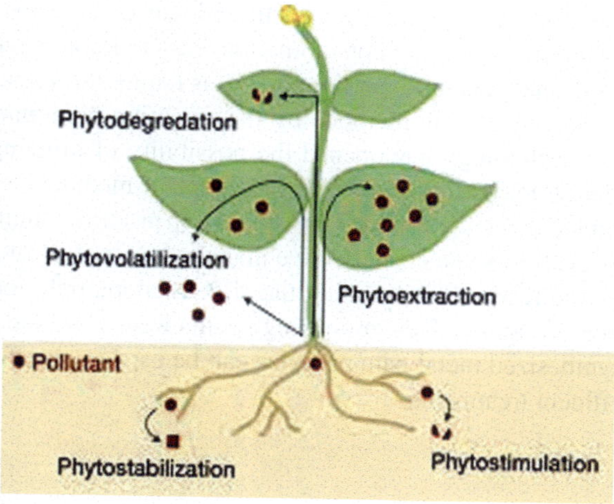


**Fig. 9** Different phytoremediation process

Plants which are used for this phytostabilization technique should have extreme root system to adsorption, absorption and accumulation of pollutants/contaminants in the root and high tolerance to varying pH (Ismail 2012; Materac et al. 2015; Sri Lakshmi et al. 2017; https://www.unep.or.jp/Ietc/Publications/Freshwater/FMS2/2.asp; Cunnigham and Berti 1993).

Different phytoremediation process (Pilon-Smith 2005) is shown in Fig. 9.

## 4 Conclusion

In earth, overall quality and standard of life of every living organisms is decided based on the growth supporting system for its survival. Among the various growth supporting system, water has been considered one of the important source for their survival. In simple, on earth no life without water. Advancement and industrialization of the urbanization, more amount of the polluted water is discharged from various industries. Among the various industrial pollutants of our environment, textile dye and its effluent has been identified as most predominant pollutant. Treatment techniques are classified as physical, chemical and biological methods, but each treatment technique has its own limitations in the dye removal. In physical method (ion exchange, adsorption and membrane filtration), dye molecules are simply transferred to another phase rather than removing the pollutant also this method is applicable only to the small volume of effluent. In chemical method (electrochemical degradation, chemical oxidation and ozonation), an effective pretreatment sludge production is required. When compared to the physical and chemical methods, biological method (phytoremediation) has many advantages and can adopt itself for its survival, grow in the high polluted area and detoxify the contaminants over the period of time. Major advantages of the biological method (phytoremediation) are environmental friendly, low cost (cost effective), in situ and can be engaged at the infected site without disturbing the site of contamination, easy to implement, no secondary pollution and no skilled person (expertise person) is required for handing and implementation.

Search of new methods by the scientific community in the past few decades, nanotechnology has opened the possibility of utilizing in different area due to its unique properties. Green synthesis (plant mediated synthesis) of nanoparticle has gained great interest due to its single step process, elimination of harmful reagents for its synthesis in a cost effective manner. Various investigations have been performed by the researchers utilizing the silver nanoparticle for effective catalytic degradation of organic dyes, promising results were observed. Hence, in near future, green synthesized metal nanoparticles can be used as an effective technique in the textile effluent treatments.

## References

Abbas SH, Ali WH (2018) Electrocoagulation technique used to treat wastewater: a review. Am J Eng Res 7(10):74–88

Ahmad A, Siti S, Sing C, Asma K, Waseem AW, Rajeev K, Mohd R (2015) Recent advances in new generation dye removal technologies: novel search of approaches to reprocess waste water. RSC Adv 39:1–58

Aksu Z, Karabayir G (2008) Comparison of biosorption properties of different kinds of fungi for the removal of gryfalan black RL metal-complex dye. Bioresour Technol 99:7730–7741

Aksu Z, Tezer S (2005) Biosorption of reactive dyes on the green alga Chlorella vulgaris. Process Biochem 40:1347–1361

Alhassani H, Muhammad R, Ashraf S (2007) Efficient microbial degradation of toluidine blue dye by Brevibacillus sp. Dyes Pigm 75:395–400

Ali H (2010) Biodegradation of synthetic dyes-a review. Water Soil Pollut 213:251–273

Alkio M, Tabuchi TM, Wang X, Colon-Carmona A (2005) Stress responses to polycyclic aromatic hydrocarbons in *Arabidopsis* include growth inhibition and hypersensitive response-like symptoms. J Exp Bot 56:2983–2994

Arulmurugan A, Chithra K, Thilakavathi R, Balasubramanian N (2007) Degradation of textile effluent by electro coagulation technique. Bull Electrochem 23:247–252

Base Exchange (Ion Exchange) Water Softening Process. https://textilelearner.blogspot.com/2012/05/base-exchange-ion-exchange-water.html

Bharathiraja B, Jayamuthunagai J, Praveenkumar R, Iyyappan J (2018) Phytoremediation techniques for the removal of dye in wastewater. In: Varjani SJ et al. (eds), Bioremediation: applications for environmental protection and management, energy, environment, and sustainability. Springer Nature Singapore Pvt Ltd, pp 243–252

Burns RG, DeForest JL, Marxsen J, Sinsabaugh RL, Stromberger ME, Wallenstein MD et al (2013) Soil enzymes in a changing environment: current knowledge and future directions. Soil Biol Biochem 58:216–234

Butler E, Hung YT, Yeh RUL, Al Ahmad MS (2011) Electrocoagulation in wastewater treatment. Water 3:495–525

Cañizares P, Carmona M, Lobato J, Martínez F, Rodrigo MA (2005) Electrodissolution of aluminum electrodes in electrocoagulation processes. Ind Eng Chem Res 44(12):4178–4185

Chen HL, Burns LD (2006) Environmental analysis of textile products. Cloth Text Res J 24:248–261

Chung YC, Chen CY (2009) Degradation of azo dye reactive violet 5 by $TiO_2$ photocatalysis. Environ Chem Lett 7:347–352

Cunnigham SD, Berti WR (1993) Remediation of contaminated soils with green plants: an overview. Vitro Cell Dev Biol 29:207–212
Environmental health practitioner manual: a resource manual for environmental health practitioners working with aboriginal and torres strait islander communities. https://www.health.gov.au/internet/publications/publishing.nsf/Content/ohp-enhealth-manual-atsi-cnt-l~ohp-enhealth-manual-atsi-cnt-l-ch6~ohp-enhealth-manual-atsi-cnt-l-ch6.1
Feng Y, Yang L, Liu J, Logan BE (2016) Electrochemical technologies for wastewater treatment and resource reclamation. Environ Sci Water Res Technol 2:800
Foo KY, Hameed BH (2010) An overview of dye removal via activated carbon adsorption process. Desalin Water Treat 19:255–274
Gupta VK, Ali SI, Mohan D (2003) Equilibrium uptake and sorption dynamics for the removal of a basic dye (basic red) using low-cost adsorbents. J Colloid Interf Sci 265(2):257–264
Hassan MAA, Li TP, Noor ZZ (2009) Coagulation and flocculation treatment of wastewater in textile industry using chitosan. J Chem Nat Resour Eng 4(1):43–53
Holkar CR, Jadhav AJ, Pinjari DV, Mahamuni NM, Pandit AB (2016) A critical review on textile wastewater treatments: possible approaches. J Environ Manage 182:351–366
https://www.profilt.net/en/water-treatment-principles/
Huang D, Hu C, Zeng G, Cheng M, Xu P, Gong X et al (2017) Fenton image—combination of Fenton processes and biotreatment for wastewater treatment and soil remediation. Sci Total Environ 574:1599–1610
Ismail S (2012) Phytoremediation: a green technology. Ira J Plant Physiol 3(1):567–576
Jadhav JP, Phugar SS, Dhanve RS, Jadhav SB (2010) Rapid biodegradation and decolorization of direct orange 39 (orange TGLL) by an isolated bacterium Pseudomonas aeruginosa strain BCH. Biodegrdation 21(3):453–463
Khandare RV, Govindwar SP (2015) Phytoremediation of textile dyes and effluents: current scenario and future prospects. Biotech Adv 33:1697–1714
Kimmatkar KS, Purohit AV, Sanyal AJ (2017) Phytoremediation techniques and species for combating contaminants of textile effluents—an overview. Int J Sci Res 6(3):2174–2181
Lin SH, Chen ML (1997) Combined ozonation and ion exchange treatments of textile wastewater effluents. J Environ Sci Health Part A Environ Sci Eng Toxicol 32(7):1999–2010
Lin SH, Chen ML (1997) Purification of textile wastewater effluents by a combined Fenton process and ion exchange. Desalin 109:121–130
Macek T, Mackova M, Kas J (2000) Exploitation of plants for the removal of organics in environmental remediation. Biotech Adv 18:2334
Mahdavi Talarposhti A, Donnelly T, Anderson GK (2001) Colour removal from a simulated dye wastewater using a two-phase anaerobic packed bed reactor. Water Res 35:425–432
Martínez-Huitle CA, Andrade LS (2011) Electrocatalysis in wastewater treatment: recent mechanism advances. Quim Nova 34(5):850–858
Materac M, Wyrwicka A, Sobiecka E (2015) Phytoremediation techniques in wastewater treatment. Environ Biotech 11(1):10–13
Membrane materials. https://www.separationprocesses.com/Membrane/MT_Chp03.htm
Metcalf and Eddy (1991) Wastewater engineering: treatment, disposal and reuse, 3rd edn. McGraw Hill, New York
Mohan DR, Pramila J, Kavitha E, Nithya D, Hema Mala G, Tarun M (2014) Remediation of textile effluents by membrane based treatment techniques: a state of the art review. J Chem Pharma Sci 4:296–299
Naje AS, Chelliapan S, Zakaria Z, Abbas SA (2015) Treatment performance of textile wastewater using electrocoagulation (EC) process under combined electrical connection of electrodes. Int J Electrochem Sci 10:5924–5941
Palanivelan R, Rajakumar S, Jayanthi P, Ayyasamy PM (2013) Potential process implicated in bioremediation of textile effluents: a review. Adv Appl Sci Res 4(1):131–145
Parameswaran V, Nagarajan ER, Murugan A (2014) Treatment of textile effluents by ion-exchange polymeric materials. Int J ChemTech Res 6(6):3332–3335

Parvin F, Ferdaus Z, Tareq SM, CHoudhur TS, Islam JMM, Khan MA, et al (2014) Effect of gamma-irradiated textile effluent on plant growth. Int J Recycl Org Waste Agric 4(1):23–30
Patel H (2018) Charcoal as an adsorbent for textile wastewater treatment. Sep Sci Tech 53(17):2797–2812
Pera-Titus M, García-Molina M, Banos MA, Gimenez J, Esplugas S (2004) Degradation of chlorophenols by means of advanced oxidation processes: a general review. Appl Catal B Environ 47(4):219–256
Pesoutova R, Hlavinek P, Matysikova J (2011) Use of advanced oxidation processes for textile wastewater treatment—a review. Food Environ Safety J Faculty Food Eng Stefancel Mare University, Suceava 10(3):59–65
Phytoremediation: an environmentally sound technology for pollution prevention, control and redmediation. https://www.unep.or.jp/Ietc/Publications/Freshwater/FMS2/2.asp
Pilon-Smith E (2005) Phytoremediation. Annual Rev Plant Biol 56:15–39
Pinherio HM, Touraud E, Tomas O (2004) Aromatic amines from azo dye reduction: status review with emphasis on direct UV spectrophotometric detection in textile industry wastewaters. Dye Pigment 61:121–139
Pooja M, Jyotsna K, Arun U, Jasdev B (2019) Assessment of phytoremediation potential of *Chara vulgaris* to treat toxic pollutants of textile effluent. J Toxicol Article ID 8351272
Popuri AK, Guttikonda P (2015) Treatment of textile dyeing industry effluent using activated carbon. Int J Chem Sci 13(3):1430–1436
Principles of flocculation dispersion selective flocculation. In: Principles of water treatment: coagulant and flocculant—profit.http://www.columbia.edu/~ps24/PDFs/Principles%20of%20Flocculation%20Dispersion%20Selective%20Flocculation.pdf
Radjenovic J, Sedlak DL (2015) Challenges and opportunities for electrochemical processes as next-generation technologies for the treatment of contaminated water. Environ Sci Technol 49:11292–11302
Rahman Bhuiyan MA, Mizanur Rahman M, Shaid A, Khan MA (2014) Application of gamma irradiated textile wastewater for the pretreatment of cotton fabric. Environ Ecol Res 2(3):149–152
Rani B, Maheshwari R, Yadav RK, Pareek D, Sharma A (2013) Resolution to provide safe drinking water for sustainability of future perspectives. Res J Chem Environ Sci 1:50–54
Raskin I, Smith RD, Salt DE (1997) Phytoremediation of metals using plants to remove pollutants from the environment. Curr Opins Biotech 8:221–226
Robinson T, McMullan G, Mardhant R, Nigam P (2001) Remediation of dyes in textile effluent: a critical review on current treatment technologies with a proposed alternative. Bioresource Technol 77(3):247–255
Lakshmana Prabu S, Suriyaprakash TNK, Ruckmani K, Thirumurugan (2016) Effective waste water treatment and its management. In: Rathoure AK, Dhatwalla VK (eds) Toxicity and waste management using bioremediation. IGI Global, pp 312–334
Sabur MA, Khan AA, Safiullah S (2012) Treatment of textile wastewater by coagulation precipitation method. J Sci Res 4(3):623–633
Schnoor JL, Licht LA, McCutcheon SC, Wolfe NL, Carreira LH (1995) Phytoremediation of organic and nutrient contaminants. J Environ Sci Tech 29:318A-323A
Shah MP (2018) Azo dye removal technologies. Austin J Biotechnol Bioeng 5(1):1090
Shivangi R (2012) Microbial degradation and treatment studies on textile waste water of Bagru region (Rajasthan). In: Doctoral thesis, Jayoti Vidyapeeth Women's University, Jaipur (Rajasthan)
Shriram B, Kanmani S (2014) Ozonation of textile dyeing wastewater—a review 15(3):46–50. https://www.researchgate.net/publication/281776011_Ozonation_of_Textile_Dyeing_Wastewater_-_A_Review
Sri Lakshmi K, Sailaja H, Anji Reddy M (2017) Phytoremediation—a promising technique in waste water treatment. Int J Scientific Res Manag 5(6):5480–5489
Susarla S, Medina VF, McCutcheon SC (2002) Pytoremediation: an ecological solution to organic chemical contamination. Ecol Eng 1818:647–658

Ukiwe LN, Ibeneme SI, Duru CE, Okolue BN, Onyedika GO, Nweze CA (2014) Chemical and electro-coagulation techniques in coagulation-floccculation in water and wastewater treatment—a review. J Adv Chem 9(3):1988–1999

UNEP (2002) Phytoremediation: an environmentally sound technology for pollution prevention control and remediation. In: An introductory guide to decision makers

Newsletters and Technical publications freshwater management series No. 2. United Nations Environment Programmed Dg of Technology, Industry and Economics

Vasanthy M, Santhiya M, Swabna V, Geetha A (2011) Phytodegradation of textile dyes by water hyacinth (*Eichhornia Crassipes*) from aqueous dye solutions. Int J Environ Sci 1(7):1702–1717

Wawrzkiewicz M, Hubicki Z (2015). Anion exchange resins as effective sorbents for removal of acid, reactive, and direct dyes from textile wastewaters, ion exchange—studies and applications, Ayben Kilislioglu, IntechOpen, pp 37–72. https://doi.org/10.5772/60952. Available from https://www.intechopen.com/books/ion-exchange-studies-and-applications/anion-exchange-resins-as-effective-sorbents-for-removal-of-acid-reactive-and-direct-dyes-from-textil

What is membrane separation? https://www.asahi-kasei.co.jp/membrane/microza/en/kiso/kiso_1.html

Why is water so important to life on Earth. https://sciencing.com/about-6384365-water-important-life-earth-.html

Xia H, Wu L, Tao Q (2003) A review on phytoremediation of organic contaminants. China J Appl Ecol 14:457–460

# Bioremediation of Dye Effluent Using *Dodonae Viscosa* Seed Powder

**B. Sivasathya, A. Kalaicelvi, and K. K. Kavitha**

**Abstract** Large-scale production and extensive application of synthetic dyes can cause considerable environmental pollution and are serious health-risk factors. Although, the growing impact of environmental protection on industrial development promotes the development of eco-friendly technologies. Release of considerable amounts of synthetic dyes to the environment causes public concern, legislation problems and is a serious challenge to environmental scientists. In the present study the plant seed powder has been used as a substance for dye effluent treatment. The degradation efficiency of seed powder has been tested and compared with the chemical treatment. The *Dodonae viscosa* seed powder is taken in the following five different concentrations 1, 2, 3, 4 and 5 g/l by standard jar test method. The raw dye industry effluent is treated with the seed powder. The effluent colour is degraded and changed to light yellow colour after a period of 13 h. The colour, pH, turbidity and total suspended solids were greatly reduced when compared with the raw effluent. The optimum dosage for decolourisation and the optimum dosage for the reduction in pH, turbidity and total suspended solids were found to be 5 g/l. This is highly effective when compared with the chemical counterparts. The COD of the effluent has also reduced slightly when compared with the hypochlorite treatment. Although the application of plant based natural seed powder in the degradation of dye effluent is still at its infancy, they are technically promising in terms of efficiency for further research.

**Keywords** Decolorization · *Dodonae* · Seed powder · Dye effluent · Degradation

## 1 Introduction

Manufacturing rebellion is a great benefit to mankind but there is a wide range of environmental impact created by industries. The textile industry is one of the major industries of the world which provides enormous employment opportunities to the

B. Sivasathya · A. Kalaicelvi · K. K. Kavitha (✉)
Department of Environmental and Herbal Science, Tamil University, Thanjavur 613010, India
e-mail: kavikiruthiga@gmail.com



M. Prashanthi et al. (eds.), *Bioremediation and Green Technologies*, Environmental Science and Engineering, https://doi.org/10.1007/978-3-030-64122-1_10

people thereby increasing the economy of a country. It is a vast diverse sector in terms of raw materials, processes, products and equipment and has a very complicated industrial chain (Irina-Isabella et al. 2008).India accounts for about 14% of the world's production of textile fibres and yarns. This includes jute, of which it is the largest producer. India is the second largest producer of silk, cellulose fibre and yarn and the fifth largest producer of synthetic fibre and yarn. Many countries of the world bloom because of this sector. Majority of these industries are water based. Over 3/4$^{th}$ of the freshwater drawn by the domestic and industrial effluents unavoidably end up in surface water bodies or in the groundwater affecting water quality. In India, (Tamil Nadu) main contributors to the surface and ground water pollution are the by-products of the variety of industries such as the textile and dyeing (Kaushik et al. 2005) sugar (Thamizhiniyan et al. 2009), processing chemicals, pesticides, fertilizer, pulp and paper (Malla and Mohanty 2005) distilleries, food processing dairy and sago (Dhanam 2009) mining, electroplating and others (Sah et al. 2000).

The industrial effluents are generally considered harmful but sometimes used for irrigating various crops (Malaviya et al. 2007 and Nath et al. 2009). High levels of pollutants in the water bodies cause an increase in Biological Oxygen Demand (BOD), Chemical Oxygen Demand (COD), Total Dissolved Solids (TDS), Total Suspended Solids (TSS), toxic metals such as Cd, Cr, Ni and Pb and fecal coliform and the presence of heavy metals in the environment causes deleterious effects to plants and human beings, particularly at certain levels of exposure and absorption. The industrial use of water is very low as compared to agricultural use, the disposal of industrial effluents on land or surface water bodies make water assets inappropriate for other uses (Buechler and Mekala 2005). It is also one of the industries which consumes colossal amount of portable and industrial water.

The low cost adsorbents obtained from various plant sources have been found to be more economical and sometimes more effective than commercially available activated carbon for dye removal from the industrial wastes. Dyed effluents can be treated by using low cost adsorbents such as fly ash, bagasse pith, neem leaf powder, babul bark, sewage treatment bio solids (sludge) and coconut coir (Sivamani and Lena Grace 2009). The adsorption characteristics of Eriochrome Black-T dye on the activated Nilgiri leaves, was an alternative low cost adsorbent to the commercially available activated carbon for the decolorization of dyed textile effluents (Ishtiyak and Chhipa 2015). Based on above reference, in the present study plant seeds were used for the treatment of the dye effluents.

The Objectives of the study are to collect dye industry effluents and analyze the physicochemical parameters as per standard methods (APHA 2005) and to prepare the seed powder from plant seeds (*Dodonae viscosa)* for treatment of dye effluents. We aim to produce a bio filter column to increase the treatment effect on the dye industry effluents.

## 2 Materials and Methods

### 2.1 Sample Collection

The effluent samples were collected from the Common Effluent Treatment Plant (CETP) treated at Angeripalayam CETP, in Tirupur. Effluents from 58 dyeing units are treated in this CETP.

### 2.2 Preparation of Dodonae Viscosa Seed Powder

The *Dodonae* ***viscosa*** seeds were collected from recently harvested dry seed. The seed coat and wings were removed and the seeds were ground to powder using domestic blender. Then the powder was sieved through 600 μm stainless steel sieve. Degradation processes are depended on a multitude of variable interrelated factors: Temperature, Color, Alkalinity, natural coagulant and intensity and duration of stirring, duration mixing and flocculation. The optimal dose of the seed powder cannot be found by analyzing the raw water. Rather, it must be determined by an experiment on laboratory scale. Such a test ought to follow this procedure. Previous researchers had tried to utilize maize seed as coagulant/ coagulant aid (Raghuwanshi et al. 2002; Mandloi et al. 2004). Previously removal of dye, surfactant and other contaminations using *Moringaoleifera* seed was studied by some scientists (Beltran-Heredia and Martin 2008; Bhatia et al. 2006; Kwaambwa et al. 2010). The chemical composition and other properties like active agent of seeds of *Moringa oleifera* was demonstrated by Ndabigengesere et al. (1995) and Gassenschmidt et al. (1995). Based on the previous researches we tried indigenous plant *Dodonae viscosa* seed which is available in Tamil University campus.

The ***Dodonae viscosa*** seed powder was taken in the following doses (1, 2, 3, 4 and 5 g/l) for treating 1 L of dye effluent samples. High speed stirring initially for 1 min and low speed for 10 min. The stirring is done with the help of a Jar Test Apparatus. Then it was allowed until the water colour was removed or degraded.

### 2.3 Preparation of Filter Column

The filter column is prepared by using sand, activated carbon and dry egg shell. Initially all the ingredients of the filter column are washed using HCl in order to remove any dust particles attached to it. The sand is then washed with distilled water and 50 ml of the sand is packed into the filter column. Activated charcoal is prepared by using dry Amla seed powder. The seed is crushed into crystals by using a blender. It is burnt at 250 °C for few minutes using a Muffle furnace. The charcoal thus obtained is then washed with HCl until the pH is neutralized. It is further washed by using distilled water and dried in an oven. This is then packed into the filter column for about 50 ml. The column is then packed with 75 ml of egg shells which is pretreated with HCl and heated in an oven for about 24 h.

### 2.4 Physico Chemical Parameters

Degradation of natural resources like water is occurring at unprecedented rates in India and around the world in developing as well as in developed countries (Buechler and Mekala 2005). The effluent generated by the textile industry is one of the major sources of pollution. Characterization of textile process effluent streams is very important to develop strategies for water treatment and reuse (Yusuff and Sonibare 2004).The effluent was characterized and the Table 1 shows the values of various parameters like pH, TDS, TSS, Chloride, Alkalinity, Turbidity, and Colour, Total Hardness, Calcium hardness, Magnesium hardness, free residual chloride and COD. The effluent characteristics are shown in Table 1.

## 3 Results and Discussion

Textile effluent collected from Tirupur, exhibited higher level of water quality parameters and above permissible limits of ISI and FAO standard. The effect of natural decolorizing agent for textile effluent treatment by using seed of *Dodonae viscosa* was studied. The chosen seed powder is taken in the following five different concentrations 1, 2, 3, 4 and 5 g/l by standard jar test method. The raw dye industry effluent is treated with the seed powder in neutral pH by adding HCL. The effluent colour is degraded and changed to light yellow colour after a period of 13 h. The colour of the raw effluent was found to be 5180 pt. co whereas the effluent treated with the seed powder at an optimum dose of 5 g/l was found to be 8.49% Table 1.

The present study found that *Dodonae viscosa* was effective for colour removal when followed by filtration through the biofilter colour—86.87%, TSS—81.06%, and Turbidity—59.48% was achieved Fig. 1. Khader et al. (2018) studied natural coagulants for removal of COD, oil and turbidity from produced waters in the petroleum industry. The natural coagulant of eggplant seed, when used as a coagulant, the highest removal of oil, COD were 99.42%, 88.2% and 95.91% with eggplant, radish and Cicer arietinum seed, respectively. It showed that a maximum of 86.87% colour removal was achieved when compared to sodium hypochlorite (77.03%). When compared to previous works, a maximum percentage colour reduction of 62.8% was reported by Patel and Vashi (2013); Kalaicelvi and Kavitha (2019), was reported colour 92% and TSS-85.33%, reduction was achieved using *M.oleifera* seed coagulant and biofilter to treat the dye effluent. The effect of sodium hypochlorite, seed powder and biofilter column on decolorization and COD of dye effluent is given in Fig. 2.

**Table 1** Decolorization effect of *Dodonae viscosa* plant seed powder and sodium hypochlorite to treat dye effluent

| S. No. | Parameters | Raw effluent | Sodium hypo chlorine solution 2.3 ml | Concentration of *Dodonae viscose* seed powder | | | | | Filtration of 5 g/l treated effluent |
|---|---|---|---|---|---|---|---|---|---|
| | | | | 1 g/l | 2 g/l | 3 g/l | 4 g/l | 5 g/l | Eggshell, sand, and activated carbon (amla seed) |
| 1 | Color (pt co) | 5180 | 1190 | 6480 | 6180 | 5540 | 5120 | 4740 | 680 |
| 2 | pH | 9.48 | 9.72 | 7.14 | 7.28 | 7.16 | 7.19 | 7.23 | 7.04 |
| 3 | Turbidity (NTU) | 64.9 | 50.4 | 168.3 | 169.4 | 158.8 | 136.4 | 135.8 | 26.3 |
| 4 | Total dissolved solids (mg/l) | 7242 | 7328 | 7312 | 7298 | 7324 | 7318 | 7296 | 7312 |
| 5 | Total suspended solids (mg/l) | 169 | 97 | 312 | 294 | 268 | 204 | 158 | 32 |
| 6 | Total hardness (mg/l) | 430 | 460 | 450 | 440 | 440 | 450 | 430 | 470 |
| 7 | Calcium hardness (mg/l) | 240 | 250 | 240 | 240 | 250 | 240 | 250 | 290 |
| 8 | Magnesium hardness (mg/l) | 190 | 210 | 210 | 200 | 190 | 210 | 180 | 180 |
| 9 | Free residual chloride (mg/l) | – | 44.6 | – | – | – | – | – | – |
| 10 | Chloride (mg/l) | 3318 | 3489 | 3412 | 3378 | 3356 | 3444 | 3393 | 3408 |
| 11 | Total alkalinity (mg/l) | 1400 | 1180 | 1060 | 1130 | 1100 | 1050 | 1080 | 1050 |
| 12 | COD (mg/l) | 1132 | 1224 | 2938 | 2946 | 2816 | 2918 | 2864 | 1212 |
| 13 | Colour (visual) | Dark purple | Dark yellow | Dark green | Dark green | green | Brownish green | Dark yellow | Lightly yellow |

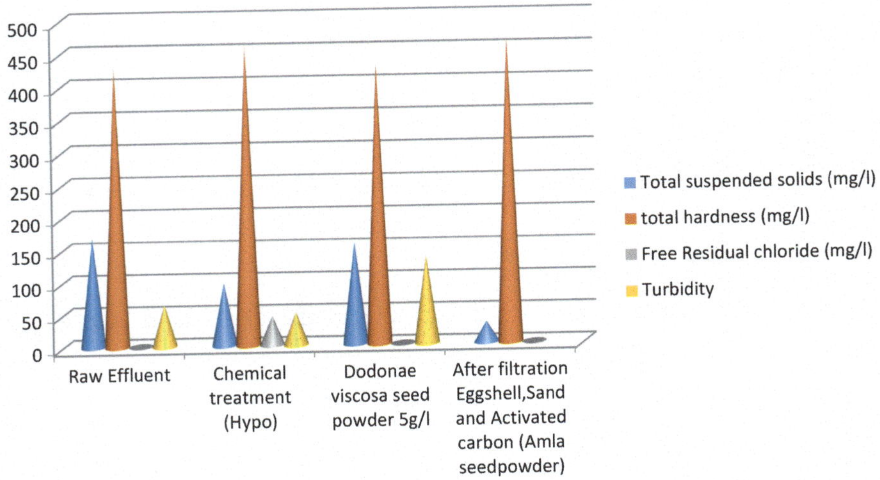


**Fig. 1** Effect of TSS, total hardness, free residual chloride and turbidity of dye effluent treatment by chemical, *Dodonae viscosa* seed powder and biofilter column

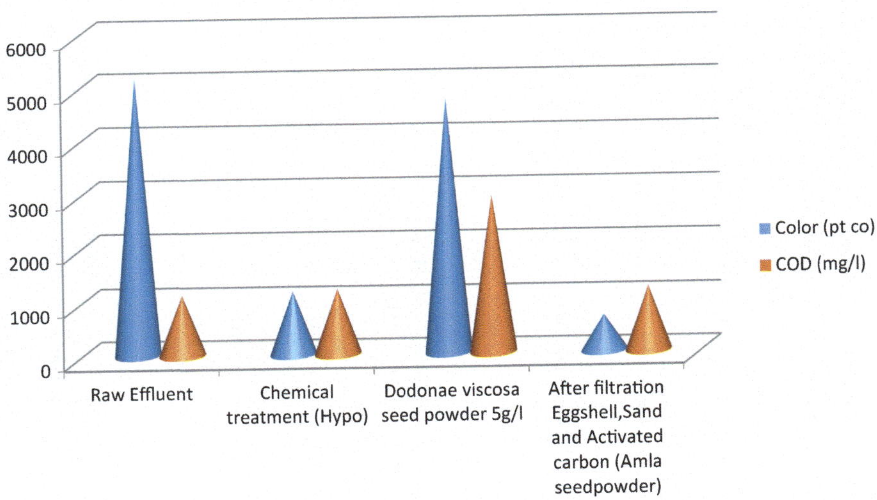


**Fig. 2** Effect of sodium hypochlorite, seed powder and biofilter column on decolorization and COD of dye effluent

The colour, turbidity and total suspended solids were reduced greatly to more than 50% using the filter column. This is highly effective when compared with their chemical counterparts. The effective dosage for decolorization and the optimum dosage and time duration for the reduction in pH, turbidity and total suspended solids were found to be 5 g/l and 13 h, respectively. As these plant materials are easily available in nature, and because of their relative abundance in nature they are cost effective. It helps in reducing the crisis in pollution. The method of operation is simple and requires little or no maintenance.

## 4 Conclusion

Industrialization is considered to be the key for the development in economic terms. At the same time, it is also recognized to be the root cause for environmental pollution. Due to different types of industries, environmental pollution is one of the vital problems presently facing in India and the World. The environmental pollution caused by the textile wastewater poses a worldwide threat to public health and it gives rise to new initiatives for environmental restoration for both economic and ecological reasons. The feasibility for the treatment of the dye effluent using the natural seed powder and chemically based hypochlorite solution were tested. It was found that the natural seed powder act as an excellent alternative to the chemical treatment. Because of its relative abundance in nature and as it functions effectively this method of treatment is economical and does not cause any damage to the environment. Utilization of these plant based substances represents important progress in sustainable environmental technology as they are renewable resources and their application is directly related to the improvement of the quality of life for underdeveloped communities. In the present study it can be clearly found that the colour removal in the dye effluent using the *Dodonae viscosa* seed powder was highly effective and shows a percentage removal of more than 86.87%. This decolorisation by itself shows a tremendous reduction in pollution, which is followed by the reduction in pH and turbidity. Hence our current study proves to be a promising hope for future research and also to implement the technique used on a commercial basis.

## References

Beltrán-Heredia J, Sánchez Martín J (2008) Azo dye removal by Moringaoleifera seed extract coagulation. Color Technol 124(5):310–317

Bhatia S, Othman Z, Ahmad AL (2006) Palm oil mill effluent pretreatment using Moringaoleifera seeds as an environmentally friendly coagulant: laboratory and pilot plant studies. J Chem Technol Biotechnol 81(12):1852–1858

Buechler S, Mekala GD (2005) Local responses to water resource degradation in India: groundwater farmer innovations and the reversal of knowledge flows. J Environ Devel 14(4):410–438

Dhanam S (2009) Effect of dairy effluent on seed germination, seedling growth and biochemical parameter in Paddy. Bot Res Int 2(2):61–63

Gassenschmidt U, Jany KD, Bernhard T, Niebergall H (1995) Isolation and characterization of a flocculating protein from Moringaoleifera Lam. Biochimicaet Biophysica Acta (BBA) General Subjects 1243(3):477–481

Kalaicelvi, Kavitha (2019) Bioremediation of *Moringa oleifera* seed using biofilter for decolourisation of textile dye effluent. RJLBPCS 5(6):1–10

Kaushik P, Garg VK, Singh B (2005) Effect of textile effluent on different cultivar of wheat. Biores Technol 96:1189–1193
Khader EH, Mohammed THJ, Mirghaffari N (2018) Use of natural coagulants for removal of cod, oil and turbidity from produced waters in the petroleum industry. J Pet Environ Biotechnol 9(3):374
Khadikar PV, Joshi S, Bajaj AV, Mandloi D (2004) Correlations between the benzene character of acenes or helicenes and simple molecular descriptors. Bioorg Med Chem Lett 14(5):1187–1191
Kumar M, Raghuwanshi NS, Singh R, Wallender WW, Pruitt WO (2002) Estimating evapotranspiration using artificial neural network. J Irrigation Drain Eng 128(4):224–233
Kwaambwa HM, Hellsing M, Rennie AR (2010) Adsorption of a water treatment protein from Moringaoleifera seeds to a silicon oxide surface studied by neutron reflection. Langmuir 26(6):3902–3910
Ladhe UV, Wankhede SK, Patil VT, Patil PR (2011) Adsorption of erichrome black t from aqueous solutions on activated carbon prepared from mosambi peel. J Appl Sci Environ Sanitation 6(2):149–154
Malaviya P, Kour R, Sharma N (2007) Growth and yield responses of Capsicum annum L. to distillery effluent irrigation. Ind J Environ Ecoplan 14:643–646
Mohanty A, Herrington JD, Koven NS, Fisher JE, Wenzel EA, Webb AG, Miller GA (2005) Neural mechanisms of affective interference in schizotypy. J Abnorm Psychol 114(1):16
Nath K, Singh D, Shyam S, Sharma YK (2009) Phytotoxic effects of chromium and tannery effluent on growth and metabolism of Phaseolusmungo Roxb
Ndabigengesere A, Narasiah KS, Talbot BG (1995) Active agents and mechanism of coagulation of turbid waters using Moringaoleifera. Water Res 29(2):703–710
Njoku VO, Hameed BH (2011) Preparation and characterization of activated carbon from corncob by chemical activation with H3PO4 for 2, 4-dichlorophenoxyacetic acid adsorption. Chem Eng J 173(2):391–399
Patel H, Vashi RT (2013) Comparison of naturally prepared coagulants for removal of cod and color from textile wastewater. Global NEST J 15(4):522–528
Rajeshkannan R, Rajasimman M, Rajamohan N (2011) Decolourization of malachite green using tamarind seed: optimization, isotherm and kinetic studies. Chem Industr Chem Eng Quart/CICEQ 17(1):67–79
Sah SK, Sah JP, Lance VA (2000) Industrial effluents and their use in agriculture along the Narayani River, Nawalparasi, Nepal. Environ Agric At the cross road of the New Millennium, 456–466
Sivamani S, Leena GB (2009) Removal of dyes from wastewater using adsorption—a review. Ijbst 2(4):47–51
Thamizhiniyan P, Sivakumar PV, Lenin M, Sivaraman M (2009) Sugar mill effluent toxicity in crop plants. J Phytol 1(2)
Yin CY (2010) Emerging usage of plant-based coagulants for water and wastewater treatment. Process Biochem 45(9):1437–1444
Yusuff RO, Sonibare JA (2004) Characterization of textile industries' effluents in Kaduna, Nigeria and pollution implications. Global Nest Int J 6(3):212–221

# Optimization of Culture Conditions for Enhanced Decolorization of Azo Dyes by *Aspergillus flavus* Isolated from Dye Contaminated Soil

**M. Dexilin, B. Gowri Manogari, and Thamaraiselvi Kaliannan**

**Abstract** Dyes released by the textile industries pose a threat to the environmental safety. Recently, dye decolorization through biological means has gained impetus as these are cheap and can be applied to wide range of dyes. The main focus of the present investigation is to evaluate the bioremediation potential of *Aspergillus flavus* isolated from textile dye contaminated soil for the decolorization of synthetic acid dyes. Among the isolated four fungal strains namely *Aspergillus flavus, Fusarium oxysporum, Rhizopus sp.* and *Aspergillus terrus, Aspergillus flavus* was capable of decolorising 96 and 97% for Acid Blue (AB) 193 and Acid Violet (AV) 90, respectively under optimum conditions of pH 6.0 and temperature 25 °C. Hence from the results, it can be concluded that this fungus can be used as eco-friendly and economically effective tool to decolorize textile dye effluents.

**Keywords** Acid blue 193 · Acid violet 90 · *Aspergillus flavus* · Decolorization · Effluent

## 1 Introduction

Synthetic dyes are widely used in textile dyeing, paper printing, color photography, pharmaceutical, food, cosmetic and leather industries (Rafii et al. 1990; Kuhad et al. 2004; Rodriguez Couto 2009). Since 1856, over 105 different dyes have been produced worldwide with an annual production of over $7 \times 10^5$ metric tons (Chen et al. 2003). Of all the familiar dyestuffs in the world, azo dyes make up about a half, making them the largest group of synthetic colorants that are released into the environment (Zhao and Hardin 2007; Sen et al. 2016). Effluents from the textile industries containing dye are highly coloured and are therefore visually identifiable (Kılıç et al. 2007) and are visible even at 1 mg/ml concentration (Pandey et al. 2007). The discharge of colored effluents into the environment is detrimental, not

M. Dexilin · B. Gowri Manogari · T. Kaliannan (✉)
Laboratory of Molecular Bioremediation and Nanobiotechnology, School of Environmental Sciences, Department of Environmental Biotechnology, Bharathidasan University, Tiruchirappalli - 620 024, Tamil Nadu, India
e-mail: kthamaraiselvi@hotmail.com



M. Prashanthi et al. (eds.), *Bioremediation and Green Technologies*, Environmental Science and Engineering, https://doi.org/10.1007/978-3-030-64122-1_11

only because of their colour but due to the lethal and carcinogenic properties of dye compounds and their end products (Weisburger 2002).

Different approaches are available for the remediation of dye effluents which include a number of physical, chemical and biological methods namely, oxidation and reduction, adhesion, photocatalysis, coagulation, microbial degradation methods, etc., (Fu and Viraraghavan 2001). Nevertheless, these technologies are usually ineffectual in the removal of color, costly and less adaptable to an extensive range of dye waste waters (Banat et al. 1996). Biological methods are generally considered environmentally approachable as they can lead to complete mineralization of organic pollutants at low cost (Pandey et al. 2007).

Among the biological methods, bioremediation is promising tool for the elimination of toxic pollutants and it is the application of microorganisms (fungi, bacteria, actinomycetes, yeasts and algae). Microorganisms which are naturally having the tolerant ability to utilise dye compounds as their sole energy source are of extraordinary interest for scientific community to eliminate the environmental contaminants that results from the industrial sources (Ali et al. 2010). Dye degradation by bacteria predominantly depends on the enzymatic transformation where the azo dyes are cleaved with the enzyme azo-reductase along with its coenzyme (Zimmermann et al. 1982; Moutaouakkil et al. 2003; Saroj et al. 2015), whereas fungal degradation of azo dyes are catalyzed by extracellular enzymes ie., ligninolytic peroxidases (Young and Yu 1997; Selvam et al. 2003; Máximo et al. 2003). Although bacterial dye degradation have been widely studied, in recent days fungal have gained importance in dye degradation due to their unique extra cellular enzyme systems. Fungi are capable of degrading the complex dye structures when compared with bacteria (Forss and Welander 2009). Fungal mycelia have an additive advantage over single cell organism by solubilizing the insoluble substrates by producing extracellular enzymes (Gajera et al. 2015). Higher concentration of dye toxicants can be easily degraded by fungi due to their advantageous enzyme arrangements and improved cell to surface ratio. Various studies have reported that fungi like white rot decolorize the synthetic dyes and the processes are mediated by lignin peroxidase (Ollikka et al. 1993), manganese peroxidase (Heinfling et al. 1998) and laccase (Abadulla et al. 2000; Soares et al. 2001; Murugesan and Kalaichelvan 2003). Based on the fact that the fungi have the capacity of degrading wide variety of organic and inorganic dye components, the present study was carried out to isolate and screen fungal strains for decolorization of azo dyes (AB 193 and AV 90) and optimize the pH and temperature for better understanding the process.

## 2 Materials and Methods

### *2.1 Dye Stuffs and Chemicals*

Acid Blue 193 (AB 193) and Acid violet 90 (AV 90) commonly used in local textile processing industries were obtained from Rajasthan Dye Chemical Ltd., Chennai, India. Dyes were dissolved in double distilled water to prepare a stock concentration (1000 mg $L^{-1}$) and solutions of the desired concentrations were obtained by successive dilutions. Media components and chemicals were purchased from Himedia Labs, Bombay, India.

### *2.2 Sample Collection*

Soil samples were collected from the textile industry at Tirupur, Tamil Nadu, India and were stored at 4 °C until further process.

### *2.3 Screening and Identification of AB 193 and AV 90 Tolerant Fungi*

The soil samples were serially diluted using sterile distilled water and from the $10^{-2}$ to $10^{-7}$ dilutions, 0.1 mL of sample was transferred onto Czapek-Dox agar plate containing AB 193 and AV 90 (50 mg $L^{-1}$). Spread plate method was performed and the plates were incubated at 30 ± 1 °C for 4 days. Fungal colonies were identified using steriobinocular microscope. Fungi which showed decolorization in Czapek-Dox agar medium were screened further by transferring in a fresh Czapek-Dox agar plates containing different dye concentrations (50 and 100 mg $L^{-1}$). Colonies were selected on the basis of their ability to decolorize the dye on the plates. Isolated colonies were then subsequently transferred to Czapek-Dox agar for further study. Fungal strains with greatest dye tolerance capacity were identified by its different biochemical characteristics and structural arrangements. Wet mount was prepared by suspending culture in a few drops of lacto phenol cotton blue solution. The slides were examined under high and low power magnification with the aid of a dissecting microscope (Cappuccino and Sherman 2004).

# 3 Decolorization Experiments

## 3.1 *Optimization of Incubation Time on Decolorization of Azo Dyes*

The isolated fungus was tested for its ability to decolorize AB 193 and AV 90 (100 mg $L^{-1}$). About 50 mL of C-limited Czapek-Dox medium was amended separately with the dyes and subsequently inoculated with 5 ml spore suspension ($10^8$ spore/ml) from a slant culture. The flasks were incubated at 30 ± 1 °C for 12 days on a rotary shaker (150 rpm) and the samples were withdrawn at 2 days intervals and centrifuged at 10,000 rpm for 10 min. Supernatant collected was assessed for the decolorization of AB 193 and AV90 in UV–Vis spectrophotometer (UV–Vis 1700, Shimadzu, Japan) at its maximum absorbence wave length of 586 and 526 nm, respectively. Two control flasks (i) dye + medium without inoculum and (ii) medium with inoculum without dye were maintained. The percentage decolorization was calculated according to the following formula Bergsten-Torralba et al. (2009)

Decolorization (%) = Initial OD—Final OD *X* 100/Initial OD... 0.11.1

## 3.2 *Optimization of pH and Temperature on Decolorization of Azo Dyes*

For optimization of physical parameters, the fungal inoculated shake flasks were incubated for 10 days at different initial pH values from 2 to 10 (2, 4, 6, 8, 10) to determine the optimal pH for dye decolorization. For investigating the effect of temperature, the shake flasks inoculated with fungal inoculum were incubated at varying temperatures from 20 to 35 °C (20, 25, 30 35 °C) at optimum pH for 10 days. The experimental conditions were maintained as mentioned earlier. Samples were withdrawn at regular intervals and analyzed for color removal.

## 3.3 *Optimization of Initial Dye Concentration on Decolorization of Dyes*

Fungal isolates were added into a 250 mL medium containing varying concentrations (25, 50, 75, 100 and 125 mg $L^{-1}$) of AB 193 and AV 90 individually. At optimum pH and temperature, the flasks were incubated in a rotary shaker (Orbitek) at 150 rpm.

**Table 1** Dye decolorization pattern of fungal isolates

| Organisms | Acid blue 193 (AB 193) | | Acid violet 90 (AV 90) | |
|---|---|---|---|---|
| | 50 mg $L^{-1}$ | 100 mg $L^{-1}$ | 50 mg $L^{-1}$ | 100 mg $L^{-1}$ |
| *Fusarium oxysporum* | − | − | + | + |
| *Rhizopus* sp. | + | + | − | − |
| *Aspergillus terreus* | + | − | − | − |
| *Aspergillus flavus* | + | + | + | + |

+ decolorization zone observed
−decolorization zone not observed

# 4 Results and Discussion

## 4.1 Screening and Selection of Dye Tolerant Fungi

In the present study one fungal strain was selected among the four isolated strains, *Aspergillus flavus, Fusarium oxysporum, Rhizopus* sp. and *Aspergillus terrus, Aspergillus flavus* for the decolorization of AB 193 and AV 90 based on its high dye decolorization ability (Table 1). This selected strain was identified as *Aspergillus flavus* based on various cultural characteristics, morphology and by the presence of conidiospores (Table 2). It was observed that the *Aspergillus flavus* and *Rhizopus* species tolerated upto 100 mg $L^{-1}$ of AB 193 and in case of AV 90, *Aspergillus flavus* and *Fusarium oxysporum* tolerated upto 100 mg $L^{-1}$, respectively. Above results reveals that decolorization by *Aspergillus flavus* was found to be high (100 mg/L) for both AB 193 and AV 90. Hence for further studies, *Aspergillus flavus* was selected as an efficient strain for decolorization.

In present study *A. flavus* decolorized dye by forming clear zone on agar plates, which may due to their extracellular, nonspecific and nonstereoselective enzyme system (Hofrichter 2002). Fungal decolorization capacity can be confirmed by their ability to form a clear halo zone on agar plates containing dye AB 193 or AV 90 (Yamini et al. 2012). Previous studies disclose that fungal strains were able to decolorize dyes on agar plates (Jarosz-Wilkolazka et al. 2002; Eichlerova et al. 2006).

## 4.2 Effect of Incubation Time on AB 193 and AV 90 Decolorization by Aspergillus Flavus

Initially, the decolorization was very low on the second day that is 42 and 43% for AB 193 and AV 90, respectively. Then it gradually increased with increase in incubation time. The decolorization efficiencies of *Aspergillus flavus* were 48 and 45% for AB

**Table 2** Identification of fungal isolates by macroscopic and microscopic appearance

| Organisms | Macroscopic appearance | Microscopic appearance |
|---|---|---|
| *Aspergilus flavus* | White colonies become greenish—blue as culture matures | Single-celled spores (Conidia) in chains developing at the end of the sterigma arising from the terminal bulb of the conidiophore, the vesicle; long conidiophores arise from a septate mycelium |
| *Fusarium oxysporum* | Wolly, white, fuzzy colonies changing colour to pink | Multicelled spores (conidia) are oval or crescent-shaped and attached to conidiophores arising from a septate mycelium |
| *Aspergillus terreus* | White colonies become greenish—blue as cultures matures | Single-celled spores (Conidia) in chains developing at the end of the sterigma arising from the terminal bulb of the conidiophore, the vesicle; long conidiophores arise from a septate mycelium |
| *Rhizopus* sp. | White colonies become gray—yellowish brown as cultures matures | Single-celled spores (Sporangia) are globose to ovoid shaped and nonseptate or sparsely septate broad hyphae. They are rhiozoids and unbranched usually |

193 and AV90, respectively and it slowly increased and reached maximum of 96 and 97% within 10 days of incubation time as represented in Figs. 1 and 2.

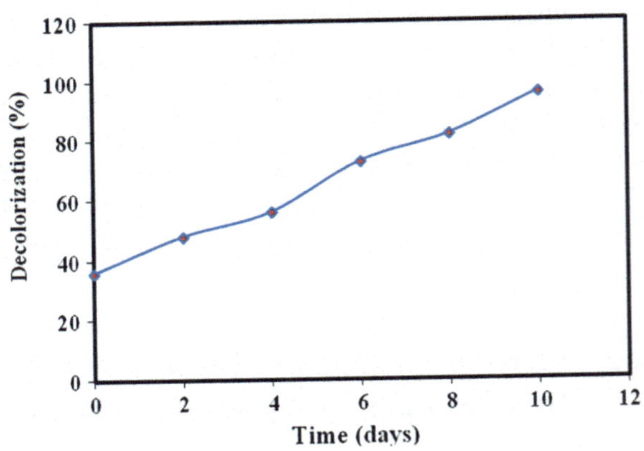


**Fig. 1** Effect of incubation time on decolorization of AB 193 *by Aspergillus flavus*

**Fig. 2** Effect of incubation time on decolorization of AV 90 *by Aspergillus flavus*

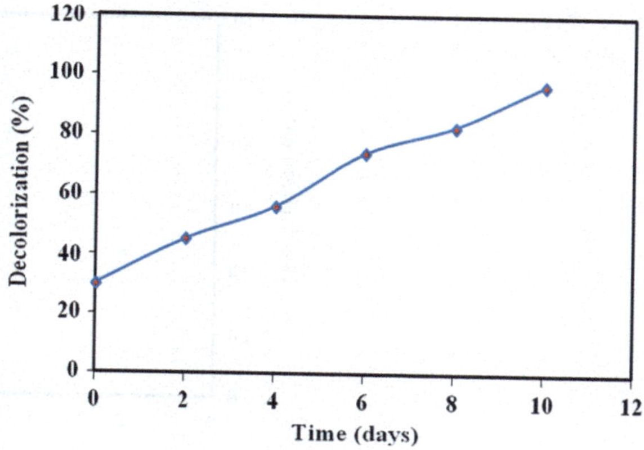


## 4.3 *Effect of pH and Temperature for AB 193 and AV 90 Decolorization by* Aspergillus Flavus

At initial pH 2, the decolorization of both dyes was low. Maximum decolorization of AB 193 (96%) and AV 90 (97%) were observed around pH 6 by *Aspergillus flavus* (Fig. 3 and 4). When the pH was increased to 10 the decolorization of dyes was decreased in both cases. It is observed that the fungus showed higher decolorization efficiency in acidic condition for both the dyes.

The pH of the medium plays a vital role in decolorization of dyes. The effects of pH may be related to the transport of dye molecule across the cell membrane (Kodam et al. 2005). Study of Asgher et al. (2008) shows *Schizophyllum commune* can work in acidic media, and decolorize maximum (77%). Murugesan and Kalaichelvan (2003) also reported that G. lucidum laccase prefers acidic orange for higher decolorization extent. The maximal decolorization was observed at pH 4.0 for both RB-5 and Remazol Brilliant Blue R (RBBR). However Yesilada et al. (2003) studied that *Funalia troggii* had effective decolorization capacity in broad pH range of 6–11 and

**Fig. 3** Effect of pH on decolorization of AB 193 by *Aspergillus flavus*

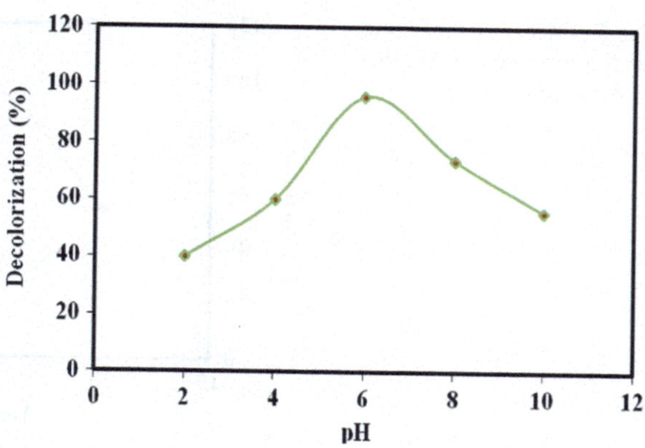

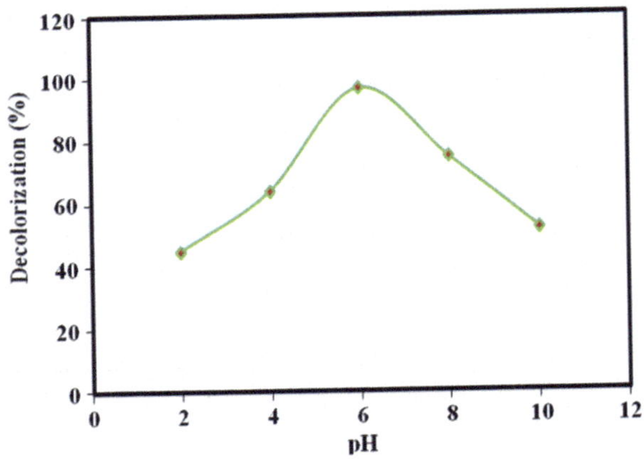


**Fig. 4** Effect of pH on decolorization of AV 90 by *Aspergillus flavus*

this results reveals that the optimum pH of fungus may be affected by its chemical nature.

Dye decolorization was evaluated at varying temperatures (20–35 °C) for AB 193 and AV 90 and the results are shown in Figs. 5 and 6. The fungus exhibited better dye decolorization at 25 °C for both the dyes in 10 days of incubation time. At 20 °C, decolorization of dyes was 50 and 60% for AB193 and Av 90, respectively. The decreasing trend of dye decolorization of 42 and 74% were observed at higher temperatures 30 °C for AB 193 and 35 °C AV 90, respectively.

The decolorization of dye varies at different temperaures, since the temperature changes lead to a sudden alteration in the activation energy which is concerned with the cell viability (Yu et al. 2001). Decolorization of *A. flavus* reached maximum within 10 days at lower concentrations of dye which indicates that increasing the dye concentration gradually decreases the decolorization rate, probably due to the toxic effect of dyes with regard to the fungal biomass concentration (Tony et al. 2009).

The dye decolorization is inversely related to the concentration of the dye due to which it shows inhibitory effects of high dye stuffs in the solution. Similar results

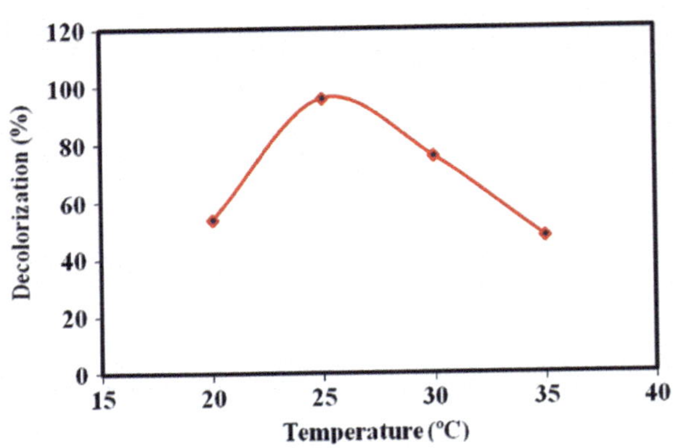


**Fig. 5** Effect of temperature on decolorization of AB 193 by *Aspergillus flavus*

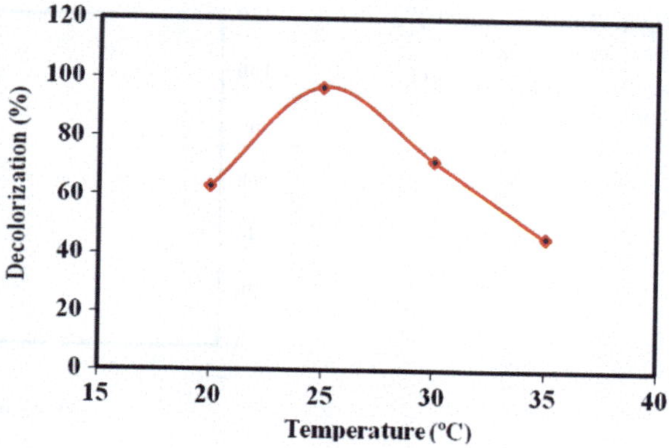


**Fig. 6** Effect of temperature on decolorization of AV 90 by *Aspergillus flavus*

has been observed in decolorization of acid fuchsin, acid violet and evans blue up to 83–91% within 2 h whereas it slowly decreased to 45% for bromophenol blue within 8 h. The difference in the decolorization efficiencies for the dyes is from the different redox potentials of the colored substrates (Chivukula and Renganathan 1995).

## 4.4 *Effect of Initial Dye Concentration on AB 193 and AV 90 Decolorization*

The decolorization efficiencies of *Aspergillus flavus* against AB 193 and AV 90 reached 100% till 100 mg/L whereas when the concentration of dye was increased slowly, decreased decolorization was observed which is shown in Figs. 7 and 8. It was observed that at concentrations 100 mg/L the decolorization was quite efficient, and it reached up to 96% (AB 193) and 97% (AV 90) respectively. When the concentration was gradually increased to 125 mg/L, the decolorization of AB 193 and AV 90 decreased up to 65 and 56%, respectively.

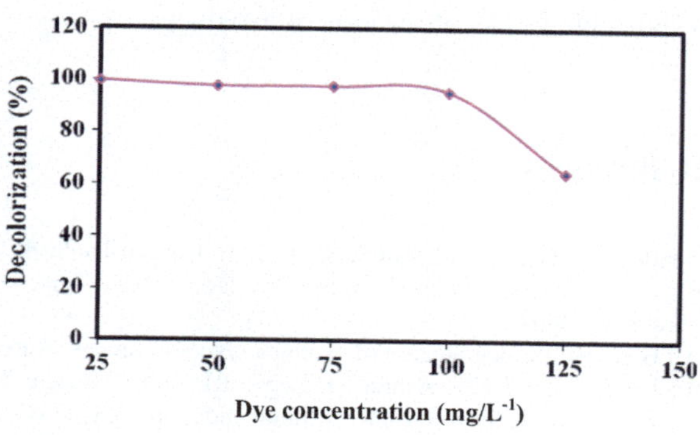


**Fig. 7** Effect of initial dye concentration on decolorization of AB 193 by *Aspergillus flavus*

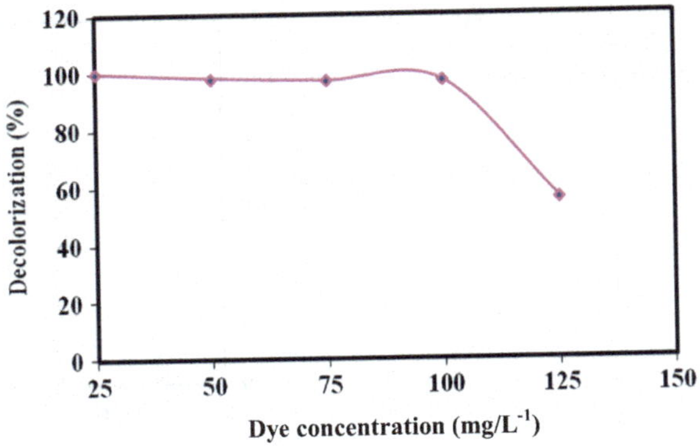


**Fig. 8** Effect of initial dye concentration on decolorization of AV 90 by *Aspergillus flavus*

The growth of the fungi may be affected by the presence of dyes at toxic concentrations and in turn shows effect on decolorization efficiency. Also, the structure of the dye is also influential in deciding the extent to which dye is decolorized. As reported earlier (Kapdan et al. 2000) that at lower dyestuff concentration 0.05% complete decolorization of everzol turquoise blue G by *C. versicolor* was observed whereas only partial decolorization could be achieved for dyestuff concentrations above 0.05%. *Phanerochaete chrysosporium* showed a color removal efficiency of 95–100% on direct blue 15 in repeated batches (Pazarlioglu et al. 2005) following first order kinetics with respected to initial concentrations of dye.

## 5 Conclusion

In summary, the present study indicates that the fungal isolate *Aspergillus flavus* is an excellent strain for the decolorization of azo dyes. The optimum pH and temperature for enhanced decolorization is at 6 and 25 °C, respectively. This implies the fungus *Aspergillus flavus* serve as a potential candidate for the decolorization of textile effluents containing the dyes. In future, further approach has to be focused on enzyme responsible for decolorization of azo dyes.

## References

Abadulla E, Tzanov T, Costa S, Robra K-H, Cavaco-Paulo A, Gübitz GM (2000) Decolorization and detoxification of textile dyes with a laccase from *Trametes hirsuta*. Appl Environ Microbiol 66:3357–3362

Ali H (2010) Biodegradation of synthetic dyes—a review. Water Air Soil Pollut 213:251–273

Asgher M, Bhatti HN, Ashraf M, Legge RL (2008) Recent developments in biodegradation of industrial pollutants by white rot fungi and their enzyme system. Biodegrad 19:771

Banat IM, Nigam P, Singh D, Marchant R (1996) Microbial decolorization of textile-dyecontaining effluents: a review. Bioresour Technol 58(3):217–227

Bergsten-Torralba LR, Nishikawa MM, Baptista DF, Magalhães DP Silva da M (2009) Decolorization of different textile dyes by *Penicillium simplicissimum* and toxicity evaluation after fungal treatment. Braz J Microbiol 40:808–817
Cappuccino J, Sherman N (2004) In Cummings B (ed) Microbiology: a laboratory manual, 7 edn. San Francisco, California, USA
Chen K-C, Wu J-Y, Liou D-J, Hwang S-CJ (2003) Decolorization of the textile dyes by newly isolated bacterial strains. J Biotechnol 101:57–68
Chivukula M, Renganathan V (1995) Phenolic azo dye oxidation by laccase from *Pyricularia oryzae*. Appl. Environ. Microbiol. 61:4374–4377
Eichlerová I, Homolka L, Nerud F (2006) Synthetic dye decolorization capacity of white rot fungus Dichomitus squalens. Bioresour Technol 97:2153–2159
Forss J, Welander U (2009) Decolourization of reactive azo dyes with microorganisms growing on soft wood chips. Int Biodeterior Biodegrad. 63:752–758
Fu Y, Viraraghavan T (2001) Fungal decolorization of dye wastewaters: a review. Bioresour Technol 79:251–262
Gajera HP, Bambharolia RP, Hirpara DG, Patel SV, Golakiya BA (2015) Molecular identification and characterization of novel *Hypocrea koningii* associated with azo dyes decolorization and biodegradation of textile dye effluents. Process Safe Environ Protec 98:406–416
Heinfling A, Martínez MJ, Martínez AT, Bergbauer M, Szewzyk U (1998) Transformation of industrial dyes by manganese peroxidases from *Bjerkandera adusta* and *Pleurotus eryngii* in a manganese-independent reaction. Appl Environ Microbiol 64:2788–2793
Hofrichter M (2002) Review: lignin conversion by manganese peroxidase (MnP). Enzyme Microb Technol 30:454–466
Jarosz-Wilkołazka A, Malarczyk E, Pirszel J, Skowroński T, Leonowicz A (2002) Uptake of cadmium ions in white-rot fungus trametes versicolor: effect of Cd (ii) ions on the activity of laccase. Cell Biol Int 26:605–613
Kapdan IK, Kargi F, McMullan G, Marchant R (2000) Biological decolorization of textile dyestuff by coriolus versicolor in a packed column reactor. Environ Technol 21:231–236
Kılıç NK, Nielsen JL, Yüce M, Dönmez G (2007) Characterization of a simple bacterial consortium for effective treatment of wastewaters with reactive dyes and Cr(VI). Chemosphere 67:826–831
Kodam KM, Soojhawon I, Lokhande PD, Gawai KR (2005) Microbial decolorization of reactive azo dyes under aerobic conditions. World J Microbiol Biotechnol 21:367–370
Kuhad RC, Sood N, Tripathi KK, Singh A, Ward OP (2004) Developments in microbial methods for the treatment of dye effluents. In: Microbiology. Academic Press, p. 185–213.
Máximo C, Amorim MTP, Costa-Ferreira M (2003) Biotransformation of industrial reactive azo dyes by *Geotrichum* sp. CCMI 1019. Enzyme Microb Technol 32:145–151
Moutaouakkil A, Zeroual Y, Zohra Dzayri F, Talbi M, Lee K, Blaghen M (2003) Purification and partial characterization of azoreductase from *Enterobacter agglomerans* a. Biochem Biophys 413:139–146
Murugesan K, Kalaichelvan PT (2003) Synthetic dye decolourization by white rot fungi. Indian J Esp Biol 41(09)
Ollikka P, Alhonmäki K, Leppänen V-M, Glumoff T, Raijola T, Suominen I (1993) Decolorization of Azo, Triphenyl Methane, Heterocyclic, and Polymeric Dyes by Lignin Peroxidase Isoenzymes from *Phanerochaete chrysosporium*. Appl Environ Microbiol 59:4010–4016
Pandey A, Singh P, Iyengar L (2007) Bacterial decolorization and degradation of azo dyes. Int Biodeterior Biodegrad 59:73–84
Pazarlıoğlu NK, Sariişik M, Telefoncu A (2005) Laccase: production by *Trametes versicolor* and application to denim washing. Process Biochem 40:1673–1678
Rafii F, Franklin W, Cerniglia CE (1990) Azoreductase activity of anaerobic bacteria isolated from human intestinal microflora. Appl Environ Microbiol 56:2146–2151
Rodríguez Couto S (2009) Dye removal by immobilized fungi Biotechnol Adv 27:227–235

Saroj S, Dubey S, Agarwal P, Prasad R, Singh RP (2015) Evaluation of the efficacy of a fungal consortium for degradation of azo dyes and simulated textile dye effluents. Sustain Water Resour Manag 1:233–243

Selvam K, Swaminathan K, Chae K-S (2003) Decolourization of azo dyes and a dye industry effluent by a white rot fungus *Thelephora* sp. Bioresour Technol 88:115–119

Sen SK, Raut S, Bandyopadhyay P, Raut S (2016) Fungal decolouration and degradation of azo dyes: a review. Fungal Biol. Rev. 30:112–133

Soares GMB, de Amorim MTP, Costa-Ferreira M (2001) Use of laccase together with redox mediators to decolourize Remazol Brilliant Blue R. J Biotechnol Biotechnol Textile Indus Perspect New Millenn 89:123–129

Tony BD, Goyal D, Khanna S (2009) Decolorization of textile azo dyes by aerobic bacterial consortium. Int Biodeterior Biodegrad 63:62–469

Weisburger JH (2002) Comments on the history and importance of aromatic and heterocyclic amines in public health. Mutatation res/fundamental and molecular mechanisms of mutagenesis. In: 8th international conference on carcinogenic/mutagenic n-substituted aryl compounds 506–507, 9–20

Yamini D, Sivakami V, Soundarya M, Kumar DM (2012) Biodegradation of textile dyes, direct brilliant violet and direct greenish blue by *Aspergillus* sp. J Acad Indus Res 1(6):329–331

Yesilada O, Asma D, Cing S (2003) Decolorization of textile dyes by fungal pellets. Process Biochem 38:933–938

Young L, Yu J (1997) Ligninase-catalysed decolorization of synthetic dyes. Water Res 31:1187–1193

Yu J, Wang X, Yue PL (2001) Optimal decolorization and kinetic modeling of synthetic dyes by Pseudomonas strains. Water Res 35(15):3579–3586

Zhao X, Hardin IR (2007) HPLC and spectrophotometric analysis of biodegradation of azo dyes by *Pleurotus ostreatus*. Dyes Pigm 73:322–325

Zimmermann T, Kulla HG, Leisinger T (1982) Properties of purified orange II Azoreductase, the enzyme initiating azo dye degradation by pseudomonas KF46. Eur. J. Biochem. 129:197–203

# Potential Utilization of a Weed *Prosopis Juliflora* Leaf Extract Nanoparticle for Dye Degradation and Antibacterial Activity

**S. Sahithya and C. Krishnaveni**

**Abstract** The present study was conducted to evaluate the silver nanoparticle synthesized using leaf extract of *Prosopis juliflora.* For possible dye degradation and antibacterial activity on Gram positive and gram-negative bacteria (*Bacillus subtilis* and *Pseudomonas putida*). The acetone extract showed good anti-bacterial activity against both the organisms, the inhibitory effect increased with increase in concentration of the extract. The presence of Silver Nanoparticles in the leaf extract was observed by reduction of silver ions into silver nanoparticles. Fourier Transform Infra-Red (FTIR) analysis was done to detect the functional groups which will promote the antibacterial and dye degradation capabilites.

## 1 Introduction

Plants have served human beings as natural source for the treatments and therapies from ancient times, amongst them medicinal herbs have gained attention because of their wide use and less side effects. Higher plants serve as reservoirs of large number of organic compounds that are highly beneficial as they possess curative properties which are treated as secondary metabolites. The knowledge of chemical compounds of a plant is very essential to identify new resources to meet the ever-increasing demand on medicinal plants. Weed species interfere with our endeavors, such as agriculture or animal farming, recreational pursuits, including gardening, transport, bush walking and water sports etc. Globally, the utilization of weeds has been patchy over the past few decades. Utilization has been recognized as an effective means of weed management. Weeds have been used for long time as sources of food, fiber, dye, medicines, etc. *Prosopis juliflora* is a shrub or a small tree belonging to the family Fabaceae, a kind of mesquite. Its common uses include forage, firewood and environmental management (Ruto et al. 2018). But it has become an invasive weed in several countries where it was introduced. Though it is considered as an invasive weed, it has been used as a folk remedy for diarrhea, dysentery, flu, inflammation, measles,

S. Sahithya · C. Krishnaveni (✉)
Department of Botany, PSGR Krishnammal College for Women, Coimbatore, Tamil Nadu, India
e-mail: c.krishnaveni@hotmail.com



M. Prashanthi et al. (eds.), *Bioremediation and Green Technologies*, Environmental Science and Engineering, https://doi.org/10.1007/978-3-030-64122-1_12

sore throat and in healing of wounds. Several alkaloids have been isolated from the leaf extracts possessing pharmacological properties (Cronk and Fuller 2014). view of therapeutic importance of this plant, the present research work aims at screening the silver nanoparticle leaf extracts exploited for dye degradation and antimicrobial activity. In this way it can easily be eliminated and used in a more meaningful way (Abiyot and Getachew 2006).

## 2 Materials and Methods

### 2.1 Plant Material

The leaves of *Prosopis juliflora* were collected from Peelamedu, Coimbatore, Tamil Nadu, India. The leaves of *Prosopis juliflora* were washed thoroughly, shade dried and powdered mechanically. The dried leaf material is extracted with different solvents like acetone, ethanol and distilled water in a Soxhlet apparatus (Abdul et al. 2013).

### 2.2 Synthesis of Silver Nanoparticles

2 ml of extract was added to the aqueous solution of 2 ml of 2 mM Silver nitrate. Then the sample was incubated in dark for 24 h. After 24 h, the appearance of reddish-brown colour indicates the presence of silver nanoparticles. The formation and completion of silver nanoparticles was characterized by spectrophotometer. The analysis was carried out in a Visible Spectrometer-Deep vision 2305 with a resolution of 1 nm between 300 and 600 nm (Raja et al. 2012).

### 2.3 Dye Degradation Analysis

About 10 mg of methylene blue dye was added to 1000 mL of double distilled water (stock solution). About 10 mg of the bio-synthesized silver nanoparticles was added to 100 mL of methylene blue dye solution, this suspension was well mixed and was put under the sunlight. The absorbance spectrum of the supernatant was subsequently measured using UV–Vis spectrophotometer at the different wavelength. Concentration of dye during degradation was calculated by the absorbance value at 660 nm (Vanaja et al. 2014).

Percentage of dye degradation was estimated by the following formula:

$$\%\text{degradation} = (\text{Ci} - \text{Co}/\text{Ci}) * 100 \tag{12.1}$$

where Ci represents the initial dye concentration and Co represents the final dyeion concentration.

### 2.4 Anti-bacterial Activity

The Acetone, Ethanol and Distilled water extracts of Prosopis juliflora, were analyzed for their anti-bacterial activity for *Pseudomonas putida, Bacillus subtilis* (Ahmad et al. 1988; Jeevan et al. 2012).

### 2.5 Fourier Transform Infra-Red Analysis (FTIR)

Fourier Transform-Infrared Spectrophotometer (FTIR) is a tool to identify types of chemical compounds in the sample. Dried leaf powder (ethanolic extract) of test plant was used for FTIR analysis. 1 mg of dried powder was encapsulated in 10 mg of KBr pellet, in order to prepare translucent sample disc. The powdered sample of the pellet was loaded in FTIR Spectroscope (Shimadzu, Japan) with a scan range from 400 to 4000 $cm^{-1}$ with a resolution of 4 $cm^{-1}$ and the results were recorded (Arya et al. 2018).

## 3 Results

### 3.1 Synthesis of Silver Nanoparticles

The presence of silver nanoparticles in the leaf extract was observed by reduction of silver ions into silver nanoparticles. The color change of Ethanol extract from green to reddish brown was observed visually.

### 3.2 FTIR Analysis of Ethanol Leaf Extracts

Fourier Transform Infra-Red (FTIR) analysis was performed to detect the molecules present in the ethanol extracts of the leaves. A typical FTIR spectrum of the compounds is shown in (Fig. 1). The absorption bands were seen at 3367.86, 1634.74, 1393.63, 1057.04, 703.08 and 400.25 $cm^{-1}$. The intense band at 3367.86 corresponds to C = C–H Stretching, the band at 1634.74 corresponds to OH (Absorbed water), 1393.63 corresponds to t-butyl stretching, band at 1057.04 corresponds to

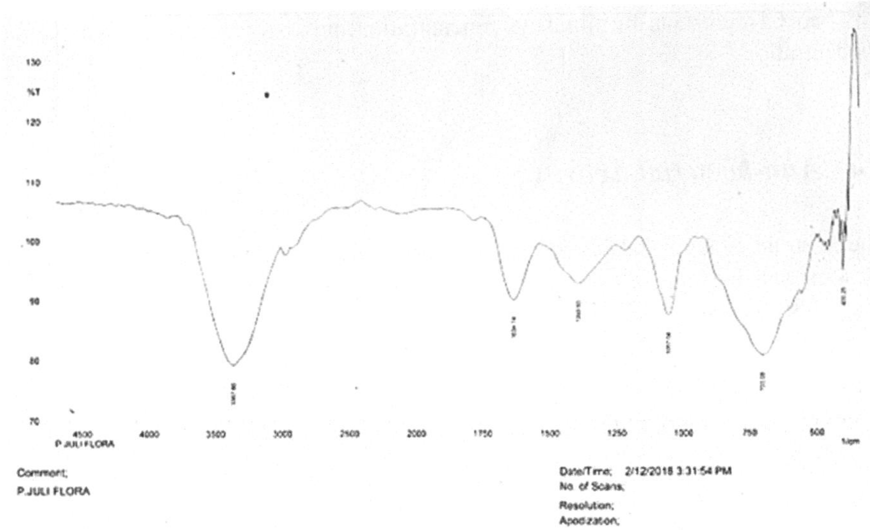


**Fig. 1** FTIR analysis of the ethanol extract

N–S Stretching, band at 703.08 corresponds to C–C Deformation and the band at 400.25 corresponds to S–S Stretching (Fig. 1) (Thakur et al. 2014).

## 3.3 Anti-bacterial Activity

The anti-bacterial activity of the Acetone, Ethanol and Dist. water extracts of *Prosopis juliflora* was tested against bacteria such as *Bacillus subtilis* and *Pseudomonas putida.* Acetone extract showed good anti-bacterial activity against both the organisms tested than ethanol and dist. water extracts. The zone of inhibition was higher against *Pseudomonas putida.* The inhibitory effect increased with increase in concentration of the extract. The positive control Streptomycin showed the inhibitory effect against the bacteria (Patil et al. 2012). The measured results of the Zone of inhibition are given in the Table 1.

## 3.4 Dye Degradation

Photocatalytic degradation of methylene blue was carried out by using *Prosopis juliflora* leaf extract silver nanoparticles under solar light. Dye degradation was identified by color change (also measured via spectrophotometer at 660 nm). Initially, the color of dye (blue color) changed after 1 h of exposure to sunlight with silver nanoparticles and the intensity decreased with prolonged exposure to light. Finally,

**Table 1** Antimicrobial activity of the weed

| S. No. | Organisms | Zone of inhibition (mm) | | | | | | | | | |
|---|---|---|---|---|---|---|---|---|---|---|---|
| | | Acetone extract (µl/ml) | | | Ethanol extract (µl/ml) | | | Dist. water extract (µl/ml) | | | Streptomycin (50 µl/ml)—Control |
| | | 25 | 50 | 75 | 25 | 50 | 75 | 25 | 50 | 75 | |
| 1 | *Bacillus subtilis* | 24 | 26 | 31 | 09 | 10 | 12 | – | – | 08 | 08 |
| 2 | *Pseudomonas putida* | – | – | 12 | 25 | 28 | 30 | – | – | – | – |

the degradation process was terminated at 48 h (colorless state) (Vanaja et al. 2014; Raina et al. 2020).

## 4 Discussion

Weeds are one of the most aggressive, troublesome and undesirable elements of the world's vegetation (Mahgoub et al. 2005). In the present study an attempt was made to explore the potential application of the leaf extract of the weed to explore its viability as an antimicrobial agent and make silver nanoparticles for dye degradation.

Present study shows that the acetone extract showed good anti-bacterial activity against *Bacillus subtilis* and ethanol extract showed good anti-bacterial activity against *Pseudomonas putida* than dist. water extract. The zone of inhibition was higher against *Pseudomonas putida* (30 mm). The inhibitory effect increased with increase in concentration of the extract. The positive control Streptomycin showed the inhibitory effect against the bacteria (Nakano et al. 2004; Nutter et al. 2016). Among the different solvent extracts from leaves of *P. juliflora* tested for antimicrobial activity only methanol extract showed significant anti-bacterial activity based on the zone of inhibition in a diffusion assay (Zainal et al. 1988). The results of ethanol fractions obtained from various parts of *P. juliflora* showed that the leaf, pod and flower extracts were efficient in inhibiting growth of bacteria (Mohamedsham et al. 2010).

The change of the colour of the ethanol extract from green to reddish-brown was observed in the present study. Colour of methanol extract changing from green to reddish-brown was observed in the similar study carried out by (Arya et al. 2018) on the synthesis of silver nanoparticles from the leaf extracts. The absorption bands at 3367.86, 1634.74, 1393.63, 1057.04, 703.08 and 400.25 $cm^{-1}$ shows the presence of C = C–H Stretching, OH (water), t-butyl stretching, N–S Stretching, C–C Deformation and S–S Stretching. The FTIR analysis revealed about 15 peaks representing the available functional groups within the Acetone extract of *P. juliflora*. The major functional groups are alcohols, phenols, aromatics, alkyl halides, alkenes and primary amines (Arya et al. 2018; Picariello et al. 2017).

The photocatalytic degradation potential of the nanoparticle can be observed from the subsequent decrease of the absorbance value of dye. The percentage of degradation efficiency of silver nanoparticles was observed as 95.6% at 48 h (Table 2). The degradation percentage was increased as increasing the exposure time of dye silver nanoparticles (Vanaja et al. 2014; Raina et al. 2020).

## 5 Conclusion

The present study supports the view that, even a weed can be used for activities which might make it a useful plant. The leaves extract of *P. juliflora* were found to be a

**Table 2** Percentage of dye degradation of *Prosopis juliflora* leaf extract silver nanoparticles

| Time (h) | % degradation |
|---|---|
| 1 | 9 |
| 2 | 16 |
| 4 | 27 |
| 8 | 36 |
| 16 | 48 |
| 24 | 72 |
| 48 | 96 |

decent antibacterial agent and the leaf extract sourced silver nanoparticles was also proved to be a potential agent for dye degradation. Hence this invasive weed can be used for more potential application apart from its use as timber and firewood; the weed can easily be eliminated and used in a different and a meaningful way.

**Acknowledgements** We extend my wholehearted thanks to our Managing Trustees and administrators of our college, and the Director, South Indian Textile Research Institute (SITRA).

## References

Abdul W, Mehreen G, Syed BJ, Muhammad N (2013) Phytochemical analysis of medicinal plants occurring in local area of Mardan. Biochem Anal Biochem 2(4):1000144. https://doi.org/10.4172/2161-1009.1000144

Abiyot B, Getachew T (2006) The Prosopis dilemma, impacts on dryland biodiversity and some controlling methods. J Drylands 1(2):158–164. https://doi.org/10.4172/2161-1009.1000144

Ahmad A, Khan KA, Ahmad VU, Qazi S (1988) Antibacterial activity of an alkaloidal fraction of *Prosopis juliflora*. Fitoterapia 59(6):481–484

Arya G, Kumari RM, Gupta N, Kumar A, Chandra R, Nimesh S (2018) Green synthesis of silver nanoparticles using *Prosopis juliflora* bark extract: reaction optimization, antimicrobial and catalytic activities. Artif Cells Nanomed Biotechnol 46(5):985–993. https://doi.org/10.1080/21691401.2017.1354302

Cronk QC, Fuller JL (2014) Plant invaders: the threat to natural ecosystems. Routledge

Jeevan P, Ramya K, Rena AE (2012) Extracellular biosynthesis of silver nanoparticles by culture supernatant of Pseudomonas aeruginosa. Ind J Biotechnol 11(1):72–76

Mahgoub O, Kadim IT, Forsberg NE, Al-Ajmi DS, Al-Saqry NM, Al-Abri AS, Annamalai K (2005) Evaluation of Meskit (*Prosopis juliflora*) pods as a feed for goats. Anim Feed Sci Technol 121(3–4):319–327. https://doi.org/10.1016/j.anifeedsci.2005.01.016

Mohamedsham SH, Hansipriscilla D, Kavitha T (2010) Antimicrobial activity and phytochemical analysis of selected Indian folk medicinal plants. Int J Pharma Sci Res 1(10):430–434

Nakano H, Nakajima E, Hiradate S, Fujii Y, Yamada K, Hasegawa K (2004) Growth inhibitory alkaloids from mesquite (*Prosopis juliflora* (Sw.) DC.) leaves. Phytochemistry 65:587–591. https://doi.org/10.1016/j.phytochem.2004.01.006

Nutter J, Fritz R, Iurlina MO, Saiz AI (2016) Effect of Prosopis sp. honey on the growth and fermentative ability of *Pediococcus pentosaceus* and *Lactobacillus fermentum*. LWT Food Sci Technol 70:309–314

Patil RS, Kokate MR, Kolekar SS (2012) Bioinspired synthesis of highly stabilized silver nanoparticles using Ocimum tenuiflorum leaf extract and their antibacterial activity". Spectrochimica Acta A 91:234–238

Picariello G, Sciammaro L, Siano F, Volpe MG, Puppo MC, Mamone GF (2017) Comparative analysis of C-glycosidic flavonoids from *Prosopis* spp. and *Ceratonia siliqua* seed germ flour. Food Res Int 99:730–738

Raja K, Saravanakumar A, Vijayakumar R (2012) Efficient synthesis of silver nanoparticles from *Prosopis juliflora* leaf extract and its antimicrobial activity using sewage. Spectrochim Acta a Mol Biomol Spectrosc 97:490–494

Raina S, Roy A, Bharadvaja N (2020) Degradation of dyes using biologically synthesized silver and copper nanoparticles. Environ Nanotechno Monit Manag 13:100278. https://doi.org/10.1016/j.enmm.2019.100278

Ruto MC, Ngugi CM, Kareru PG, Cheruiyot K, Rechab SO, Madivoli ES, Maina EG (2018) Antioxidant activity and antimicrobial properties of *Entada leptostachya* and *Prosopis juliflora* extracts. J Med Plants Econ Dev 2:1–8

Thakur R, Singh R, Saxena P, Mani A (2014) Evaluation of antibacterial activity of *Prosopis juliflora* leaves. Afr J Tradit Complement Altern Med 11(3):182–188

Vanaja M, Paulkumar K, Baburaja M, Rajeshkumar S, Gnanajobitha G, Malarkodi C, Sivakavinesan M, Annadurai G (2014) Degradation of methylene blue using biologically synthesized silver nanoparticles. Bioinorg Chem App 2014:742346. https://doi.org/10.1155/2014/742346

Zainal AS, Abdel-Rahim AM, Abu-Ali RM, Radwan SS (1988) Antimicrobial substance(s) in the leaf litter of the xerophyte *Prosopis juliflora*. Zentral Blattfür Mikrobiologie 143:375–381

# Experimentation and Evaluation of Biodegradative Kinetic Parameters of Nitrate Removal from Drinking Water

**Rajakumar Sundaram, P. M. Ayyasamy, S. Umadevi, Y. C. Song, and P. Lakshmanaperumalsamy**

**Abstract** Nitrate ($NO_3^-$) contamination in drinking water is regulated by environmental agencies around the world since it causes health problems at higher concentrations. $NO_3^-$removal in drinking water has been a challenge for a long time. Microbial denitrification is one of the promising processes for remediation of $NO_3^-$ from contaminated samples. In the present study, the effect of various carbon sources, temperatures, pH and inoculum concentrations on the removal of $NO_3^-$ was first investigated using assimilatory $NO_3^-$ reducing bacterial consortium (*Pseudomonas* sp. KW1 and *Bacillus* sp. YW4) in Mineral Salts Medium (MSM) containing 100 mg/Lof $NO_3^-$. Starch at 1%, temperature at 30°Cand pH at 7 and 1% of inoculum concentration were found to be optimum. Monod and Haldane kinetic models were applied to evaluate the rate of nitrate removal by bacteria. Further, Friedman test was used to determine the role of each factor on degradation process. Second phase of experiments was studied in drinking water using a pilot scale treatment plant and the results showed that the consortium (KW1 and YW4), reduced nitrate to a level at 92% (100–8 mg/L) and verified with statistical test. Further, boiling, ultraviolet radiations, filtration by adsorption and chlorination methods were evident for consecutive elimination of microorganisms from the treated water.

**Keywords** Nitrate · Carbon source · Denitrification · Coagulants · Pilot scale plant

R. Sundaram (✉) · S. Umadevi
Department of Marine Biotechnology, Bharathidasan University, Tiruchirappalli 620024, Tamil Nadu, India
e-mail: kodairaj@gmail.com

P. M. Ayyasamy
Department of Microbiology, Periyar University, Salem 636011, Tamil Nadu, India

Y. C. Song
Department of Environmental Engineering, Korea Maritime and Ocean University, Busan 49112, South Korea

P. Lakshmanaperumalsamy
Department of Environmental Science, Bharathiyar University, Coimbatore 641046, India



M. Prashanthi et al. (eds.), *Bioremediation and Green Technologies*, Environmental Science and Engineering, https://doi.org/10.1007/978-3-030-64122-1_13

# 1 Introduction

Nitrate ($NO_3^-$)in drinking water often causes methemoglobinemia and other health disorders such as hypertension, increased infant mortality, goiter, stomach cancer, thyroid disorder, cytogenetic defects and birth defects (WHO 2003). During the last two decades, extensive attention has been paid to the management and control of nitrate contamination. As nitrate is stable and highly soluble ion and has a low potential for adsorption or co-precipitation, it is difficult to remove $NO_3^-$ by conventional water treatment processes including lime softening and filtration (Kapoor and Viraraghavan 1997).Although, reverse osmosis (Schoeman and Steyn 2003), ion exchange (Matosicet al., 2000; Kim and Benjamin 2004),catalytic reduction (Reddy and Lin 2000; Sunet al., 2010), electrodialysis (Elmidaouiet al., 2001), electroccoagulation (Jeonget al., 2014), adsorption process (Hamoudi and Belkacemi 2013), chemical denitrification (Mortazaviet al. 2011) and phytoremediation (Ayyasamyet al., 2009) are effective in removing $NO_3^-$ from contaminated water, but they are very expensive with a limited potential application (Kesseruet al., 2002). The most versatile and widely used technology in the removal of $NO_3^-$ from water and/or wastewater is microbiological denitrification (Shin and Cha 2008).

Biological denitrification performed with bacteria can be involved in autotrophic and heterotrophic ways. However, the biological system needs essential electron donors to reduce$NO_3^-$ from drinking water. Many workers have used a variety of carbon sources as an electron donor toenhance the microbial growth for theremoval of $NO_3^-$. Gomez et al. (2000) found that ethanol was the suitable carbon source among studied the effect of sucrose, ethanol and methanol for the removal of $NO_3^-$ from contaminated groundwater. Shanthi et al. (2005) reported 95 to 100% reduction when methanol was used as a carbon source. Akunnaet al. (1994) showed that glucose induced a very high degree of $NO_3^-$ removal in aaqueous mediumunder anaerobic condition. Kim et al. (2002) reported thatnitrogen removal efficiency of 99.5% at a hydraulic residence time of 1 h while starch was supplemented to groundwater. In a study conducted by Fernandez-Nava et al. (2010), three different carbon sources namely, wastewater from sweet manufacturing, saccharose-rich residue from the production of soft drinks and lactic acid-rich residue from a dairy plant were investigated for their potential use in nitrate reduction from wastewater. Authors found that there was maximum nitrate removal efficiency using the wastewater of sweet production and saccharose-rich residue in short time. Thus, in this study, commercially available carbon sources such as glucose, starch, cellulose, sucrose and acetic acid were tested for their involvement in the removal of $NO_3^-$from drinking water.

The major issue in biological methodis the removal of bacterial cells and suspended particles. Coagulating agents have been widely applied to remove chemical ions, colloidal particles and microorganisms (Jiang et al. 2006; Ayyasamy et al. 2007). Several authors have studied biological and chemical treatment separatelyfor the removal of $NO_3^-$ and microorganisms from the water system. However, combined treatment system of microorganisms and coagulants on the removal of $NO_3^-$was not reported. Hence, this study was designed to remove $NO_3^-$and heterotrophic bacteria from drinking water through biological and chemical treatment processes.

# 2 Material and Methods

## 2.1 Isolation of Nitrate Reducing Bacteria

The water and sediment samples were collected under aseptic conditions in sterile bottles from Kodaikanal and Yercaud lakesin Tamil Nadu, India. Bacterial populations were estimated by pour plate technique using nutrient agar (per litre containing 3 g of beef extract, 3 g of yeast extract, 5 g of peptone, 5 g of NaCl,20 g of agar, pH 7.0 ± 0.2). Well-defined colonies were isolated on the basis of morphological characteristics, transferred to nutrient agar and were identified up to the generic level after purification (Buchanan and Gibbons 1974). All the strains were tested for nitrate reduction using potassium nitrate broth (per litre containing 5 g of peptone, 3 g of beef extract, 5 g of NaCl, 5 g of $KNO_3$, pH7.0 ± 0.2). The ability of the isolates to reduce $NO_3^-$ to nitrite ($NO_2^-$) and ammonium ($NH_4^+$) was determined by the addition of Nessler's reagent to the cultures. The appearance of yellow orange showed that $NO_3^-$and/or$NO_2^-$ has been reduced to $NH_4^+$. Based on the intensity of the color, the isolates were categorized ($^+$ low, $^{++}$ moderate and $^{+++}$ high $NO_3^-$ reduction). The bacterial species, *Pseudomonas* sp. (KW1) and *Bacillus* sp. (YW4) were found to be most efficient for nitrate reduction and used as consortium for further studies of nitrate reduction. After screening, the selected isolates were purified by repeated streaking on nutrient agar and stored at 4 °C.

## 2.2 Inoculum Preparation

A loopful of cultures (KW1 and YW4)were inoculated in presterilized 100 ml nutrient broth separately. The flasks were kept in a shaker at 120 rpm for 12 h at 30 °C. The culture broths were centrifuged at 10,000 rpm for 20 min. Cell suspensions were prepared using sterile distilled water and adjusted to 1OD ($10^4$ CFU/mL) using UV–Visible Spectrophotometer (model: Hitachi U 3210) at 600 nm. One ml containing $10^4$ CFU/mL of the cell suspension was used as inoculum.

### 2.3 *Optimization of Carbon Source, Temperature, pH and Inoculum Concentration*

Nitrate removal was evaluated under batch mode condition using 100 ml of synthetic medium (MSM) (per litre containing 0.1 g of $KH_2PO_4$, 1 g of $K_2HPO_4$, 0.005 g of $CaCl_2$, 0.1 g of $MgSO_4$, 0.05 g of $Na_2SiO_3$, pH 7 ± 0.2) supplemented with 100 mg/L of $NO_3^-$and 1% offive different carbon sources such as glucose, starch, cellulose, sucrose and acetic acid. To this, 1 mL ($10^4$ CFU/mL) each of KW1 and of YW4 (1:1 v/v) was aseptically addedand incubated for 48 h in a mechanical shaker (120 rpm). The control was also maintained with the same concentration of $NO_3^-$ but without the bacterial inoculum. Every 6 h, the $NO_3^-$level was determined by phenol disulphonic acid method (Rao 2000). Starchwas found to enhance the$NO_3^-$ removal and hence it was selected as a carbon source for further study. The above experiment was repeatedsubsequently to find out the optimum conditions of temperature (among 25, 30, 35, 40 and 45 °C), pH (among 6, 7, 8 and 9), starch (among 0.2, 0.4, 0.6, 0.8, 1, 1.2 and 1.4%) and inoculum concentrations (among 1, 2, 3, 4 and 5%) for the removal of $NO_3^-$in the synthetic medium.

### 2.4 *Growth Kinetics of Nitrate Removal*

To investigate the biodegradation kinetics of nitrate, several kinetic models have been evaluated. The expression of substrate utilization in a batch reactor with respect to time can be correlated by the first-order differential equation describes in Eq. (13.1).

$$-\frac{dS}{dt} = KsS \tag{13.1}$$

where, *S* is the substrate concentration (mg/l), *Ks* is the first order rate constant ($h^{-1}$) and *t* is the incubation time (h). Also, the Monod equation (Eq. 13.2) was applied for nitrate biodegradation with different concentrations of starch, temperature, pH and cell inoculum.

$$\mu = \frac{\mu m S}{Ks + S} \tag{13.2}$$

where, $\mu_m$ is the maximum specific degradation rate ($h^{-1}$), *S* is the substrate concentration (mg/l) and *Ks* is the Monod constant (mg/l). In some cases, the system disturbs due to high concentration of inhibitory compounds. In this case, Haldane's model has a well-fitted model for determination of the kinetic parameters and it has the similar form as described in Eq. (13.3).

$$\mu = \frac{\mu \max S}{Ks + S + \frac{S^2}{Ki}} \tag{13.3}$$

where, $K_i$ is the inhibition coefficient (mg/l). The biokinetic parameters in this model were estimated by the Lineweaver–Burk plot (1/μ vs. 1/S) (Najafpour 2007). Also, the most important factor in biodegradation process was determined by Friedman test by use of the statistical package SPSS.

## 2.5 Lab Scale Study for Removal of Nitrate in Drinking Water

The lab scale setup consists of reservoir, reactor tank, settling tank, filtration tank and collection tank (Fig. 1). The reactor and settling tanks were fitted with stirrer. The filtration tank was made of rectangular glass vessel with dimension of 55 cm in height, 60 cm in length and 35 cm in width. To this, large pebbles (2.0 to 2.5 cm) were packed in bottom of the vessel followed by small pebbles (0.7–1.5 cm), gravel (0.4–0.6 cm), coarse sand (0.05–0.1 cm), fine sand (0.15–0.3 mm) and activated carbon (0.75–1.0 μm). The filtration tank was washed thoroughly with sterile distilled water to remove the adhering particles in the filter bed. The bottom of the vessel was connected with an outlet to a collection tank.

About 10 L of fresh drinking water with 100 mg/L of $NO_3^-$ was added to the reactor tank. Further, 1% of soluble starch and 10 mL of KW1 and YW4 (1:1 v/v) containing $10^4$ CFU/mL were added and allowed to react for 48 h at an optimum temperature of 30 °C and pH 7. The speed of stirrer in both reactor and settling tank was adjusted to 120 rpm. After bacterial treatment, the water sample was transferred to settling tank and mixed thoroughly with commercial grade alum [$(Al_2(SO_4)_3$] and lime [$CaCO_3$], at normal dosage (alum—25 mg/L and lime—6 mg/L) prescribed by Central Pollution Control Board, India, for 15 min at 120 rpm and 10 min at 15 rpm and allowed to settle for 30 min (Ayyasamy et al. 2007). Further, the bacterial biomass and coagulant particles were removed through the outlet of settling tank followed by filtration. The bacterial growth, amount of $NO_3^-$, $NO_2^-$ and $NH_4^+$ were estimated in the treated water sample for every 6 h (APHA 2005).

## 2.6 Pilot Scale Study for Removal of Nitrate in Drinking Water

The pilot scale plant consists of reservoir (1000 L), inoculum tank (10 L), reactor tank (1200 L), settling tank (1200 L), sand filter (1200 L) (mode of packing was similar as mentioned in Sect. 2.4) and collection tank (1200 L). Each tank was interconnected with pipelines as shown in the Fig. 2. The reactor tank was fitted with peddler to mix bacterial inoculum with water. The settling tank was also fitted with peddler for mixing of coagulating agents. The treatment plant was designed and constructed by IVC laboratory and Environmental Services, Chennai, Tamil Nadu, India.

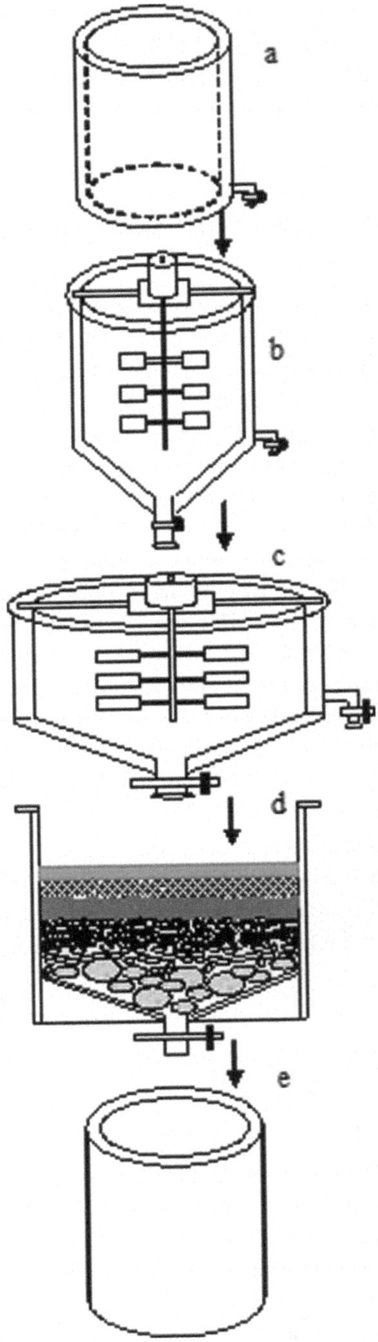


**Fig. 1** Lab scale plant for the removal of nitrate from drinking water. **a** Reservoir **b** bioreactor **c** settling tank **d** sand filter **e** collection tank

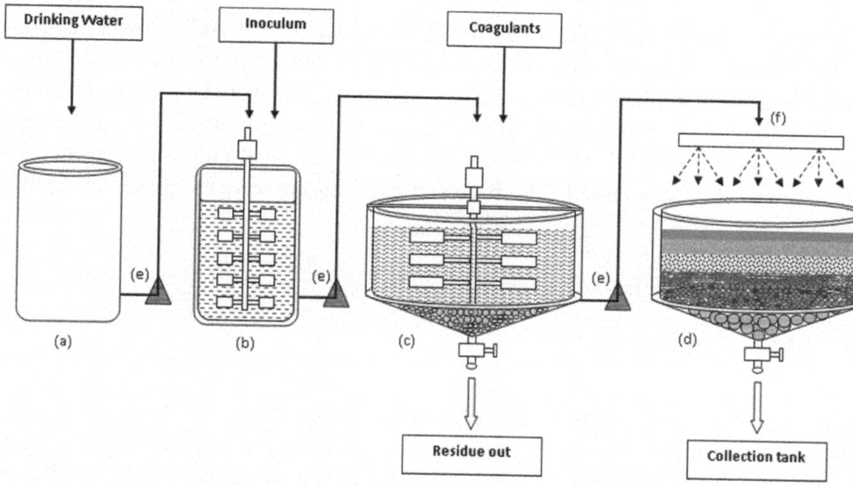


**Fig. 2** Pilot scale plant for the removal of nitrate from drinking water. **a** Reservoir **b** bioreactor **c** settling tank **d** filtration tank **e** pump **f** shower

About 1000 L of drinking water containing 100 mg/L of $NO_3^-$ and 1% starch as a carbon source was prepared in a large reservoirand was transferred to bioreactor. One hundred mL ($10^{4\,CFU}$/mL) of bacterial consortium (KW1 and YW4) (1:1 v/v) was added to the bioreactor and agitated mechanically at 120 rpm for 6, 12, 18, 24, 30, 36, 42 and 48 h.The temperature and pH was maintained at an optimum level of 30 °C and 7 °C respectively. After the retention time the bacterially treated water was pumped in to settling tank. The commercial grade alum (25 mg/L) and lime (6 mg/L) were added and mixed thoroughly for 15 min with the help of rotating peddler (120 rpm). Then, the samples were allowed to settle for 30 min and the supernatant was pumped in to sand filter. In each process, the samples were collected and analyzed for bacterial population and physico-chemical parameters such as pH, electrical conductivity (EC), turbidity, total solids (TS), total suspended solids (TSS), total dissolved solids (TDS), total hardness, chloride, $NO_3^-$, $NO_2^-$, $NH_4^+$, sulphate and phosphate (APHA 2005).

## 2.7 *Disinfection Methods for the Removal of Bacteria from Treated Water Sample*

Disinfection studies were carried out for the treated water samples using various disinfection methods such as boiling, ultraviolet (UV) radiation, filtration by adsorption and chlorination. The sample of 100 ml was taken in a sterile conical flask and boiled for 10 min. In ultraviolet technique, 20 ml sample was taken in sterile Petri dish and subjected to UV radiation at a distance of 20 cm for various time intervals (10, 20, 30, 40, 50 and 60 min).The sample was passed through Eureka Forbes water filter cum purifier (Model—Aquaguard—Nova, Model type—AGWPNOVA, Serial No: BG03-005,504, Dimension—395 × 300 × 255 mm). After the above disinfection methods,the samples were collected and bacterial cells were enumerated by pour plate technique. For chlorination, about 100 ml of the sample was taken in separate conical flasks and 0.1, 0.2, 0.3, 0.4 and 0.5 mg/Lof residual chlorine (commercial bleaching powder) was added. The bacterial populations were estimated at regular intervals.

### 2.8 Statistical Analysis

The relationship between the bacterial growth, concentration of nitrate, nitrite and ammonium was determined by Pearson correlation coefficient in terms of percentage difference using SPSS software package. Further, the mean and standard deviation of the physic-chemical parameters of water samples were calculated using the statistical package within Microsoft® Excel Version 2010.

## 3 Results

### 3.1 Nitrate Reducing Bacteria in Water and Sediment Samples

The bacterial species of *Pseudomonas, Bacillus, Moroxella, Micrococcus, Alcaligenes, Corynebacterium*and members of Enterobacteriaceae were isolated from water and sediment samples of Kodaikanal and Yercaud Lake. The ability of $NO_3^-$ reduction was tested using potassium nitrate broth and the results were given in Table 1. Among them,*Pseudomonas*sp. (KW1) and *Bacillus* sp. (YW4) were found to be the most efficient bacteria in removing $NO_3^-$. The consortium of KW1 and YW4 was further tested for the removal of $NO_3^-$.

### 3.2 Batch Mode Study for the Removal of Nitrate in the Synthetic Medium

The maximum removal of $NO_3^-$ from 100 to 0.6 mg/L was recorded in the synthetic medium supplemented with starch (Fig. 3a) followed by glucose (100–18.35 mg/L). The $NO_3^-$ reduction in the presence of acetic acid, cellulose and sucrose was 100–27.7, 100–34.8 and 100–44.9 mg/L respectively. The study conducted in the synthetic medium with starch at various concentrations showed that at 1% starch the consortium KW1 + YW4 reduced maximum level of $NO_3^-$(100–0.58 mg/L) at 48 h of

**Table 1** Selection of nitrate reducing bacteria (Nitrate reduction test)

| S.No | Strain No | Genera | Nitrate/ Nitrite / Ammonia |
|---|---|---|---|
| 1 | *KW 1* | *Pseudomonas* | +++ |
| 2 | *KW 2* | *Bacillus* | + |
| 3 | KW 3 | *Moroxella* | ++ |
| 4 | KW 4 | *Bacillus* | ++ |
| 5 | KW 5 | *Micrococcus* | ++ |
| 6 | KW 6 | *Bacillus* | ++ |
| 7 | KW 7 | *Pseudomonas* | ++ |
| 8 | KW 8 | *Alcaligenes* | ++ |
| 9 | KS 1 | *Bacillus* | ++ |
| 10 | KS 2 | *Alcaligenes* | + |
| 11 | KS 3 | *Pseudomonas* | + |
| 12 | KS 4 | *Bacillus* | ++ |
| 13 | KS 5 | *Pseudomonas* | + |
| 14 | YW 1 | *Bacillus* | ++ |
| 15 | YW 2 | *Bacillus* | + |
| 16 | YW 3 | *Corynebacterium* | + |
| 17 | YW 4 | *Bacillus* | +++ |
| 18 | YW 5 | *Corynebacterium* | + |
| 19 | YW 6 | *Bacillus* | ++ |
| 20 | YW 7 | *Bacillus* | ++ |
| 21 | YW 8 | *Micrococcus* | - |
| 22 | YW 9 | Enterobacteriaceae | + |
| 23 | YW 10 | *Moroxella* | + |
| 24 | YS 1 | *Bacillus* | ++ |
| 25 | YS 2 | *Alcaligenes* | + |

$^{+}$less nitrate reduction$^{++}$ moderate nitrate reduction$^{+++}$ high nitrate reduction
$^{-}$no reduction

incubation (Fig. 3b). Whereas the starch at 0.2, 0.4, 0.6 and 0.8% concentrations, the $NO_3^-$ reduction was less compared to 1% concentration. The above results showed that starch at 1% was found to be optimum for $NO_3^-$ reduction by the bacterial consortium.

The bacterial consortium reduced $NO_3^-$ from 100 to 0.6 mg/Lin the synthetic medium amended with 1% starch at 30 °C incubation at 48 h. The $NO_3^-$ reduction at other temperatures (25, 35, 40 and 45 °C)was comparatively less (Fig. 3c). The maximum $NO_3^-$reduction (100–0.61 mg/L) was noticed in pH 7. The $NO_3^-$ reduction was very negligible at pH 6, 8 and 9 (Fig. 3d). There was no significant variation in the reduction of $NO_3^-$ at different dosage of inoculum (Fig. 3e).

In the batch mode experiment conducted with various carbon sources, $NO_3^-$ was removed in the synthetic medium supplemented with 1% starch as a sole carbon source. The highest $NO_3^-$ reduction rate was obtained because the bacterial cultures were amylolytic (starch degraders) and were capable of utilizing starch as a carbon source. These results are in good agreement with the results of Kim et al. (2002) where the use of starch as a carbon sources in the bioremediation of ground water

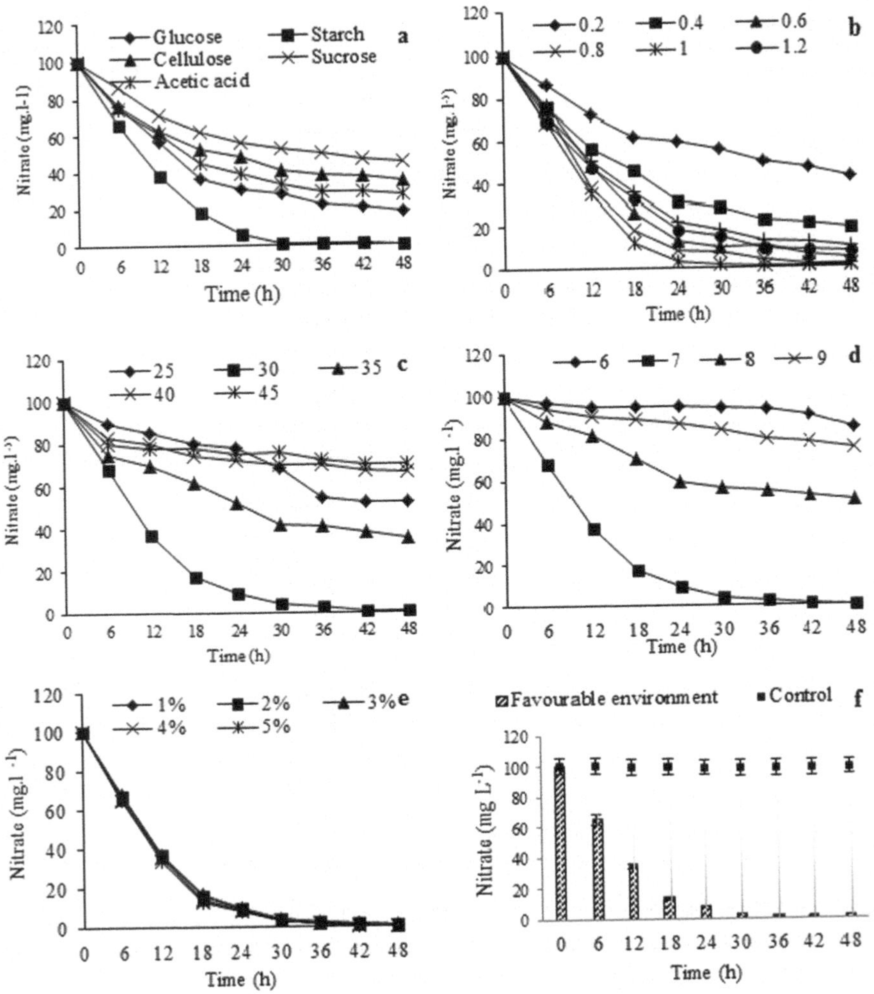


**Fig. 3** Removal of nitrate in the synthetic medium under batch mode condition **a** Effect of carbon sources **b** effect of starch concentration **c** effect of temperature **d** effect of pH **e** effect of cell concentration **f** effect of all favourable conditions

$NO_3^-$. Correspondingly, Ayyasamy et al. (2007) attained 89.08% of $NO_3^-$ reduction in ground water amended with 1% starch by *Pseudomonas* sp. (RS 7).

The temperature effect on the denitrification rate is another important feature in the design of a denitrification process (Oh and Silverstein 1999; Orhon et al. 2004). Thermophilic and psychrophilic denitrifying bacteria are known to have different temperature optima than the mesophiles. The bacterial consortium (KW1 and YW4) which were mesophiles, reduced maximum percentage of $NO_3^-$ (99.4%) in the synthetic medium. Ayyasamy et al. (2007) reported a maximum of 89.08% of $NO_3^-$ at 30 °C. In the present study also, higher (99.4%) $NO_3^-$ removal was observed at

30 °C. The lower and higher temperature affected the bacterial growth in the synthetic medium containing $NO_3^-$.

The $NO_3^-$ removal by bacterial consortium was severely influenced by various pH. The maximum level 99.4% of $NO_3^-$ was reducedat pH 7 and at 30 °C. The nature of alkaline and acidic conditions may interfere in $NO_3^-$ reduction. Carrera et al. (2003) studied hydrogenotrophic denitrification using a fluidized bed sand reactor and showed that the optimum pH for $NO_3^-$ removal was 7.5. The consortium functions to its maximum at neutral pH and since pH of drinking water is normally near neutral the consortium might have effectively reduced $NO_3$.

## 3.3 *Growth Kinetics of Nitrate Removal*

The first-order kinetic model of substrate utilization (Eq. 13.1) and $R^2$, the determinant coefficient was demonstrated (Table 2). It was observed that, the increasing level of starch concentration invers the first-order rate constant, unless it reaches equilibrium. As the favourable starch concentration was fixed with the consecutive test with different temperature, pH and cell mass, there was no correlation observed in rate constant.

The maximum specific removal rate ($\mu$), calculated by Lineweaver–Burk plot using the concept of Monod model for different starch concentration, temperature, pH and cell inoculum was 7.6923, 0.0009, 0 and 0.0617 respectively. From this result, it is obvious that increasing the starch concentration increases the bacterial growth and leads to higher removal of nitrate and pH makes the effective impact even in bit variation in neutral condition. These results were further, verified with considering the variables as independent by Friedman test. Table 3 describes the results of difference between mean ranks towards factors on role description. At, 5% significant level, cell concentration is the most important factor followed by temperature in par with pH and starch concentration.

The specific growth rate (degradation rate of nitrate) parameter is Haldane's kinetic model was illustrated in Fig. 4. In case of starch, temperature and pH, the specific growth rate increases along with the initial concentration up to a certain maximum value, then the rate decrease.

## 3.4 *Removal of Nitrate from Drinking Water in Lab Scale Study*

Water sample containing 100 mg/Lof $NO_3^-$amended with 1% starch was treated with bacterial consortium(KW1 + YW4) in lab scale was reduced 99.4% of $NO_3^-$and also increased in numbers (85 × $10^4$ CFU/mL) after 48 h. During the process of $NO_3^-$ reduction, 3.2 mg/L and 8.4 mg/L of $NO_2^-$ and $NH_4^+$ respectively were formed in

**Table 2** Rate kinetics of nitrate removal

| Environmental condition | Rate of nitrate removal ($h^{-1}$) | $R^2$ |
|---|---|---|
| *Carbon source* | | |
| Glucose | 0.0143 | 0.9409 |
| Starch | 0.0526 | 0.9546 |
| Cellulose | 0.0079 | 0.9464 |
| Sucrose | 0.0064 | 0.9359 |
| Acetic acid | 0.0105 | 0.9220 |
| *Concentration of starch (%)* | | |
| 0.2 | 0.0576 | 0.8259 |
| 0.4 | 0.1209 | 0.8507 |
| 0.6 | 0.2140 | 0.8938 |
| 0.8 | 0.2928 | 0.9230 |
| 1.0 | 0.3789 | 0.9017 |
| 1.2 | 0.2317 | 0.7587 |
| 1.4 | 0.2237 | 0.6869 |
| *Temperature (°C) + starch (1%)* | | |
| 25 | 0.0066 | 0.9325 |
| 30 | 0.0508 | 0.9929 |
| 35 | 0.0085 | 0.0965 |
| 40 | 0.0023 | 0.9343 |
| 45 | 0.0015 | 0.9311 |
| *pH + starch (1%) at 30 °C* | | |
| 6 | 0.0012 | 0.7432 |
| 7 | 0.0508 | 0.9929 |
| 8 | 0.0060 | 0.9201 |
| 9 | 0.0023 | 0.9890 |
| *Cell concentration (%) + pH 7 + starch (1%) at 30 °C* | | |
| 1 | 0.0459 | 0.9889 |
| 2 | 0.0483 | 0.9874 |
| 3 | 0.0523 | 0.9957 |
| 4 | 0.0576 | 0.9953 |
| 5 | 0.0609 | 0.9941 |

**Table 3** Role of environmental conditions on nitrate removal

| Factors | Mean rank | $\chi^2$ | P |
|---|---|---|---|
| Starch | 1.61 | 8.143 | 0.043 at 5% significant level |
| Temperature | 2.78 | | |
| pH | 2.78 | | |
| Cell concentration | 2.83 | | |

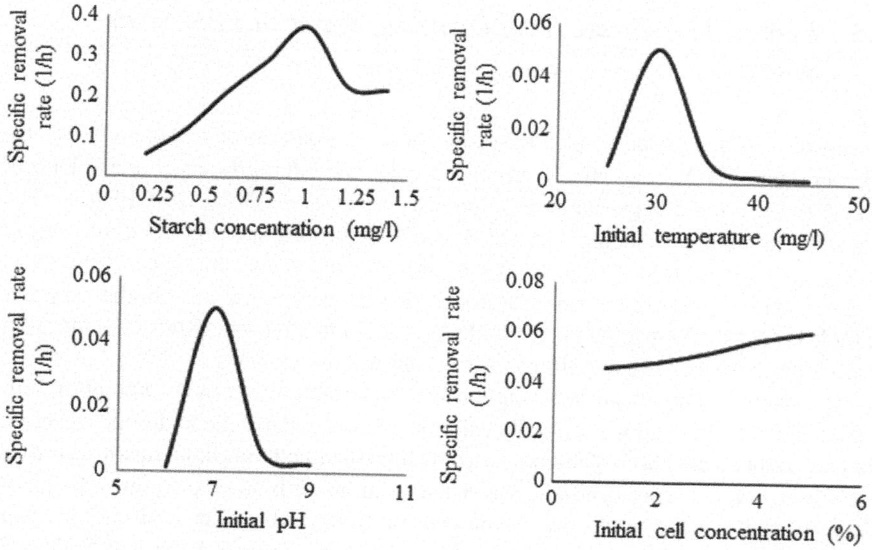


**Fig. 4** Haldane's growth kinetic model

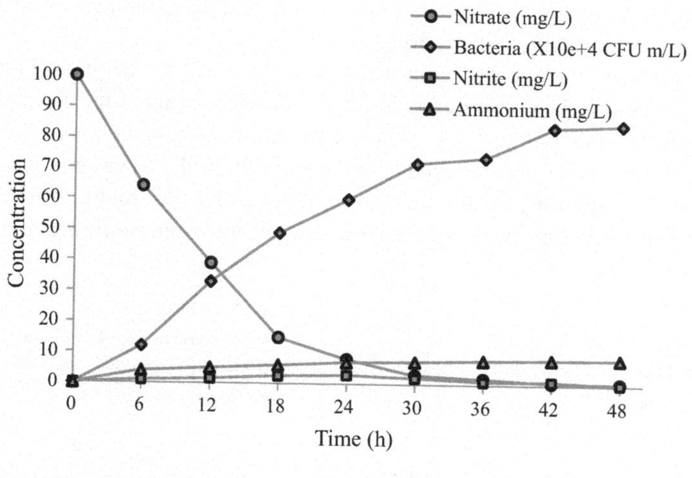


**Fig. 5** Effect of lab scale study of nitrate removal on parameters

the reactor. Complete removal of $NO_3$, $NO_2^-$ and $NH_4^+$ was noted in the bacterially treated water after coagulation followed by filtration (Fig. 5).

### 3.5 Removal of Nitrate from Drinking Water in Pilot Scale Study

Removal of $NO_3^-$ by bacterial consortium showed maximum (92%) at 48 h (Fig. 6). The amount of $NO_2^-$ and $NH_4^+$ formed were 2.1 and 7.6 mg/L respectively. During $NO_3^-$ reduction the bacterial population also increased to 74 × $10^4$ CFU/mL. The physico-chemical parameters were analysed and showed less amount compared to untreated water samples (Table 4). Due to presence of bacterial biomass, the amount of total solids (2880 mg/L) and turbidity (25 NTU) were found to be higher in water sample before sand filtration (Table 4). The values of treated water after sand filtration were with in the permissible limits of drinking water standards.

Autotrophic denitrification was attained from potable water by *Thiobacillusdenitrificans* using three laboratory-scale column reactors under three hydraulic retention times (Darbi et al. 2003). Dhamole et al. (2009) studied simultaneous removal of carbon and $NO_3^-$ from synthetic water using an airlift bioreactor. Similarly, laboratory and pilot scale set up was developed for $NO_3^-$ removal from drinking water containing 100 mg/L of $NO_3^-$ with the bacterial consortium of KW1 and YW4 and 99.4% and 92% of $NO_3^-$ was removed at 48 h. However, in pilot scale treatment the percentage of $NO_3^-$ removal was slightly lesser when compared to laboratory study. This might be due to some natural environmental factors that interfere in the bacterial growth and $NO_3^-$ utilization.

After treatment, the bacterial population was 74 × $10^4$ CFU/mL and an attempt was made to remove the microorganisms using chemical coagulants (alum and lime). The coagulants after reduced the bacterial biomass by quicker precipitation. The population in the water was reduced to 16 × $10^1$ CFU/mL. Ayyasamy et al. (2007) used different coagulating agents such as alum, lime and PAC and found that significant levels of microbial biomass removal through precipitation process. Dziubek

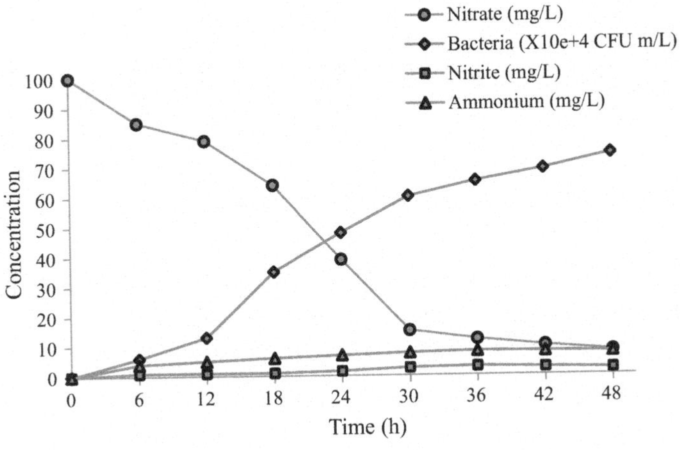


**Fig. 6** Effect of pilot scale study of nitrate removal on parameters

**Table 4** Physico-chemical parameters of water sample before and after pilot scaletreatment

| Parameters | Untreated water | Treated water | | | ISI drinking water standard |
|---|---|---|---|---|---|
| | | After bacterial treatment | After coagulation treatment | After sand filtration | |
| pH | 7.1 ± 0.5 | 7.1 ± 0.2 | 7.3 ± 0.2 | 7.3 ± 0.2 | 6.5 - 8.5 |
| Conductivity | 11 ± 0.75 | 15 ± 2 | 10 ± 1.55 | 10 ± 1.62 | NQ |
| Turbidity (NTU) | 7 ± 0.55 | 25 ± 3.78 | 2 ± 0.14 | 2 ± 0.09 | 5 |
| Total solids | 455 ± 5.44 | 2880 ± 34 | 422 ± 8.55 | 240 ± 7.35 | 1000 |
| TSS | 312 ± 2.88 | 2821 ± 31 | 409 ± 7.46 | 226 ± 5.26 | 600 |
| TDS | 143 ± 2.07 | 59 ± 0.85 | 13 ± 0.54 | 14 ± 0.63 | NQ |
| Total hardness | 108 ± 6.23 | 72 ± 8.5 | 15 ± 0.96 | 15 ± 0.74 | 300 |
| Chloride | 13 ± 0.85 | 9 ± 0.854 | 6 ± 0.15 | 6 ± 0.23 | 250 |
| Nitrate | 100 ± 0.58 | 8 ± 0.25 | 0 | 0 | 45 |
| Sulphate | 0.4 ± 0.01 | 0.05 ± 0.01 | 0.05 ± 0.02 | 0.05 ± 0.02 | 200 |
| Phosphate | 12 ± 0.15 | 8 ± 0.22 | 0.4 ± 0.13 | 0.4 ± 0.21 | NQ |
| THB (CFU/mL) | $2 \pm 0.14 \times 10^2$ | $74 \pm 0.51 \times 10^4$ | $5 \pm 0.58 \times 10^4$ | $16 \pm 0.65 \times 10^1$ | NQ |

values are expressed in mg/L except pH, EC, turbidity and THB
*NQ* not quoted

and Kowal (1989) reported that lime act as an effective precipitant for phosphates, trace metals and bacteria, and act as a coagulant for the removal of suspended and colloidal materials in municipal wastewater. In other investigation, coagulation and flocculation processes removed 90–99% of viruses and protozoan cysts from water (Bitton 1980) and 74–99.4% of *E. coli* and coliforms (Bitton 1994).

After coagulation process, sand filtration was used for the removal of remaining bacterial cells. The efficiency of sand filtration depends on the filter medium, concentration and type of coagulated solids matter to be filtered out, and the operation of the filter. The complete removal of *Salmonella* sp., *Shigella* and *Giardia* cysts using rapid sand filter and slow sand filtration (Bitton 1980; Bellamy et al. 1985).

## 3.6 Disinfectant Study

Bacterial cells were completely removed in boiled and aqua guard filtered samples. In the UV treatment, the population gradually decreased when the exposed time increased from 10 to 50 min (Fig. 7). Application of residual chlorine at 0.5 mg/L after 120 min exposure, bacterial cells were completely eliminated; whereas in the concentration of 0.1 mg/L,the population reduced minimally from 16 to 11 × $10^1$ CFU/mL (Fig. 7). The population was gradually eliminated when the chlorine concentration increased from 0.2 to 0.5 mg/L. Among the disinfectant methods, the techniques like boiling, UV treatment and water filtration using commercial units were found to be very effective.

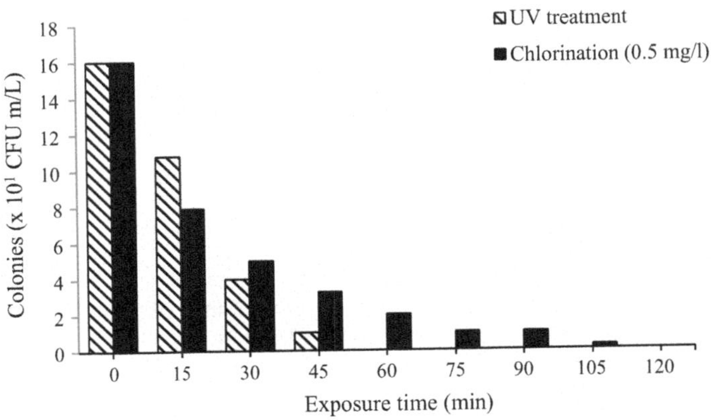


**Fig. 7** Comparison of disinfection rate of optimal chlorination with UV treatment

In a disinfection study by chlorination, the bacterial populations were completely eliminated at 0.5 mg/Lchlorine concentration after 120 min. In the case of 0.1 mg/Lof chlorine the population reduced only from 16 to 11 × $10^1$ CFU/mL. This may be due to lower concentration of chlorine that was not effective against all bacterial cells. Chlorination was effective against many pathogenic bacteria (Bitton 1994; Wagenet and Lemley 1998). But at normal dosage rates (0.2 mg/L), it does not kill all viruses, cysts or worms. Complete elimination of bacterial species by chlorination was proved at a dosage of 0.5 mg/Lat 120 min (Dziubek and Kowal 1989). In the present study, the populations were gradually eliminated when the chlorine concentration was increased from 0.2 to 0.5 mg/L. The disinfectant study such as boiling, UV and adsorption by Eureka forbes water filter completely eliminated the bacterial cells in water samples. Among the disinfectant method, boiling, UV treatment and adsorption filter method proved to be very effective. The UV wavelength ranging from 240 to 280 nm is an optimum dosage to deactivate, or effectively kill microorganisms by damaging their DNA and/ or prevent the DNA, organisms from replication (Harm 1980). Similarly boiling kills all the bacterial cells and is also direct and safe method of disinfection.

## *3.7 Statistical Analysis of Nitrate Removal*

In pilot scale study, Pearson correlation coefficient between bacterial growth and nitrate; nitrate and ammonium; bacterial growth and ammonium were observed as 95.4%, 98.10% and 95.10% respectively at 1% significant level. In large scale nitrate removal, there was a significant correlation among all the variables at 1% level. But in pilot scale study there was a notable correlation found among all the variables (Table 5). There was a positive relationship found for bacterial growth with the concentration of nitrate, nitrite and ammonium by 90.4%, 90.4% and 94.8% respectively.

**Table 5** Pearson correlation coefficient of variables in nitrate removal process

| | Nitrate | Bacteria | Nitrite | Ammonium |
|---|---|---|---|---|
| Nitrate Pearson correlation<br>Sig. (1-tailed) | – | 0.904**<br>0.000 | 0.892**<br>0.001 | 0.989**<br>0.000 |
| Bacteria Pearson correlation<br>Sig. (1-tailed) | – | – | 0.904**<br>0.000 | 0.948**<br>0.000 |
| Nitrite Pearson correlation<br>Sig. (1-tailed) | – | – | – | 0.898**<br>0.000 |
| Ammonium Pearson correlation<br>Sig. (1-tailed) | – | – | – | – |

**Correlation is significant at the 0.01 level (1-tailed)

The correlation relationship with nitrate concentration to nitrate was 89.2% and to ammonium was 98.9%, also nitrite and ammonium concentrations positively 89.8% relationship at 1% significant level.

## 4 Conclusion

The genera of *Pseudomonas* sp.,*Bacillus* sp., *Moroxella* sp., *Micrococcus* sp., *Alcaligenes* sp., *Corynebacterium* sp. and members of Enterobacteriaceaewere isolated from water and sediment samples. Among them, *Pseudomonas*sp. (KW1) and *Bacillus* (YW4) were found to be the most efficient in terms of $NO_3^-$ reduction. From the above results, it was noticed that the$NO_3^-$ reduction by KW1 and YW4 was influenced by various environmental factors. The rate of $NO_3^-$ reduction was higher in the synthetic medium supplemented with 1% starch as asole carbon source compared with those in the media supplemented with glucose, cellulose, sucrose and acetic acid under aerobic conditions. The removal of $NO_3^-$was optimum at30 °C, pH 7 and inoculum concentration of 1%. The statistical analysis reveals the role of bacterial growth in nitrate removal and the starch concentration to the bacterial growth. In the lab scale study, the bacterial consortium of KW1 and YW4 removed maximum level of $NO_3^-$(99.5%) at 48 h. In the pilot scale experiment, the natural environment highly influenced the growth of bacteria and removal of $NO_3^-$. In this study the $NO_3^-$ was removed to a level of 92% in water samples containing 100 mg/Lof $NO_3^-$ supplemented with 1% starch at 48 h. After filtration, all the parameters of treated samples were with in the permissible limits of drinking water standards. The treated water samples were further subjected to various disinfection processes and boiling, UV and membrane filter methods and were found to be very effective for removal of microbes. Demand for quality and safe drinking water make the researcher to find alternate method to treat the $NO_3^-$ contaminated water. The biological treatment under laboratory and pilot scale process using *Pseudomonas* sp. (KW1) and *Bacillus* sp. (YW4) could be effective in the removal of $NO_3^-$ from contaminated drinking water.

## References


Akunna JC, Bizeau C, Moletta R (1994) Nitrate reduction by anaerobic sludge using glucose at various nitrate concentrations: ammonification, denitrification and methanogenic activities. Environ Technol 15:41–49

APHA (2005) Standard methods for the examination of water and wastewater, 21st edn. American Public Health Association, Baltimore, Maryland, USA.

Ayyasamy PM, Shanthi K, Lakshmanaperumalsamy P, Lee SJ, Choi NC, Kim DJ (2007) Two-stage removal of nitrate from ground water using biological and chemical treatment. J Biosci Bioeng 104:129–134

Ayyasamy PM, Rajakumar S, Sathishkumar M, Swaminathan K, Shanthi K, Lakshmanaperumalsamy P, Lee S (2009) Nitrate removal from synthetic medium and groundwater with aquatic macrophytes. Desalination 242:286–296

Bellamy WD, Silverman GP, Hendricks DW, Logsdon GS (1985) Removing Giardia cysts with slow sand filtration. J Am Water Works Assn 77:52–60

Bitton G (1980) Introduction to environmental virology. Wiley, New York

Bitton G (1994) Waste water microbiology. Wiley, New York

Buchanan RE, Gibbons NE (eds) (1974) Bergey's manual of determinative bacteriology, 8th edn. The Williams and Wilkins Co., Baltimore

Carrera J, Vicent T,BLafuenteBFJ (2003) Influence of temperature on denitrification of an industrial high-strength nitrogen wastewater in a two-sludge system.Water SA 29:11–16

Darbi A, Viraraghavan T, Butler R, Corkal D (2003) Column studies on nitrate removal from potable water. Water Air Soil Pollut 150:235–254

Dhamole PB, Nair RR, D'Souza SF, Lele SS (2009) Simultaneous removal of carbon and nitrate in an airlift bioreactor. Bioresour Technol 100:1082–1086

Dziubek AM, Kowal AL (1989) High-pH coagulation-adsorption: a new technology for water treatment and reuse. Water Sci Technol 21:511–517

Elmidaoui A, Elhannouni F, Menkouchi MA, Chay L, Elabbassi H, Hafsi M, Largeteau D (2001) Pollution of nitrate in Moroccan ground water: removal by electrodialysis.Desalination 136:325–332

Fernandez-Nava Y, Marannon E, Soons J, Castrillon L (2010) Denitrification of high nitrate concentration wastewater using alternative carbon sources. J Hazard Mater 173:682–688

Gomez MA, Lopez JG, Garcıa EH (2000) Influence of carbon source on nitrate removal of contaminated groundwater in a denitrifying submerged filter. J Hazard Mater 80:69–80

Hamoudi S, Belkacemi K (2013) Adsorption of nitrate and phosphate ions from aqueous solutions using organically-functionalized silica materials: Kinetic modeling. Fuel 110:107–113

Harm W (1980) Biological effects of ultraviolet radiation. International Union of Pure and Applied Biophysics, Biophysics series, Cambridge

Jeong JY, Song YH, Kim JH, Park JY (2014) Simultaneous removal of nitrate, phosphate, and fluoride using a ZVI-packed bed electrolytic cell. Desalinat Water Treat 52:737–743

Jiang JQ, Wang S, Panagoulopoulos A (2006) The exploration of potassium ferrate (VI) as a disinfectant/coagulant in water and wastewater treatment. Chemosphere 63:212–219

Kapoor A, Viraraghavan T (1997) Nitrate removal from drinking water—review. J Environ Eng 123:371–380

Kesseru P, Kiss I, Bihari Z, Polyak B (2002) Investigation of the denitrification activity of immobilized Pseudomonas butanovora cells in the presence of different organic substrates. Water Res 36:1565–1571

Kim J, Benjamin MM (2004) Modeling a novel ion exchange process for arsenic and nitrate removal.Water Res38:2053–2062

Kim YS, Nakano K, Lee TJ, Kanchanatawee S, Matsumura M (2002) On-site nitrate removal of groundwater by an immobilized psychrophilic denitrifier using soluble starch as a carbon source. J Biosci Bioeng 93:303–308

Matosic M, Mijatovic L, Hodzic E (2000) Nitrate removal from drinking water using ion exchange—comparison of chloride and bicarbonate form of the resins. Chem Biochem Eng 14:141–146
Mortazavi SB, Ramavandi B, Moussavi G (2011) Chemical reduction kinetics of nitrate in aqueous solution by Mg/Cu bimetallic particles. Environ Technol 32:251–260
Najafpour GD (2007) Biochemical engineering and biotechnology. Elsevier Science Ltd., Amsterdam
Oh J, Silverstein J (1999) Oxygen inhibition of activated sludge denitrification. Water Res 33:1925–1937
Orhon D, Genceli EA, Sozen S (2004) Experimental evaluation of the nitrification kinetics for tannery wastewaters. Water SA 26:43–50
Rao MVS (ed) (2000) Water and water analysis. Environmental Research Academy International, Viaskhapatnam, Andra Pradesh, India
Reddy KJ, Lin J (2000) Nitrate removal from groundwater using catalytic reduction. Water Res 34:995–1001
Schoeman JJ, Steyn A (2003) Nitrate removal with reverse osmosis in a rural area in South Africa. Deasalination 155:15–26
Shanthi K, Ayyasamy PM, Rajakumar S, Ramasamy K, Lakshmanaperumalsamy P (2005) Influence of carbon sources on nitrate reduction with sludges. Asian J Microbiol Biotechnol Environ Sci 7:431–438
Shin KH, Cha DK (2008) Microbial reduction of nitrate in the presence of nanoscale zero-valent iron. Chemosphere 72:257–262
Sun D, Yang J, Li J, Yu J, Xu X, Yang X (2010) Novel Pd-Cu/bacterial cellulose nanofibers: preparation and excellent performance in catalytic denitrification. Appl Surf Sci 256:2241–2244
Wagenet L, Lemley A (1998) Chlorination of drinking water. Water treatment notes, Cornell Cooperative extension, New York State College of Human Ecology, https://www.cce.cornell.edu/factsheets/wq-fact-sheets/fact5.html
WHO (2003) Guidelines for drinking water quality, 3rd edn. Geneva.

# Clean Technology and Promising Microbial Technological Findings for Environmental Mitigation

# Economic and Environmental Benefits of Cleaner Technology in Industrial Pollution Control: Case Study of Select Sugar Industry in Tamil Nadu

**X. Agnello J. Naveen, S. Boopathi, A. Arivoli, and A. Kannan**

**Abstract** Sugar industry pollutes air, water and soil that different types of advanced pollution control technologies were used to reduce the pollution levels to permissible limits. But still Sugar industry manages to recycle and reuse its by-products based on the "concept of industrial ecology" on its own production premises in a holistic positive environmental management approach. Due to the production totally based on agricultural products using "biomass", which is organic from the starting till the end of the product, there is a total life cycle assessment (LCA). Recycling of products like water recycling in a closed loop water saving system, molasses reuse, Co-generation (energy conservation), variable frequency drives (VFD), cane cutting, bagasse, press muds, composting using sludge are few methods followed in this unit. Sugar industry cost variables Economic and environmental variables, older (Conventional) and newer (Cleaner) technology and their negative and positive advantages were compared. Variables like capital cost, variable cost, viability period of the equipment, depreciation cost, buy back cost, benefit cost and environmental benefits like energy in (kWh) per year, water in liters per year and other recycling process like "Add-on" and "Process change" technologies are taken in consideration. The main objective is to focus on the cost aspects between the two technologies, conventional and cleaner technology in pollution control. This was carried out by comparing cost benefit analysis and Return on Investment (ROI) for the old and clean technologies. The other parameter compared was cost benefit liter per year using cleaner recycling leading to environmental advantage. There are nine technologies used in this industry that has been analyzed.

X. Agnello J. Naveen (✉)
Department of Enivronmental Management, Bharathidasan University, Tiruchirappalli, India
e-mail: agnello.naveen18@gmail.com

S. Boopathi
Department of Economics, Bharathiar University, Coimbatore, India

A. Arivoli
Department of Environmental Science, Government Arts and Science College, Ariyalur, India

A. Kannan
Department of Environmental Economics, Kamaraj University, Madurai, India



M. Prashanthi et al. (eds.), *Bioremediation and Green Technologies*, Environmental Science and Engineering, https://doi.org/10.1007/978-3-030-64122-1_14

**Keywords** Conventional technology · Cleaner technology · Return on investment · Recycling

## 1 Introduction

Sugar is one of the largest agro-based industries. Sugarcane, a major raw material, is an important crop for bio-products because it produces sugar with a by-product bagasse (Renouf et al. 2008). Bagasse is used as involvement resource in 80 sugarcane producing countries (Botha and Von Blottnitz 2006). In the world over, the top of five nations viz., India, Brazil, Thailand, Australia, and China, accounted for 40% of the total sugar production, while sugar is produced in about 115 countries in the world. Sugar is produced 70% from sugarcane and 30% from sugar beet and cassava, etc. (Contreras et al. 2009). In the year of 2015–2016, 526 mills are operated in India, which produced 33.90 million tons of sugar from Andhra, Pradesh, Karnataka, Madhya Pradesh, Tamil Nadu and Uttar Pradesh (ISMA 2012). Sugar industry is mainly seasonal and operates only for 150–210 days in a year (November–May) (Kolhe et al. 2009). A huge volume of waste is generated during the operation of sugar production and has a huge amount of pollution load; classified in terms of suspended solids, organic matter, and press mud, bagasse and air pollutants another important is wastewater. The mills generate wastewater in the ratio of 1:2 (Jadhav et al. 2013; Vinish 2014). Each process in the sugar industry with crushing capacity of 1500–5000 tons per day requires 1500–10,000 $m^3$/day of water. Sugar industries are used for chemical and coagulation of impurities and refining of end products. These entire chemicals, one-way or another, are contributing towards increased water pollution level. Sugar mills account in the industries which discharge huge amount of effluent per day without any or partly treatment during the crushing season (Trivedy 1998). Sugar industry is categorized under the highly polluting water industries (Red category industry) (CERP 1989). To drop this the mandatory laws of pollution control board stipulated to the sugar industry to install ETP with CPCB prepared in accordance to the guidelines dated in 19-01-2015, On techno-Economic feasibility for implementation of Zero liquid discharged (ZLD) mechanism for reuse and recycle the effluent water conservation and irrigation protocol as alternate to ZLD water(CPCB 2015). To install continuous effluent online monitoring system (CEMS) for all the ETP for the measurement of parameters like flow pH, COD, BOD and TSS (CPCB 2015) corresponding by the flow meter.

## 2 Study Area

The study of select sugar industry is situated in Erode is an "Ultra red" category industry as the pollution load is very high. The production of sugar is 4750 TCD (tonne capacity per day).

### 2.1 *Materials and Methods*

**Economic Parameters**

**Cost variables**: In this cost analysis basic cost parameters like capital investment, variable cost, Buy back cost and viability of the mechanism (Life time of the mechanism in years) of the two treatment process are Elicited by the environmental engineer has a secondary data this are necessary cost to find (Benefit cost per liter).

**VC = Variable cost, FC = Fixed cost, BB = BuyBack cost, Viability Period of the mechanism.**

## 3 Steps Involved

1. **The total Buyback cost** is equal to capital investment in Rupees. Multiply with buyback cost in percent, is divide by 100 is equal to Rupees. (Total buyback cost).
2. **Actual capital Investment** equals to capital Investment in Rupees minus Total buyback cost in Rupees is equal to fixed cost per year in Rupees,
3. **The Fixed cost** is given fixed cost equals Actual capital investment in rupees divided by viability period of the mechanism in years, it's given Rupees,
4. **Total cost (TC)** equal to fixed cost in Rupees per year plus variable cost in Rupees, gives Rupees minus Depreciation cost per year.
5. **For Return on Investment (ROI)** Profit is equal to benefited amount—Total cost + Depreciation cost per year.

## 4 Cost Variables for Return on Investment (ROI)

In this cost analysis basic cost parameters like capital Investment, variable cost, Buy back cost and viability of the mechanism (Life time of the mechanism in years) of the two technology are Elicited from the environmental engineer and energy auditor has a secondary data this are necessary for cost variables to find Return on investment (ROI) (Phillips and Philips 2006).

$$\textbf{Profit} = \textbf{Total Revenue} - \textbf{Total operational cost}.$$
$$\textbf{Return on Investment (R.O.I)} = \textbf{Profit/Total cost} * \textbf{100} \qquad (1)$$

Cost variable for Cost Benefit Ratio is (Total Revenue) and Total cost (Fixed cost + Variable cost + Depreciation cost + Pollution and operational cost (Siva 2016).

***Note:*** Cost benefit Ratio is equal to Total Benefit value divided by Total cost.

$$\textbf{Cost benefit Ratio} = \textbf{Total Benefited Value or Total Revenue/Total Cost} \qquad (2)$$

Figure 1. shows the operational flowchart of select sugar industrial unit production process, this flow chart describes a classification of pollution control technologies in water (closed-loop), air, solid waste (co-processing) and energy conserve (co-generation and inbuilt technology), this industry has adopted conventional (older), cleaner (newer) technologies which is described detail below in Tables 1 and 2 with a detail explanation of concepts.

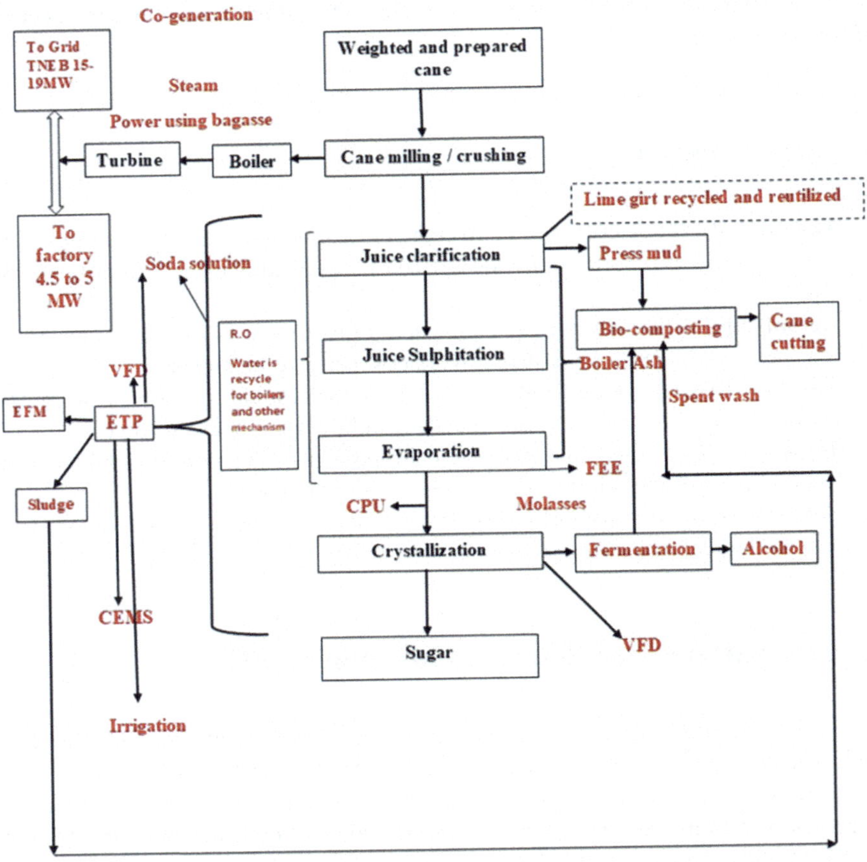


**Fig. 1** The operational process and cleaner technology used in select sugar industry

**Table 1** Comparison between cleaner and conventional technology using return on investment and cost benefit ratio (ROI and CBR) in select Sugar Industry

| S. No. | Technology | Cost variables | Conventional technology parameters | Cleaner technology parameters | Return on investment (ROI) | | Cost-Benefit Ratio (CBR) | |
|---|---|---|---|---|---|---|---|---|
| | | | | | Conventional Technology (%) | Cleaner technology (%) | Conventional technology | Cleaner technology |
| 1. | **Condensate polishing unit (CPU*)** | 1. Capital Investment<br>2. Variable Cost<br>3. Viability period<br>4. BuyBack cost<br>5. Depreciation cost<br>**6. Profit**<br>**7. Benefit**<br>8. TC | Rs.500 lakh<br>Rs.7.33 lakh p.a.<br>10 years<br>6%<br>Rs.0.35 lakh p.a.<br>**Rs.33.75 lakh p.a**<br>**22,500,000 liter p.a**<br>Rs. 54.33 lakh p.a. | Rs.700 lakh<br>Rs.10.55 lakh p.a.<br>20 years<br>7.5%<br>Rs.0.695 lakh p.a.<br>**Rs.81 lakh p.a**<br>**54,000,000 liter p.a.**<br>Rs. 42.92 lakh p.a. | −38.5 | **87.1** | **0.6121** | 1.887 |

(continued)

**Table 1** (continued)

| S. No. | Technology | Cost variables | Conventional technology parameters | Cleaner technology parameters | Return on investment (ROI) | | Cost-Benefit Ratio (CBR) | |
|---|---|---|---|---|---|---|---|---|
| | | | | | Conventional Technology (%) | Cleaner technology (%) | Conventional technology | Cleaner technology |
| 2. | **Cooling tower technology** | 1. Capital Investment<br>2. Variable Cost<br>3. Viability period<br>4. BuyBack cost<br>5. Depreciation cost<br>**6. Profit**<br>**7. Benefit**<br>**8. TC** | Rs. 20 lakh<br>Rs. 1.7 lakh p.a.<br>8 years<br>5%<br>Rs. 0.25 lakh p.a<br>**Rs. 108 lakh p.a.**<br>**72,000,000 liter p.a.**<br>Rs. 4.075 lakh p.a. | Rs. 34 lakh<br>Rs. 7.06 lakh p.a.<br>20 years<br>6.5%<br>Rs. 0.5 lakh p.a.<br>**Rs. 202.5 lakh p.a.**<br>**135,000,000 liter p.a**<br>Rs. 8.64 lakh p.a. | 1881.5 | **2237.9** | **26.50** | 23.43 |

(continued)

**Table 1** (continued)

| S. No. | Technology | Cost variables | Conventional technology parameters | Cleaner technology parameters | Return on investment (ROI) | | Cost-Benefit Ratio (CBR) | |
|---|---|---|---|---|---|---|---|---|
| | | | | | Conventional Technology (%) | Cleaner technology (%) | Conventional technology | Cleaner technology |
| 3. | **Effluent treatment plant with (ASP*) and RO* UF*and MGF*.** | **1.** Capital Investment<br>2. Variable Cost<br>3. Viability period<br>4. BuyBack cost<br>5. Depreciation cost<br>6. **Profit**<br>**7. Benefit**<br>8. TC | Rs. 435 lakh<br>Rs. 24.77 lakh p.a.<br>25 years<br>12.5%<br>Rs. 1 lakh p.a.<br>**Rs. 283.5 lakh p.a.**<br>**189,000,000 liter p.a**<br>Rs. 40.99 lakh p.a. | Rs. 1250 lakh<br>Rs. 97.3 lakh p.a.<br>10 years<br>14%<br>Rs. 8.54 lakh p.a.<br>**Rs. 208.88 lakh p.a.**<br>**52,200,000 liter p.a.**<br>Rs. 204.8 lakh p.a. | **605.8** | −2.17 | **6.916** | 1.019 |

(continued)

**Table 1** (continued)

| S. No. | Technology | Cost variables | Conventional technology parameters | Cleaner technology parameters | Return on investment (ROI) | | Cost-Benefit Ratio (CBR) | |
|---|---|---|---|---|---|---|---|---|
| | | | | | Conventional Technology (%) | Cleaner technology (%) | Conventional technology | Cleaner technology |
| 4. | **Co-generation** | **1.** Capital Investment<br>2. Variable Cost<br>3. Viability period<br>4. BuyBack cost<br>5. Depreciation cost<br>6. **Profit**<br>**7. Benefit**<br>8. TC | Rs. 550 lakh<br>Rs. 9 lakh p.a.<br>4 years<br>6.5%<br>Rs. 2.35 lakh p.a.<br>**Rs. 9.24 lakh p.a.**<br>**N.A**<br>Rs. 137.56 lakh p.a. | Rs. 9500 lakh<br>Rs. 156 lakh p.a.<br>20 years<br>6.5%<br>Rs. 10 lakh p.a.<br>**Rs. 162.45 lakh p.a.**<br>**3,420,000 p.a.**<br>Rs.5 97.4 lakh p.a. | −94.99 | **−73.8** | 0.0671 | **0.2719** |

(continued)

**Table 1** (continued)

| S. No. | Technology | Cost variables | Conventional technology parameters | Cleaner technology parameters | Return on investment (ROI) | | Cost-Benefit Ratio (CBR) | |
|---|---|---|---|---|---|---|---|---|
| | | | | | Conventional Technology (%) | Cleaner technology (%) | Conventional technology | Cleaner technology |
| 5. | **Variable frequency drive (VFD)** | 1. Capital Investment<br>2. Variable Cost<br>3. Viability period<br>4. BuyBack cost<br>**5.** Depreciation **cost**<br>**6. Profit**<br>**7. Benefit**<br>8. TC | Rs. 20 lakh<br>Rs. 0.1 lakh p.a.<br>5 years<br>5%<br>0.1<br>**Rs. 0.1025 lakh p.a.**<br>**7000 (kWh) p.a.**<br>Rs.3.9 lakh p.a. | Rs. 45 lakh<br>Rs. 0.5 lakh p.a.<br>10 years<br>7.5%<br>Rs. 0.33 lakh p.a.<br>**Rs. 2.58115 lakh p.a.**<br>**54,340 (kWh) p.a.**<br>Rs.4.66 lakh p.a. | −99.93 | **−51.7** | 0.026 | **0.5538** |

(continued)

**Table 1** (continued)

| S. No. | Technology | Cost variables | Conventional technology parameters | Cleaner technology parameters | Return on investment (ROI) | | Cost-Benefit Ratio (CBR) | |
|---|---|---|---|---|---|---|---|---|
| | | | | | Conventional Technology (%) | Cleaner technology (%) | Conventional technology | Cleaner technology |
| 6. | **Free flowing evaporator (FFE)** | **1.** Capital Investment<br>2. Variable Cost<br>3. Viability period<br>4. BuyBack cost<br>5. Depreciation cost<br>**6. Profit**<br>**7. Benefit**<br>8. TC | Rs. 450 lakh<br>Rs. 40 lakh p.a<br>10 years<br>5%<br>Rs. 0.5 lakh p.a<br>**Rs. 3.0875 lakh p.a.**<br>**60,720 (kWh) p.a.**<br>Rs. 80.75 lakh p.a. | NA* | −96.79 | NA* | 0.03823 | NA* |

(continued)

**Table 1** (continued)

| S. No. | Technology | Cost variables | Conventional technology parameters | Cleaner technology parameters | Return on investment (ROI) | | Cost-Benefit Ratio (CBR) | |
|---|---|---|---|---|---|---|---|---|
| | | | | | Conventional Technology (%) | Cleaner technology (%) | Conventional technology | Cleaner technology |
| 7. | **Air pollution control (Double stage Electrostatic precipitator)** | **1.** Capital Investment<br>2. Variable Cost<br>3. Viability period<br>4. BuyBack cost<br>5. Depreciation cost<br>**6. Profit**<br>**7. Benefit**<br>8. TC | Rs. 150 lakh<br>Rs. 14.55 lakh p.a.<br>10 years<br>8%<br>Rs. 10.6 lakh p.a.<br>**Rs. 0.52 lakh p.a.**<br>**11,000 (kWh) p.a.**<br>Rs. 28.3 lakh p.a. | NA* | −135.6 | NA* | 0.01837 | NA* |
| 8. | **Caustic Soda processing** | 1. Capital Investment<br>2. Variable Cost<br>3. Viability period<br>4. BuyBack cost<br>5. Depreciation cost<br>**6. Profit**<br>**7. Benefit**<br>8. TC | Rs. 35 lakh<br>Rs. 0.3 lakh p.a.<br>4 years<br>5%<br>Rs. 0.1 lakh p.a.<br>**Rs. 3.087 lakh p.a.**<br>**638,720 (kWh) p.a.**<br>Rs. 8.6125 lakh p.a. | NA* | NA* | −65.31 | 0.358 | NA* |

(continued)

**Table 1** (continued)

| S. No. | Technology | Cost variables | Conventional technology parameters | Cleaner technology parameters | Return on investment (ROI) | | Cost-Benefit Ratio (CBR) | |
|---|---|---|---|---|---|---|---|---|
| | | | | | Conventional Technology (%) | Cleaner technology (%) | Conventional technology | Cleaner technology |
| 9. | **Cane Cutting processing** | 1. Capital Investment<br>2. Variable Cost<br>3. Viability period<br>4. Buyback cost<br>5. Depreciation cost<br>**6. Profit**<br>**7. Benefit**<br>8. TC | NA* | Rs. 22 lakh<br>Rs. 7.55 lakh p.a.<br>30 years<br>10%<br>Rs. 25 lakh p.a.<br>**Rs. 5.1 lakh p.a.**<br>**1,700,000 cane cutting p.a.**<br>Rs. 14.81 lakh p.a. | NA* | **58.7** | NA* | **0.3443** |

*ASP** Activated sludge processing, *RO** Reverse Osmosis, *MGF** Multi-grade Filter, *NA* Not Available, *TC* Total cost, *p.a.* per annum, kWh- kilowatt-hour

**Table 2** Comparison between cleaner and conventional technology using cost benefit per liter and liter benefit per year in select Sugar Industry

| Technology | Cost variables | Conventional technology cost parameters | Cleaner technology cost parameters | Cost benefit liters | | Total Liters benefit per year | |
|---|---|---|---|---|---|---|---|
| | | | | Conventional technology | Cleaner technology | Conventional technology | Cleaner technology |
| Condensate polishing unit (CPU) | 1. Capital Investment<br>2. Variable Cost<br>3. Viability period<br>4. BuyBack cost<br>5. Depreciation cost<br>**6. Profit**<br>**7. Benefit**<br>8. TC | Rs.500 lakh<br>Rs.7.33 lakh p.a.<br>10 years<br>6%<br>Rs.0.35 lakh p.a.<br>**Rs.33.75 lakh p.a.**<br>**22,500,000 liter p.a**<br>Rs.54.33 lakh p.a. | Rs.700 lakh<br>Rs.10.55 lakh p.a.<br>20 years<br>7.5%<br>Rs.0.695 lakh p.a.<br>**Rs.81 lakh p.a.**<br>**54,000,000 liter p.a.**<br>Rs.42.92 lakh p.a. | **0.2414** | 0.07948 | 4.141 | **12.58** |
| **Cooling tower technology** | 1. Capital Investment<br>2. Variable Cost<br>3. Viability period<br>4. BuyBack cost<br>5. Depreciation cost<br>**6. Profit**<br>**7. Benefit**<br>8. TC | Rs.20 lakh<br>Rs.1.7 lakh p.a.<br>8 years<br>5%<br>Rs.0.25 lakh p.a<br>**Rs.108 lakh p.a.**<br>**72,000,000 liter p.a.**<br>Rs.4.075 lakh p.a. | Rs.34 lakh<br>Rs.7.06 lakh p.a.<br>20 years<br>6.5%<br>Rs.0.5 lakh p.a.<br>**Rs.202.5 lakh p.a.**<br>**135,000,000 liter p.a**<br>Rs.8.64 lakh p.a. | 0.0056 | **0.0064** | **176.6** | 156.25 |
| **Effluent treatment plant with (ASP*) and RO* UF*and MGF*.** | 1. Capital Investment<br>2. Variable Cost<br>3. Viability period<br>4. BuyBack cost<br>5. Depreciation cost<br>**6. Profit**<br>**7. Benefit**<br>8. TC | Rs.435 lakh<br>Rs.24.77 lakh p.a.<br>25 years<br>12.5%<br>Rs.1 lakh p.a.<br>**Rs.283.5 lakh p.a.**<br>**189,000,000 liter p.a**<br>Rs.40.99 lakh p.a. | Rs.1250 lakh<br>Rs.97.3 lakh p.a.<br>10 years<br>14%<br>Rs.8.54 lakh p.a.<br>**Rs.208.88 lakh p.a.**<br>**52,000,000 liter p.a.**<br>Rs.204.8 lakh p.a. | 0.0216 | **0.392** | **47.132** | 2.54 |

*ASP** Activated sludge processing, *RO** Reverse Osmosis, *MGF** Multi-grade Filter, *NA* Not Available, *TC* Total cost, *p.a.* per annum, kWh- kilowatt-hour

## 5 Explanation of Cost Analysis Using ROI, CBR, cost-benefit per year and benefit liter per year for CPU

Water requirement is enormous in the case of sugar industry production; the mode of water supply for this sugar industry to a large extent is from ground water. Condensate polishing unit (CPU) the return on investment in conventional technology is (−38.5%) and cleaner technology is (87.1%) for Rs. 1 investment. On comparing the two technologies Cleaner technology (upgraded CPU) shows a higher ROI. In terms of cost-benefit ratio conventional technology (0.6212) and cleaner technology (1.887) and so cleaner technology shows a higher benefit. The Environmental advantages of recycling water using CPU is 54,000,000 liter per year. Cost-benefit per liter in conventional technology is (0.2414) and cleaner technology is (0.0794). Benefit liter per year in conventional technology is (4.141) and cleaner technology is (12.58). The benefit liter per year in cleaner technology is higher than the conventional technology. The cost benefit per year is higher in conventional than cleaner technology due to the quantity of water is higher (recycle of hot water 3 times a day), the annual cost saving from CPU is Rs. 81 in lakh. per year, as shown in (Tables 1 and 2).

## 6 Explanation of Cost Analysis Using ROI, CBR, cost-benefit per liter and benefit liter per year for Cooling Tower

Cooling tower the return on investment in conventional technology (1881.5 percent) and (2237.9 percent) for Rs. 1 investment. On comparing the two technologies Cleaner technologies (upgraded cooling tower) shows a higher ROI and In terms of cost-benefit ratio conventional technology (26.50) and cleaner technology (23.43) in term of cost-benefit ratio conventional technology shows a higher benefit. The Environmental advantages are recycling of water using in cooling tower is 135,000,000 liters per year, Cost-benefit per liter in conventional technology is (0.0056) and cleaner technology is (0.0064) the cleaner technology shows slightly higher in value. Benefit liter per year in conventional technology is (176.6) and cleaner technology is (156.25) conventional technology shows a higher liter of benefits per year than cleaner technology. The annual cost saving from water saving is Rs. 202.5 per year in lakh. Though cleaner technology showed lesser cost-benefit ratio than conventional technology, the environmental advantage offsets this through water saving, as shown in (Tables 1 and 2).

## 7 Cost Analysis Using ROI, CBR, cost-benefit per liter and liter benefit per year using for ETP and Advance Treatment Technology (MGF, RO, UF)

As shown in (Tables 1 and 2), Effluent treatment plant (ETP) the return on investment in conventional technology (Activated sludge process) is (605.8%) and cleaner technology (MGF (Multi grade filter), RO (Reverse osmosis) and UF (Ultra filtration) is (−2.17%) for Rs.1 investment. Comparing the two technologies, conventional technology is gain and cleaner technology is loss and the operational and maintenance cost is very high comparing with the conventional technology cost. In terms of cost-benefit ratio conventional technology (6.916) and cleaner technology (1.019), leading to conventional technology showing a higher benefit. Cost-benefit per liter in conventional technology is (0.0216) and cleaner technology is (0.393) and liters benefit per year in conventional technology is (46.10) and cleaner technology is (2.53). Conventional technology shows a higher value in the cost-benefit per liter. The amount of water used is very less in cleaner technology due to the recycled water being used only for boiler feed purpose with of 52,000,000 liter per year and ETP the advantage are 70 Tons per year (70,000) and the water recycle is 1890 lakh liters per year and the benefited amount is Rs. 208.88 per annum in lakhs and thus the total benefit from the Effluent treatment plant (ETP) is Rs. 283.5 per annum in lakhs. CPU, Cooling tower, ETP (ASP) and MGF, RO, UF is said to be a "closed-loop system" water is recycled has an "add-on" technology and the end of recycling process from the ETP and aeration tank the water is send to irrigate 94.77 acres of sugar cane farming land.

## 8 Conventional Technology

ASP used in ETP as a secondary treatment technology in sugar industry has outcome advantage in sludge processing and can be reused as a composting material, but on comparing with the TNSPCB standards, after treatment the outcome exceeds the permissible limits in parameters like BOD, COD, TDS, Sulphate, Chloride, Sodium and oil grease high organic load, has been there since the inception of the industry in 1980s and has led to advance technology Viability of this ASP used as an "add-on" technology in this industry to screen the primary water pollutants is evident and this whole system CPU, Cooling tower, ETP with advance technology is called has "Closed-loop" water system technology without wasting of water or draining the water into surface water, as shown in (Tables 1 and 2).

# 9 Explanation of Cost Analysis Using ROI and CBR for Co-generation

Regarding Co-generation, from 1986 to 1990 industry used direct electricity from non-renewable energy source this is the conventional technology (−94.99%) and from 1990 "Igni fluid" boiler form bagasse (Co-generation) was installed as cleaner technology (-73.82%) for Rs. 1 investment. Comparing the two technologies cleaner technology (Co-generation) shows a higher ROI, but there is loss. In terms of cost-benefit ratio, conventional technology (0.0671) and cleaner technology (0.2719), (shown in Table 2) cleaner technology shows a higher benefit.

## *9.1 Advantage of Co-generation*

1. It is as follows: power export to grid during season, Sugar plant and aux consumption: 4.5–5 MW, Power export to TNEB Grid 8–10 MW 77.28 GWh; emission reduction, 62950.75 tonnes per year; and annual revenue at Rs. 270/tonne of $CO_2$ reduction, Rs. 1.7 crores.
2. About 15% of coal is used for initial ignition of the raw material, 85% biomass, has raw materials with a ratio by calculating the specific calorific value (CV) with quantity approximate usage determine per annum. For steam production-825, Enthalpy, 2081 Cv.
3. Running Period, 180–250 days; power export to grid, 10–14 MW; power cost given by Tamil Nadu electricity board (TNEB) using bagasse as a fuel, Rs. 4.52/unit; power cost given by TNEB using bagasse and coal as a fuel, Rs. 4.50–4.90/unit; total sales during off season, Benefits in Rs. 16,245,000 per year.
4. 85 percent biomass has raw materials, only 15 percent of coal is used as per government norms.
5. Power savings is 3,420,000 kWh per year.
6. Air pollution is very low as compared to other technology like nuclear power and atomic power as the dispersion of the air is too low in the stack. Double stage Electrostatic precipitator (ESP) is used to control the air pollution.
7. The farmer gets Economic benefit due to selling their raw material (biomass) for co-gen like Coconut shell, Julie flora, Pith, Chipper dust, Groundnut shell to the industry.
8. Solid waste generated by Boiler Ash-Bagasse 21.67(T/D), Boiler Ash coal 30 (T/D) Bagasse ash will used as manure and coal ash will be sold out to brick manufactures.

**Advantage**: Automated flow meter for energy calculation. No human error accurate in readings, Energy saving technology, Automated sensor are attached.

## 10 Energy Conservation Technology

### *10.1 Process Description of Variable Frequency Drive (VFD)*

Principle is Changing Direct current into Alternate current. Key areas were (Variable frequency drive) VFD used. Effluent treatment plant (ETP), Molasses tank, Membrane technology, cooling tower blowdown, washing and process condensate polishing unit (CPU), Boiler water feeding and falling-film evaporators (FFE). Conveyor belt and some of the other auxiliary instruments (Figure 1, shown in Table 1).

**Explanation of Cost Analysis using ROI and CBR for VFD**

In Variable Frequency Drive (VFD), conventional technology is (−99.93%) and energy conversation technology, 2004, where auxiliary equipment was introduced in this industry by process modification "energy saving" Variable frequency drive (VFD) is the cleaner technology showing (−51.7%) for Rs. 1 investment. Both values are negative, depicting loss in conventional and cleaner technology but cleaner better. In terms of cost-benefit ratio conventional technology (0.026) and cleaner technology (0.5538), cleaner technology shows a higher value of benefit than conventional technology. But the energy saving 54,340 kWh per year and benefit amount is Rs. 2.58115 lakh per year as shown in Table 1.

### *10.2 Falling Film Evaporator (FFE)*

**Explanation of Cost Analysis using ROI and CBR for FEE**

Free Flowing evaporator (FFE) is conventional technology (−96.79%) for Rs.1 investment it's a loss. In terms of cost-benefit ratio conventional technology (0.0382) the benefits is less in terms of cost benefit. But the Energy saving 60,720 kWh per year and benefit amount is Rs. 3.0875 lakh per year. Attaching online sensor's and automated variable frequency drive, automated flow meters, as shown in Table 1.

**Benefits**: This is used for internal cleaning purpose of the boiler and energy conserving technology.

## 10.3 Air Pollution Control (Double Stage Electrostatic Precipitator)

**Explanation of Cost Analysis Using ROI and CBR for Air Pollution Control (Double Stage Electrostatic Precipitator)**

Air pollution control (Double stage Electrostatic precipitator) is conventional technology with (−135.6 percent ) for Rs.1 investment in terms of ROI is a benefit. The cost-benefit ratio shows (0.01837) implying there is loss (shown in Table 1). But the Energy saving 11,000 kWh per annum, Rs. 0.52 lakh per annum.

## 10.4 Caustic Soda Processing

**Explanation of Cost Analysis using ROI and CBR for Caustic soda**

Caustic soda processing is a conventional technology ROI for Rs.1 investment is (−65.31 percent) which is a loss and the cost-benefit ratio of conventional technology is 0.358 the benefits is less in terms of cost-benefit analysis . But the energy saving 638,720 kWh per annum, Rs. 3.087 lakh per annum.

**Advantage**: Works on recycling process to neutralize the pH of spray pond and cleaning substance used in the Falling flow evaporator, Attaching on-line sensors and automated variable frequency drive, automated flow meters.

**Benefits-**This is used for internal cleaning purpose of the boiler and energy conserving technology.

## 10.5 Cane Cutting Technology

**Selecting Seeds**

Peeling sugarcane leaves: peeling leaf sheaths of sugarcane seeds first before the seeds are chopped after the seeds are selected, chopping off tails of lower parts from growing points for 4 to 5 cm and old stems away from the ground for 80 to 100 cm, and using most middle sugarcane stems as the sugarcane seeds, Raising nursery using single-budded chips (conventionally, 2–3 budded sets are used and normally no nursery is prepared) (Shaochun et al. 2015).

**Explanation of Cost Analysis using ROI and CBR for Cane Cutting**

Cane cutting processing conventional technology (58.7 percent) for Rs. 1 investment it's a gain. In terms of cost benefit ratio conventional technology value is (0.344), the benefits are high (shown in Table 1). Because this is directly utilized by the farmers in their farmland due to the yield is high. Cane cutting saving of Rs. 17 lakh per year and also a benefit amount is Rs. 5.1 lakh per year through bio-remediation shows the positive environmental benefit.

**Benefits of Cane Cutting**

A new and simple method of waste cane collection was implemented in an effective way to save the manpower and to create space of 150sq ft for other usage and by using this seeding framer will get a high yield in can production. Savings Rs. 0.06 lakh per month, Press mud are used as composting material (biomass fertilizer).

## 11 Recycle Products in Sugar Industry

**Raw material consumption**

**Shows the Consumption of Raw Material Per Unit of Output**

Sugar = cane crushing quantity/sugar cane quantity = 713,904.693/70950.30 = 10.062

Bagasse = cane crushing quantity/bagasse quantity = 713904.693/191607.029 = 3.726

Filter mud = cane crushing quantity/filter mud quantity = 713904.693/27811.630 = 25.669

Molasses = cane crushing quantity/molasses quantity = 713904.693/32630.23 = 21.879

Bio-compost = cane crushing quantity/bio-compost quantity = 713904.693/181.06 = 3942.9

**Shows the Product and By-Product Produced on % Cane**

Sugar recovered = sugar quantity/cane crushing quantity* 100 = 70950.30/713904.693 * 100 = 9.96

Bagasse = Bagasse quantity/cane crushing quantity * 100 = 191607.029/713904.693 * 100 = 26.84

Filter cake = Filter cake quantity/cane crushing quantity * 100 = 2780.620/713904.693 * 100 = 03.90

Molasses = Molasses quantity/cane crushing quantity * 100 = 32620.23/713904.693 * 100 = 4.57

**By Products**

Bagasse, Molasses, Press-mud, Boiler ash, Coal ash, Lime grit are three important by-products of Sugar Mill.

**Bagasse**

Bagasse, the residue after the extraction of juice from the cane, is rich in cellulose fiber, which is a major source of energy, and it is being the major substitute raw material for wood and bamboo used in the paper and pulp industry, Bagasse produced was depithed and sold as raw material for paper manufacturing only about 75% of the quantity. The remaining 25% is called the pith, is used in boiler as fuel. Bagasse is used has a raw material for energy production in Co-generation plant for Sugar industry about 119,970 tons are produced per year and the Benefit in Rs. 850 lakh per year.

**Sludge from ETP**: Press mud is added with ETP sludge to make bio-compost about 42 tons per year, Sludge thickener, used in Bio-compost the benefited amount is Rs.31 lakh per year.

**Molasses**

Molasses, a residue, subsequent to extraction of sugar from juice, is a storehouse of organic chemicals, like ethanol, ether, methanol and alcohol can be made by using molasses has a raw material. And this molasses is recycled in distillery industry to add alcohol in manufacturing of beer and other chemical products. In this Sugar Industry with a capacity of 4750 TCD operating for a period of 180 days would produce around 0.18 lakh tonnes of molasses per year; and this is recycled in cattle fodder field, Oil mills and distillery industry with a benefit amount of Rs. 446 lakh per year.

**Press Mud or Filter Muds**

Press Mud is used as fertilizer in fields, which helps in increase the cane production and used for Cane cutting purpose has a bio-compost. The production is 0.28728 tons per year and the benefit in Rs. 2.22 lakh per year.

**Lime Grits**: This is used for landfilling in cement factory and produced 0.0081 lakh tons per year and the benefit in Rs. 50 lakh per year.

**Boiler Ash and Coal Ash**: Boiler ash produced is 0.039006 lakh Tons per year and reused in cement manufacturing and the cost-benefit is Rs. 123 lakh per annum and coal ash produced is 0.054 lakh Tons per year and cost-benefit is Rs. 140 lakh per annum coal ash will be sold out to brick manufacture industry as it can also be used has a raw material.

**Belt Conveyor**: Control adsorption Cost savings Rs. 0.07 lakh per annum, Dust collector collects the dust in the Hooper, Pan Hooper Control's adsorption, Sprinklers Control adsorption, Green Belt 8522 and cost for the plantation is 5 lakhs. Coconut Plantain, Teak, Palm, Ashoka, Neem, Sobibul, Casurina and Pungan (Natural bio-remediation to control air pollution), Rain water harvesting for (recharging of rain-water).

## 12 Discussion

The sugar industry selected for study situated in Erode started in 1984, a large scale industry and Red category (Annexure B in EIA), has a Bagasse co-generation plant for paper board industry. The adverse effects of the industry like utilization off arm land for sugar cane production with Cauvery running at a distance of 0.5 km functions with a capacity of 2500 TCD (Tons capacity per day) which has now increased to 4750 TCD and this large scale red category industry has been running successfully around 180 to 200 days in a year. Presently this unit incorporates advance technologies like co-generation, co-processing, recycling, inbuilt, process-modification and closed loop water saving system technology, advanced effluent treatment plant (ETP) technology to have reusing capacity, energy conservation technology "closed-loop" of water saving technology to achieve zero liquid discharge (ZLD) connects, condensate polishing unit (CPU), cooling tower modification, cleaner technology. As a result, they procure a return on investment of (87.1) and CBR of 1.887. Cleaner technology has high benefit, Hot water is recycled 3 times a day, about 54,000,000 liter per year is recycled and Rs. 81 lakh per year cost-benefit per liter leading to 0.07948 and total benefit liters is 12.58 percent cleaner technology and In cooling tower ROI (2237.9%) for Rs. 1 investment and CBR the value is higher in conventional technology (26.50) Environmental benefit are higher in cleaner technology like water is recycled about 135,000,000 liters per year and profit amount is Rs. 202.5 lakh per year. But comparing the ASP, the older one was cheaper and slow on water treatment process, and the advantage are removing oil and grease is recycled as against the cleaner technology where maintenances is high. Connecting technology towards "closed-loop" system is a condensate polishing unit, cooling water in conventional technology was through the Activated sludge process (ASP). Cleaner technology used RO-Reverse osmosis, MGF-Multi grade filter, and UF-Ultra filtration as add-on technology. Return on investment in effluent treatment plant showed Conventional technology as (605.8%) and cleaner technology as (−2.17%) shows there is a gain by using older technology and in new by modifying the process through add-on. In the cost-benefit conventional technology shows advantage (6.916) than cleaner technology. There is also benefit on improvement of water by recycling and used in irrigation purpose where cost-benefit per liter in conventional technology is 0.02169 and cleaner technology is 3.92337. The energy saved by using the technology, Co-generation, return on investment show Conventional technology (Renewable energy) is (−94.9%) loss and cleaner technology is (−73.8%) for an investment Rs. 1. Compared to conventional technology, cleaner technology is slightly low due to operational and maintenances cost which is high and the capacity is huge. Comparing the cost benefit ratio value co-generation shows a higher value (0.2719) than conventional technology. Using Biomass as raw material, power savings is 3,420,000 kWh per year and Rs. 162.45 lakh per year is the profit. For cleaner technology, Variable frequency drive (VFD) is (−51.7%) for an investment Rs. 1, cleaner technology is slightly low as operational and maintenances cost was low than conventional technology. VFD was attached from 2004 and advantage was 54340 kWh per year electricity was saved.In controlling Air pollution (Double stage Electrostatic precipitator) is used return on investment was −1.0303% for Rs. 1 investment and energy savings is 11,000 kWh per year with Rs. 0.5225 lakh per year and cost-benefit ratio value is 0.01837 still in air pollution control conventional technology is used. Caustic soda processing return on investment was (-65.31%) for Rs. 1 investment and cost-benefit ratio value is 0.358 gave a benefit of Rs. 3.0875 lakh per year. Sugar industry has been recycling Lime grits is 810 tons per year, Boiler Ash and coal ash 3900.6 Tons per year were reused in cement manufacturing and the benefit-cost is Rs. 123 lakh per annum. Coal ash was produced 5400 Tons per year and the benefit-cost is Rs. 140 lakh per annum wherein coal ash would be sold out to brick manufacture industry to be used as raw material. Belt conveyor, Control adsorption Cost savings was Rs. 0.07 lakh per annum. Dust collector collects the dust in the Hooper, Pan Hooper controls adsorption in process itself, Sprinklers Controls air adsorption. Green Belt was created where about 8522 trees were planted and cost for the plantation was Rs. 5 lakhs. The trees planted were Coconut Plantain, Teak, Palm, Ashoka, Neem, Sobibul, Casurina and Pungan which act as natural bio-remediation to control air pollution. Rain water harvesting was adopted for recharging ground water. Recycling of molasses around 18,000 tonnes sent to cattle fodder, field, Oil mills and distillery industry with a profit of Rs. 446 lakh per year. Press mud was added with ETP sludge to make bio-compost about 42 tons per year, Sludge thickener used in Bio-compost earned a profit of Rs. 31,000 per year. Cane cutting (ROI) return of investment was 2.0876% and cost-benefit ratio value is 0.3433 leading to a profit of Rs. 6000 per month.

**Acknowledgements** The author thanks the help rendered by Dr. M. Ravichandran, Professor (Retired) Department of Environmental Management, Bharathidasan University, Tiruchirappalli in correcting the final version of the paper.

## References

Botha T, Von Blottnitz H (2006) A comparison of the environmental benefits of bagasse-derived electricity and fuel ethanol on a life-cycle basis. Energy Policy 34(17):2654–2661

Contreras AM, Rosa E, Pérez M, Van Langenhove H, Dewulf J (2009) Comparative life cycle assessment of four alternatives for using by-products of cane sugar production. J Clean Prod 17(8):772–779

CPCB (1989) corporate reasonability for Environmental protection (CREP) List of Industries under 'RED' Category Industries identified by Ministry of Environment and Forests, Govt. of India as heavily polluting and covered under Central Action Plan

CPCB (2015) Guidelines on Techno-Economic Feasibility of Implementation of Zero liquid discharge (ZLD) for water polluting industries, p 14

Indian Sugar Mill Association (2012) https://www.indiansugar.com/Statics.aspx

Kolhe AS, Sarode AG, Ingale SR (2009) Study of effluent from sugar cane industry. Sodh Samiksha Mulyankan 2:303–306

Phillips PP, Phillips JJ (2006) Return on investment (ROI) basics. American Society for Training and Development

Renouf MA, Wegener MK, Nielsen LK (2008) An environmental life cycle assessment comparing Australian sugarcane with US corn and UK sugar beet as producers of sugars for fermentation. Biomass Bioenerg 32(12):1144–1155

Siva Chandran J (2016) Cost-benefit approach of Bt cotton and Non-Bt cotton seeds application in agriculture: a case evidence from Salem and Coimbatore districts of Tamil Nadu, India. p 8

Trivedy RK (1998) Advances in waste water Treatment Technologies. 1st edn. Global Sciences, p 331

# Case Experience on WIN WIN Approach Towards Industrial Pollution Control-Economic and Environmental Efficacy of Cleaner Technology in Select Cement Industry in Tamil Nadu

**X. Agnello J. Naveen, S. Boopathi, and A. Arivoli**

**Abstract** The cement industry pollutes air, water and soil to control pollution using different types of advanced pollution control technologies to reduce the pollution levels and achieve permissible limits. But still, Cement industry manages to recycle and reuse its (Co-processing) by-products based on the "concept of industrial ecology" on its own production premises in a holistic positive environmental management approach. Due to the production totally based on mining products from "fossil fuels" like calcium carbonate, Gypsum. Co-generation (energy conservation, variable frequency drives (VFD), Cement industry cost variables Economic and environmental variables, older (Conventional) and newer (Cleaner) technology and their negative and positive advantages were compared. Variables like capital cost, variable cost, viability period of the equipment, depreciation cost, buy back cost, benefit cost and environmental benefits like energy in kWh per year, water in liters per year and other recycling process like "Add-on" and "Process change" technologies are taken in consideration. The main objective is to focus on the cost aspects between the two technologies, conventional and cleaner technology in pollution control. This was carried out by comparing cost benefit analysis and Return on Investment (ROI) for the old and clean technologies. There are nine technologies used in this industry that has been analyzed.

**Keywords** Conventional technology · Cleaner technology · Industrial ecology · Return on investment · Recycling

---

X. Agnello J. Naveen (✉)
Department of Environmental Management, Bharathidasan University, Tiruchirappalli, India
e-mail: agnello.naveen18@gmail.com

S. Boopathi
Department of Economics, Bharathiar University, Coimbatore, India

A. Arivoli
Department of Environmental Science, Government Arts and Science College, Ariyalur, India



M. Prashanthi et al. (eds.), *Bioremediation and Green Technologies*, Environmental Science and Engineering, https://doi.org/10.1007/978-3-030-64122-1_15

# 1 Introduction

The cement Industrial process is a very complex high-energy consumption to run, which leads to environmental issues by using non-renewable energy resources, heavy electricity depended industries like Aluminum smelter, copper smelter, iron and steel, cement, paper and pulp, fertilizer industries (Gupta 2013; Sengupta 2016). And it is clear that steel, aluminum, cement are the largest consumers of commercial energy compared to other industrial sectors. Energy Reduction for (Perform-Achieve-Trade) PAT (Ministry of power 2012) Indian Cement industry comprises of 210 large cement plants (BEE 2018), and that accounts for 10.3% of total fuel consumption in the manufacturing sector (Khurana et al. 2002). Cement industry depends upon combusted energy level (specific heat capacity) which is used for operation can be conducted smooth by the operators Green rating project (GRP) has focused to implement the directives of European union best available technology (BAT). Which is very essential in finding the carbon emission reduction and carbon foot printing for a year or a decade? So to update technology regulatory systems are enhancing system based support like monitoring and have prior responsible for the energy audit ISO 50001, the main focus in the cement industry is on air pollution control, adsorption and energy conservation.

# 2 Study Area

The select Cement industry taken for the study situated in Ariyalur is an "Ultra red" category industry as the pollution load is very high. The production of cement is 3 MT per day.

## 2.1 *Materials and Methods*

**Economic Parameters**

**Cost variables**: In this cost analysis basic cost parameters like capital expenditures, variable cost, Buy back cost and viability of the mechanism (Life time of the mechanism in years) of the two treatment process are Elicited by the environmental engineer has a secondary data this are necessary cost to find (Benefit cost per liter).

**VC = Variable cost, FC = Fixed cost, BB = BuyBack cost, Viability Period of the mechanism**

**Steps involved**

1. **The total Buyback cost** is equal to capital investment in Rupees. Multiply with buyback cost in %, is divide by 100 is equal to Rupees (Total Buyback cost).
2. **Actual capital Investment** equals to capital Investment in Rupees minus Total buyback cost in Rupees is equal to fixed cost per year in Rupees,

3. **The Fixed cost** is given fixed cost equals Actual capital investment in rupees divided by viability period of the mechanism in years, it is given in Rupees,
4. **Total cost (TC)** equal to fixed cost in Rupees per year plus variable cost in Rupees, gives Rupees minus Depreciation cost per year.
5. **For Return on Investment (ROI)** Profit is equal to benefited amount– Total cost + Depreciation cost per year.

## *2.2 Cost Variables for Return on Investment (R.O.I)*

In This Cost Analysis Basic Cost Parameters like Capital Investment, Variable Cost, Buy Back Cost and Viability of the Mechanism (Life Time of the Mechanism in Years) of the Two Technology Are Elicited from the Environmental Engineer and Energy Auditor Has a Secondary Data This Are Necessary for Cost Variables to Find Return on Investment (R.O.I) (Phillips and Philips 2006).

$$\text{Profit} = \text{Total Revenue} - \text{Total operational cost} \quad (1)$$

$$\text{Return on Investment(R.O.I)} = \text{Profit/Total cost}^{*}100 \quad (2)$$

Cost variable for Cost Benefit Ratio is (Total Revenue) and Total cost (Fixed cost + Variable cost + Depreciation cost + Pollution and operational cost (Siva 2016).

**Note:** Cost benefit Ratio is equal to Total Benefit value divided by Total cost.

Figure 1. Shows the operational of select cement industrial unit production process the flow chart describes a classification of pollution control technologies in waste recycling technology (Co-processing), Air pollution and Energy conserve technology, this industry adopted conventional (older), cleaner (newer) technologies which describes detail given below in Table 1 with detail explanation of concepts.

$$\text{Cost benefit Ratio} = \text{Total Benefited Value or Total Revenue/Total Cost}$$

## *2.3 Primary Data*

**Vertical roller mill inlet duct Modification**

**Older technology**

The conventional process of gas flow in mill grinding system are generally consists of vertical roller mill like separator, cyclone, mill circulation fan, and electrostatic precipitator (EP) and EP fan is used to pulverize final products, called has kiln feed raw meal, are collected at the cyclone and EP. For a high pressure loss at the cyclone there is a process change in the control mill attached to gas flow, mill circulation fan

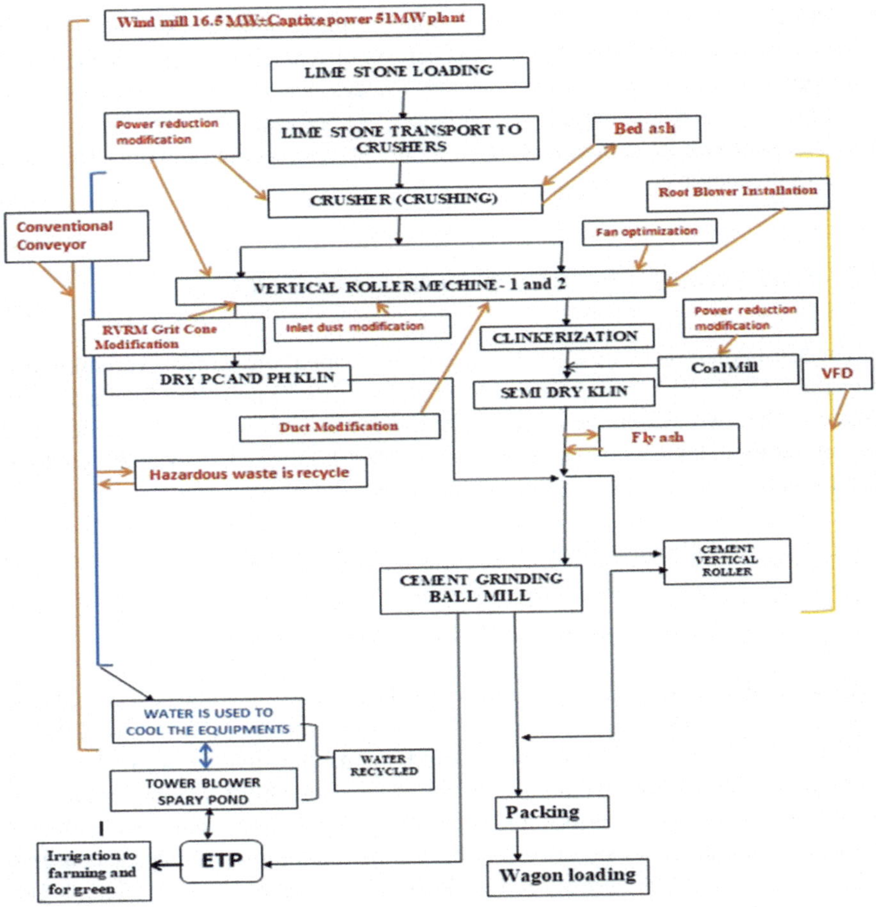


**Fig. 1** Operational process of Cement industry with Cleaner technology

and EP fan are provided after the cyclone and EP respectively. Power consumption is high and energy conservation is less in this process (Bapat 2001).

## 2.4 *Cleaner Technology Working Process of Inlet Duct Modification*

In the direct inlet dust collection system flow is simple. The raw materials are dried and ground by the mill step by step process in one-pass kiln exit gas and then done to fine product after separation is sent to EP directly, Cyclone is not installed, and system pressure loss reduces. And as the mill fan has de-dusted gas only, it's less power consumption (Jankovic et al. 2004; Chinkal et al. 2013).

**Table 1** Comparison between cleaner and conventional technology using return on investment and cost benefit ratio (ROI & CBR) in Cement Industry

| S. no | Technology | Cost variables | Conventional technology parameters | Cleaner technology parameters | Return on investment (ROI) | | Cost-Benefit ratio (CBR) | |
|---|---|---|---|---|---|---|---|---|
| | | | | | Conventional Technology (%) | Cleaner Technology (%) | Conventional technology | Cleaner Technology |
| 1. | Vertical roller mill (inlet duct Modification) | 1. CI<br>2. Variable Cost<br>3. Viability period<br>4. BuyBack cost<br>5. Depreciation cost<br>6. Profit<br>7. Benefit<br>8. TC | Rs. 9.5 lakh<br>Rs. 1.7 lakh p.a.<br>2 years<br>5.5%<br>Rs. 0.17 lakh p.a.<br>**Rs. 1.449 lakh p.a.**<br>**30,518 (kWh) p.a.**<br>6.1887 lakh p.a. | Rs. 204 lakh<br>Rs. 1.5 lakh p.a.<br>4.5 years<br>7.8%<br>Rs. 0.17 lakh p.a.<br>**Rs. 2.774 lakh p.a.**<br>**58,400 (kWh) p.a.**<br>43.297 lakh p.a. | **−79.03** | −93.3 | **0.2341** | 0.0640 |
| 2. | Vertical roller mill (Grit Cone Modification) | 1. CI<br>2. Variable Cost<br>3. Viability period<br>4. BuyBack cost<br>5. Depreciation cost<br>6. Profit<br>7. Benefit<br>8. TC | Rs. 11.5 lakh<br>Rs. 1.3 lakh p.a.<br>2 years<br>3.3%<br>Rs. 0.23 lakh p.a.<br>**2.1375 lakh p.a.**<br>**45,000 (kWh) p.a.**<br>6.86 lakh p.a. | Rs. 16 lakh<br>Rs. 3 lakh p.a.<br>5 years<br>6.7%<br>Rs. 0.4 lakh p.a.<br>**Rs. 4.561 lakh p.a.**<br>**95,900 (kWh) p.a.**<br>5.98 lakh p.a. | −72.1 | **−30.4** | 0.3115 | **0.762** |

(continued)

**Table 1** (continued)

| S. no | Technology | Cost variables | Conventional technology parameters | Cleaner technology parameters | Return on investment (ROI) | | Cost-Benefit ratio (CBR) | |
|---|---|---|---|---|---|---|---|---|
| | | | | | Conventional Technology (%) | Cleaner Technology (%) | Conventional technology | Cleaner Technology |
| 3. | Vertical roller mill (internal modification) | 1. CI<br>2. Variable Cost<br>3. Viability period<br>4. BuyBack cost<br>5. Depreciation cost<br>6. Profit<br>7. Benefit<br>8. TC | Rs. 3.6 lakh<br>Rs. 0.145 lakh p.a.<br>3 years<br>4.5%<br>Rs. 0.04 lakh p.a.<br>**Rs. 0.0285 lakh**<br>**6000 (kWh) p.a.**<br>1.291 lakh p.a. | Rs. 30.7 lakh<br>Rs. 0.32 lakh p.a.<br>4 years<br>6%<br>Rs. 0.17 lakh p.a.<br>**Rs. 5.7 lakh p.a.**<br>**120 000 (kWh) p.a.**<br>7.5345 lakh p.a. | −100.89 | **−26.5** | 0.0220 | **0.76** |

(continued)

**Table 1** (continued)

| S. no | Technology | Cost variables | Conventional technology parameters | Cleaner technology parameters | Return on investment (ROI) | | Cost-Benefit ratio (CBR) | |
|---|---|---|---|---|---|---|---|---|
| | | | | | Conventional Technology (%) | Cleaner Technology (%) | Conventional technology | Cleaner Technology |
| 4. | VRM cement girding CVRM Sp. Power Reduction in OPC | 1. CI<br>2. Variable Cost<br>3. Viability period<br>4. BuyBack cost<br>5. Depreciation cost<br>**6. Profit**<br>**7. Benefit**<br>8. TC | Rs. 5 lakh<br>Rs. 0.44 lakh p.a.<br>5 years<br>5%<br>Rs. 0.3145 lakh p.a.<br>**Rs. 0.0475 lakh**<br>Rs. 1.39 lakh p.a. | Rs. 20 lakh<br>Rs. 1.15 lakh p.a.<br>3 years<br>6%<br>Rs. 0.12 lakh p.a.<br>**Rs. 2.50 lakh p.a.**<br>**52,600 (Kwh) p.a.**<br>Rs. 3.4625 lakh p.a. | −119.19 | **−67.8** | 0.0341 | **0.337** |

(continued)

**Table 1** (continued)

| S. no | Technology | Cost variables | Conventional technology parameters | Cleaner technology parameters | Return on investment (ROI) | | Cost-Benefit ratio (CBR) | |
|---|---|---|---|---|---|---|---|---|
| | | | | | Conventional Technology (%) | Cleaner Technology (%) | Conventional technology | Cleaner Technology |
| 5. | Variable frequency drive (VFD) | 1. CI<br>2. Variable Cost<br>3. Viability period<br>4. BuyBack cost<br>5. Depreciation cost<br>**6. Profit**<br>**7. Benefit**<br>8. TC | Rs. 0.4 lakh<br>Rs. 0.05 lakh p.a.<br>2 years<br>2%<br>Rs. 0.02 lakh p.a.<br>**Rs. 0.0285 lakh p.a.**<br>**600 (kWh) p.a.**<br>Rs. 0.246 lakh p.a. | Rs. 105 lakh<br>Rs. 1.9 lakh p.a.<br>4 years<br>6%<br>Rs. 0.8 lakh p.a.<br>**Rs. 2.46 lakh p.a.**<br>**80,000 (kWh) p.a.**<br>Rs. 4.36 lakh p.a. | −96.54 | −61.9 | 0.115 | 0.564 |

(continued)

**Table 1** (continued)

| S. no | Technology | Cost variables | Conventional technology parameters | Cleaner technology parameters | Return on investment (ROI) | | Cost-Benefit ratio (CBR) | |
|---|---|---|---|---|---|---|---|---|
| | | | | | Conventional Technology (%) | Cleaner Technology (%) | Conventional technology | Cleaner Technology |
| 6. | False Air Reduction | 1. CI<br>2. Variable Cost<br>3. Viability period<br>4. BuyBack cost<br>5. Depreciation cost<br>**6. Profit**<br>**7. Benefit**<br>8. TC | Rs. 0.8 lakh<br>Rs. 0.08 lakh p.a.<br>2 years<br>3.5%<br>Rs. 0.02 lakh p.a.<br>**Rs. 0.003325 lakh p.a.**<br>**700 (kWh) p.a.**<br>Rs. 0.466 lakh p.a. | Rs. 2 lakh<br>Rs. 0.18 lakh p.a.<br>4 years<br>7.7%<br>Rs. 0.14 lakh p.a.<br>**Rs. 0.076 lakh p.a.**<br>**1600 (kWh) p.a.**<br>Rs. 0.6415 lakh p.a. | **−103.59** | −109.97 | 0.00713 | **0.1184** |

(continued)

**Table 1** (continued)

| S. no | Technology | Cost variables | Conventional technology parameters | Cleaner technology parameters | Return on investment (ROI) | | Cost-Benefit ratio (CBR) | |
|---|---|---|---|---|---|---|---|---|
| | | | | | Conventional Technology (%) | Cleaner Technology (%) | Conventional technology | Cleaner Technology |
| 7. | Compressor Power Reduction | 1. CI<br>2. Variable Cost<br>3. Viability period<br>4. BuyBack cost<br>5. Depreciation cost<br>**6. Profit**<br>**7. Benefit**<br>8. TC | Rs. 15 lakh<br>Rs. 0.11 lakh p.a.<br>2 years<br>6.7%<br>Rs. 0.15 lakh p.a.<br>**Rs. 0.2372 lakh p.a.**<br>**5000 (kWh) p.a.**<br>7.106 lakh p.a. | Rs. 18.5 lakh<br>Rs. 0.18 lakh p.a.<br>4 years<br>10%<br>Rs. 0.139 lakh p.a.<br>**Rs. 0.308 lakh p.a.**<br>**6500 (kWh) p.a.**<br>4.342 lakh p.a. | −98 | **−95.8** | 0.033 | **0.070** |

(continued)

**Table 1** (continued)

| S. no | Technology | Cost variables | Conventional technology parameters | Cleaner technology parameters | Return on investment (ROI) | | Cost-Benefit ratio (CBR) | |
|---|---|---|---|---|---|---|---|---|
| | | | | | Conventional Technology (%) | Cleaner Technology (%) | Conventional technology | Cleaner Technology |
| 8. | Reduction of specific power and fuel consumption in pyro section | 1. CI<br>2. Variable Cost<br>3. Viability period<br>4. BuyBack cost<br>5. Depreciation cost<br>**6. Profit**<br>**7. Benefit**<br>8. TC | Rs. 22 lakh<br>Rs. 7.55 lakh p.a.<br>years<br>%<br>Rs. p.a. | Rs. 20.1 lakh<br>Rs. 2.2 lakh p.a.<br>3. years<br>6%<br>Rs. 0.66 lakh p.a.<br>**Rs. 3.8 lakh p.a.**<br>**80,000 (kWh) p.a.**<br>Rs. 8.5 lakh p.a. | NA* | −63 | NA* | 0.447 |

(continued)

**Table 1** (continued)

| S. no | Technology | Cost variables | Conventional technology parameters | Cleaner technology parameters | Return on investment (ROI) | | Cost-Benefit ratio (CBR) | |
|---|---|---|---|---|---|---|---|---|
| | | | | | Conventional Technology (%) | Cleaner Technology (%) | Conventional technology | Cleaner Technology |
| 9. | Vertical roller mill (Fan optimization) | 1. CI<br>2. Variable Cost<br>3. Viability period<br>4. BuyBack cost<br>5. Depreciation cost<br>**6. Profit**<br>**7. Benefit**<br>8. TC | Rs. 4.1 lakh<br>Rs. 0.15 lakh p.a.<br>3.5 years<br>2%<br>Rs. 0.0856 lakh p.a.<br>**Rs. 0.1425 lakh p.a.**<br>**3000 (kWh) p.a.**<br>Rs. 1.298 lakh p.a. | Rs. 9.9 lakh<br>Rs. 0.37 lakh p.a.<br>1.7 years<br>5.5%<br>Rs. 0.114 lakh p.a.<br>**Rs. 1.82 lakh p.a.**<br>**38,500 (kWh) p.a.**<br>Rs. 5.873 lakh p.a. | −95.6 | **−69.7** | 0.1097 | **0.310** |

(continued)

**Table 1** (continued)

| S. no | Technology | Cost variables | Conventional technology parameters | Cleaner technology parameters | Return on investment (ROI) | | Cost-Benefit ratio (CBR) | |
|---|---|---|---|---|---|---|---|---|
| | | | | | Conventional Technology (%) | Cleaner Technology (%) | Conventional technology | Cleaner Technology |
| 10. | To stop the bag filter fan by connection of venting line with ESP inlet | 1. CI<br>2. Variable Cost<br>3. Viability period<br>4. BuyBack cost<br>5. Depreciation cost<br>**6. Profit**<br>**7. Benefit**<br>8. TC | Rs. 3.6 lakh<br>Rs. 0.245 lakh p.a.<br>3 years<br>4.5%<br>Rs. 0.04 lakh p.a.<br>**Rs. 0.1638 lakh p.a.**<br>**3450 (kWh) p.a.**<br>Rs. 1.391 lakh p.a. | Rs. 11.95 lakh<br>Rs. 0.084 lakh p.a.<br>1.5 years<br>8.5%<br>Rs. 0.02 lakh p.a.<br>**Rs. 2.622 lakh p.a.**<br>**55,200 (kWh) p.a.**<br>Rs. 7.37 lakh p.a. | −91.09 | **−64.71** | 0.1177 | **0.3555** |

(continued)

**Table 1** (continued)

| S. no | Technology | Cost variables | Conventional technology parameters | Cleaner technology parameters | Return on investment (ROI) | | Cost-Benefit ratio (CBR) | |
|---|---|---|---|---|---|---|---|---|
| | | | | | Conventional Technology (%) | Cleaner Technology (%) | Conventional technology | Cleaner Technology |
| 11. | Belt conveyor and Pipe conveyors | 1. CI<br>2. Variable Cost<br>3. Viability period<br>4. BuyBack cost<br>5. Depreciation cost<br>**6. Profit**<br>**7. Benefit**<br>8. TC | Rs. 234 lakh<br>Rs. 2.5 lakh p.a.<br>4 years<br>4%<br>Rs. 0.95 lakh p.a.<br>**Rs. 1.733 lakh p.a.**<br>**36,500 (kWh) p.a.**<br>Rs. 58.66 lakh p.a. | Rs. 400 lakh<br>Rs. 7 lakh p.a.<br>5 years<br>5%<br>Rs. 0.5 lakh p.a.<br>**Rs. 2.63 lakh p.a.**<br>**55,000 (kWh) p.a.**<br>Rs. 83 lakh p.a. | **−98.65** | −100 | 0.0295 | **0.0316** |
| 12. | Coal Mill Sp. Power Reduction by output increased | 1. CI<br>2. Variable Cost<br>3. Viability period<br>4. BuyBack cost<br>5. Depreciation cost<br>**6. Profit**<br>**7. Benefit**<br>8. TC | Rs. 10 lakh<br>Rs. 0.05 lakh p.a.<br>5 years<br>3.5%<br>Rs. 0.06 lakh p.a.<br>**Rs. 0.4275 lakh p.a.**<br>**9000 (kWh) p.a.**<br>Rs. 1.98 lakh p.a. | Rs. 15 lakh<br>Rs. 0.03 lakh p.a.<br>5 years<br>6%<br>Rs. 0.04 lakh p.a.<br>**Rs. 0.522 lakh p.a.**<br>**11,000 (kWh) p.a.**<br>Rs. 2.85 lakh p.a. | -161.25 | **-97.4** | **0.215** | 0.029 |

(continued)

**Table 1** (continued)

| S. no | Technology | Cost variables | Conventional technology parameters | Cleaner technology parameters | Return on investment (ROI) | | Cost-Benefit ratio (CBR) | |
|---|---|---|---|---|---|---|---|---|
| | | | | | Conventional Technology (%) | Cleaner Technology (%) | Conventional technology | Cleaner Technology |
| 13. | Air pollution | 1. CI<br>2. Variable Cost<br>3. Viability period<br>4. BuyBack cost<br>5. Depreciation cost<br>**6. Profit**<br>**7. Benefit**<br>8. TC | Rs. 445.8 lakh<br>Rs. 40 lakh p.a.<br>4.5 years<br>6%<br>Rs. 30.5 lakh p.a.<br>**Rs. 2.137 lakh p.a.**<br>**45,000 (kWh) p.a.**<br>Rs. 133.18 lakh p.a. | N.A | −121.2 | N.A | 0.0160 | N.A |

N.A- Not Available, TC–Total cost, CI-Capital investment, kWh- kilowatt-hour

**Explanation of Cost Analysis using ROI and CBR for (VRM) inlet duct Modification**

Vertical roller mill (VRM) machinery had a huge amount of modification in this cement industry. Considering the (VRM) inlet duct modification the return on investment in conventional technology (non-renewable energy) was (−79.3%) and cleaner technology (VRM) inlet duct Modification was (−93.3%) for Rs. 1 investment. On comparing the two technologies there is a loss in ROI. In terms of cost-benefit ratio, conventional technology (non-renewable energy) (0.2341) and cleaner technology (VRM) inlet duct Modification (0.0640) and in terms of cost-benefit ratio, conventional technology shows a higher benefit. But in cleaner technology, energy conservation was done due to less use of pressure and temperature, sensors and automated flow meters were attached for more accurate readings leading to energy saved as 58,400 Kwh per year and profit of Rs. 2.77 lakh per year, though ROI showed loss as shown in Table 1.

## 2.5 *Vertical Roller Mill Grit Cone Modification*

**Older technology**

Vertical roller mill is used directly and there were no modification done so if there is solid waste it should be recycled at the end of the process, due to this process again the operational cost can be higher due to energy use.

**Vertical roller mill Grit Cone Modification**

The process of grit cone is used in between the vertical roller mill to separate the particles in two different way coarse and fine particles High efficiency and sharpness of separation, using of different separation of materials of different grain size distribution, Easy for adjustment of the fineness of the product, Low specific power consumption, High drying efficiency within the separator, Cooling of the material with ambient air is also made has an add-on technology.

**Explanation of Cost Analysis using ROI and CBR for VRM girt cone modification**

Analysis of (VRM) grit cone Modification, the return on investment in conventional technology (non-renewable energy) was (−72.1%) and cleaner technology (VRM) grit cone Modification was (−30.4%) for Rs. 1 investment. On comparing the two technologies, both show loss in ROI. In terms of cost benefit ratio conventional technology (non-renewable energy) (0.3115) and cleaner technology (VRM) girt cone Modification (0.762) and cleaner technology showed a higher benefit (Table 1). Despite the loss in ROI and cost-benefit ratio of cleaner technology, the added benefits like energy benefit of 95,900 kWh per year and a profit of Rs. 4.5 lakhs per year was recorded. Advantage of Vertical roller mill Grit Cone Modification: Less maintenances and operational cost, High energy efficiency, recycle the fine particles

## 2.6 *Vertical Roller Mill Internal Modification*

**Process:** The vertical roller mill was modified for raw meal grinding because of the movement inside is different in terms of speed. A high speed horizontal attrition mill for dry grinding with a higher Rotation per minute (RPM) is a used to conserve energy by optimizing the electricity level from the variable frequency drive which changes the direct current (DC) to alternate current.

**Explanation of Cost Analysis using ROI and CBR for VRM internal modification**

For (VRM) internal modification, the return on investment in conventional technology (non-renewable energy) was (−100.89%) and cleaner technology (VRM) internal Modification was (−26.5%) for Rs. 1 investment. On comparing the two technologies both show loss in ROI. In terms of cost benefit ratio conventional technology (non-renewable energy) (0.0220) and cleaner technology (VRM) internal Modification (0.76) where cleaner technology showed a higher benefit. Along with loss in ROI and gain in cost-benefit ratio, cleaner technology also benefits by energy saved of 120,000 kWh per year and Rs. 5.7 lakh monetary profit per year as shown in Table 1.

## 2.7 *Vertical Roller Mill Cement Girding CVRM Sp. Power Reduction in OPC Portland Cement Grinding*

**Operational process:** The older switch is a direct current (DC) switch mode power supply change into alternate current (AC) supply modification has been done called power reduction switch.

**Explanation of Cost Analysis using ROI and CBR for VRM girt cone modification**

(VRM) Special Power Reduction in OPC showed return on investment in conventional technology (non-renewable energy) as (−119.19%) and cleaner technology (VRM) Special Power Reduction in OPC as (−67.8%) for Rs. 1 investment. On comparing the two technologies recorded loss in ROI. In terms of cost benefit ratio conventional technology (non-renewable energy) (0.0341) and cleaner technology (VRM) Special Power Reduction in OPC (0.337), cleaner technology shows a higher benefit. With loss in ROI and gain in cost-benefit ratio the cleaner technology added benefit by energy saved amount of 52,600 Kwh per year and Rs. 2.50 lakhs per year (Table 1).

## 2.8 *Variable Frequency Drive (VFD Installation)*

**Principle of working**: Direct current (DC) is changed into alternate current (AC)

**Working of Variable frequency drive (VFD)**

In Cement Industries, three phase induction motors are used because of its robust characteristics and simple maintenance. Rotating direction and speed of the three phase induction motor can be changed using Variable Frequency Drive (VFD) which is happening in recent years. The advantages of VFD are that they are energy saving, consumes less current for starting of motor, thermal and mechanical losses are less on motors, maintenance is not required often, has high power factor and a low KVA. The PLC controls and monitors VFD and VFD acts as a conciliator between 3 phase induction motor and the PLC. A conveyer is connected to the induction motor and cell sensor input is connected uniformly across the conveyer. The sensor input is connected to the PLC. This processes the input according to the ladder logic programming and initiates corresponding output to the VFD.

1. VFDs taken in line for energy conservation and optimization (More than 65 Nos of VFDs Installed for Fans, Pumps etc.)
2. VFD for Compressors (HAG–Hot generator Coal Blower)
3. Continuous monitoring of false air in each section and reduction activities (Raw mill circuit −7% including seal Air and Kiln 3.8%)
4. All major drives like Process Bag filter Fans, PH Fans, Coolers Fans; Kiln etc., are running in VFD
5. Expert Optimizer (EO) implemented for Raw Mill, Cement Mill, Coal Mill and Pyro.
6. Automated and flow meters connected to it on consumption.
7. Direct current (DC) is changed into alternative current (AC)

**Explanation of Cost Analysis using ROI and CBR for VFD**

The Table 1 shows, Variable frequency drive (VFD), the return on investment in conventional technology (non-renewable energy) was (−96.54%) and cleaner technology Variable frequency drive (VFD) was (−61.9%) for Rs. 1 investment. On comparing the two technologies recorded loss. In terms of cost benefit ratio conventional technology (non-renewable energy) (0.115) and cleaner technology (VRM) Special Power Reduction in OPC (0.564). Where the conventional technology showed higher benefit. Though loss in ROI and cost-benefit ratio, the cleaner technology would still provide benefit through energy saved to an extent of 80,000 kwh per year and Rs. 2.46 lakh per year.

## *2.9 False Air Reduction*

**Process of False air reduction**

The false air which has been escaped from the equipment can be collectively infiltrated and reused by the keeping the temperature and pressure in a minimum level of maintain if this temperature is recycled in that process is called has false air reduction (Udara et al. 2014).

For this process energy audit should be done properly 1. Optimizations of the output 2. Reduction in specific energy consumption 3. Trouble shooting in electrical, mechanical and process systems 4. Dust abatement 5. Quality assurance.

**Explanation of Cost Analysis using ROI and CBR for False air reduction**

The return on investment in conventional technology (non-renewable energy) was (−103.59%) and cleaner technology (False Air Reduction) was (−109.97%) for Rs. 1 investment. On comparing the two technologies were in loss. In terms of cost benefit ratio conventional technology (non-renewable energy) (0.00 713) and cleaner technology (False Air Reduction) (0.1184) and cleaner technology showed a higher benefit. Along with loss in ROI and higher benefit in cost-benefit ratio of cleaner technology, added benefits are energy saved as 1600 kWh per year and profit earned as Rs. 0.076 lakhs per year (Table 1).

**Operational modification done in areas:** Ball mills kiln burner, Clinker cooler, Kiln shell.

## *2.10 Compressor Power Reduction*

**Working process of compressor power reduction**

Compressed air systems has various modification depends on the capacity like Incoming air filters, inter-stage coolers, after coolers, air dryers, and moisture drain traps, receivers, piping network, filters, regulators and lubricators Air compressors can save significant amount of energy. Air compressors are used in a variety of large scale industries to supply energy requirements, to operate air tools, equipment, and instrumentation needs. Only 10–30% of energy reaches at the process end-use, and balance 70–90% of energy is being converted to useless heat energy is lost in form of sound energy, mismanagement and friction (Saidur et al. 2010).

**Explanation of Cost Analysis using ROI and CBR for Compressor Power Reduction**

The return on investment in conventional technology (non-renewable energy) was (−98%) and cleaner technology (Compressor Power Reduction) was (−95.8) for Rs. 1 investment. On comparing the two technologies showed loss in ROI. In terms of cost-benefit ratio conventional technology (non-renewable energy) was (0.033) and cleaner technology (Compressor power reduction) was (0.070) and cleaner technology showed a higher benefit. Additionally cleaner technology showed benefits in energy saved as 6500 kWh per year and profit as Rs. 0.308 lakh per year (shown in Table 1).

**Advantage of Compressor power reduction**

Energy saving technology, Operational and maintenance cost is less

## 2.11 Reduction of Specific Power and Fuel Consumption in Pyro Section

**Process**: Clinker is produced by the pyro-processing section were in this industry it has been modified to Optimize heat recovery by upgrading the clinker coolers for making in rotary kilns, Preheater/ pre-clinker kilns for clinker making in vertical shaft kilns and Low temperature heat recovery for power generation for clinker making in rotary kilns High temperature heat recovery for power generation for clinker making in rotary kilns is attached and used. (Kabir et al. 2010; Emad et al. 2013).

**Explanation of Cost Analysis using ROI and CBR for Reduction of specific power and fuel consumption in pyro section**

The return on investment of Cleaner technology (Reduction of specific power and fuel consumption in pyro section) is (−63%) for Rs. 1 investment, depicting loss. In terms of cost benefit ratio cleaner technology (Reduction of specific power and fuel consumption in pyro section) is (0.447). Though loss in ROI and cost-benefit ratio for cleaner technology, benefits in the form of energy saved 80,000 kWh per year and profit earned Rs. 3.8 lakh per year (Table 1) were offsetting the loss. Advantage: Low operational cost, Energy conserving technology.

## 2.12 Vertical Roller Mill Fan Optimization

Process of Fan optimization: High power consumption and low productivity. For stable long term operation, when the system is found occurring a large variations to control the grinding dust is accumulated in fan and when the system under small variation adjustment automatically done so the machine speed varies. Due to the decrease in the machines fan has a maximum optimization reduction occurs due to this process by temperature and pressure, energy is conserved (Danielle and Patrick 2010).

**Explanation of Cost Analysis using ROI and CBR for Vertical roller mill Fan optimization**

The return on investment in conventional technology (non-renewable energy) was (−95.6%) and cleaner technology (Vertical roller mill Fan optimization) was (−69.7%) for Rs. 1 investment, the two technologies were in loss. In terms of cost benefit ratio conventional technology (non-renewable energy) was (0.1097) and cleaner technology (Vertical roller mill Fan optimization) was (0.310) leading to cleaner technology showing a higher benefit. Despite loss in ROI and higher gain in Cost benefit ratio cleaner technology added further benefits through energy saved as 38,500 kWh per year and profit accrued as Rs. 1.82 lakh per year (Table 1).

## 2.13 To Stop the Bag Filter Fan by Connection of Venting Line with ESP Inlet

**Process**

Process Modification is done using venting line in Electrostatics precipitator (EP) inlet to conserve energy by adding automated sensor and variable frequency drive (VFD).

**Cost analysis for to stop the bag filter fan by connection of venting line with ESP inlet**

The return on investment in conventional technology (non-renewable energy) was (−91.09%) and cleaner technology (To stop the bag filter fan by connection of venting line with ESP inlet) was (−64.71%) for Rs. 1 investment, both losses. In terms of cost-benefit ratio conventional technology (non-renewable energy) was (0.1177) and cleaner technology (To stop the bag filter fan by connection of venting line with ESP inlet) was (0.3555) and cleaner technology showed higher benefits. Also cleaner technology had extra benefits in the form of energy saved as 55,200 kWh per year and profit as Rs. 2.622 lakhs per year (Table 1).

**Advantage:** Automated and sensors are available, Energy saving technology connected to variable frequency drive

## 2.14 Belt Conveyor and Pipe Conveyors

**Process**

Due to transport of materials like lime stone and calcium carbonate inside the industry conveyor systems were used. And this conveyor are classified into two, Belt and Pipe conveyor

**Belt conveyor**: Belt conveyor is an old method of transport using to rotatory poles was used.

**Pipe conveyor:** Pipe is used in a vacuum method using (air sucking) in this method there is no spillage and dispersion of air.

**Advantage:** No spillage, dust emission, Lesser area required for plant and machine, Reduced maintenance costs, Investment required for 6 Numbers of Auxiliary bag filters eliminated.

**Explanation of Cost Analysis using ROI and CBR using for Belt and Pipe conveyor**

The return on investment in conventional technology (non-renewable energy) was (−98.65%) and cleaner technology (Belt conveyor) was (−100%) for Rs. 1 investment, both losses. In terms of cost-benefit ratio conventional technology (non-renewable energy) was (0.0295) and cleaner technology (Belt conveyor) was (0.3168) where cleaner technology showed a higher benefit. Added benefits like energy saved 55,500 kWh per year and profit obtained Rs. 2.63 lakhs per year were attributed to cleaner technology (Table 1).

## 2.15 Coal Mill Sp. Power Reduction by Output Increased

**Process**

Specific heat capacity is recycled by using dry heat and increase the capacity of energy resources without using coal (non-renewable energy resource has been used less).

**Explanation of Cost Analysis using ROI and CBR using for Belt and Pipe conveyor**

The return on investment in conventional technology (non-renewable energy) was (−161.2%) and cleaner technology (Coal Mill Sp. Power Reduction by output increased) was (−97.4%) for Rs. 1 investment. On comparing the two technologies charted loss in ROI. In terms of cost-benefit ratio conventional technology (non-renewable energy) was (0.215) and cleaner technology (Coal Mill Sp. Power Reduction by output increased) was (0.029) and cleaner technology showed a higher benefit, along with energy saved as 11,000 kWh per year and profit got as Rs. 0.52 lakhs per year.

**Advantage of Compressor power reduction**

Energy saving technology, Operational and maintenance cost is high.

## 2.16 Air Pollution Control

**Process**

Air pollution control technology is used in this industry, conventional End Of Pipe (E-O-P), technology. The air is sucked by a method called as vacuum and blower is attached and then it's connected to a stack in a height for dispersion of air particulate. the CPCB has given some subsides for using Advance technology and old technology still prolong in updating technology by using senor, automated flow meters, Bag filter-cement mill, Bag filter–Coal mill, Bag filter-Cement silo, Bag filter-Fly ash silo, Bag filter-Raw mill Bag filter-Roto packers, Bag house-cement mill, Bag house-coal mill, Cooler ESP (Electrostatic precipitator) and fan, Dust suppression, Raw mill reverses air bag house and Small nuisance bag filters.

**Explanation of Cost Analysis using ROI and CBR for air pollution control**

The return on investment in conventional technology (Air pollution control) was (−121.2%) for Rs. 1 investment, loss in ROI. In terms of cost-benefit ratio conventional technology (Air pollution control) was (0.0160). Still other benefits like energy saved as 45,000 kWh per year and Profit as Rs. 2.13 lakhs per year could be seen as the positives of conventional technology. Advantage: Cost-effective and maintenance is less, Adsorption and odor is less consumption of energy is less (Table 1).

## 3 Other Environmental Management Systems in Cement Industry

1. Used oil is recycled and reused to a quantity of 120 kg per year and the method of disposal, storing in separate area and selling to authorized recyclers.
2. Quantity of waste fuel used (Tons) 11,210 per year and Equivalent of Conventional energy used (Tons of fuel) 4839 per year and Waste fuel total energy was 3.97%.
3. Fly ash is added and recycled to an extent of 42.5 ton per day and reused in manufacturing itself.
4. Bed ash is added and recycled, 7.5 ton per day with limestone.
5. Rain water harvesting pond Capacity was 40,000 $m^3$ and saved water to a value of Rs. 3 lakhs.
6. On line dust monitoring system has been attached towards vertical rolling machine for 6.5 lakhs.
7. To reduce dust insufflations in pulverized coal, High efficiency Twin Cyclone has been installed in the Coal mill circuit for a cost of Rs. 5 lakh.
8. Adsorption of air takes places in cement industry to minimize the level of adsorption various remedial measures like Sprinklers, Short sprinklers, Moisture sprinklers and Mist spray sprinklers employed.
9. High coal CV Quantity of waste fuel of 800 Tons Equivalent of conventional energy was used (Kwh of electricity or tons of fuel) converting 164.43 MT Waste to 0.77% of total energy.
10. Heat Consumption reduction by Kiln TPD increased 6500 TPD Capital Investment of Rs. 20.4 lakh, Variable cost of Rs. 1.5 lakh per year, viability period of the mechanism was 4.5 years, Benefited cost was Rs. 3.59 lakh per year, Buy back cost stood at 7.8%.
11. The Green supply chain management (GSCM) introduce new regulation norm by research and development of plants towards, Green belt (GB) management for specifically to control air pollution (absorption and adsorption) in cement industry, In this select cement industry there are 250 different type of species trees are grown predominately, this are classified into neem, Pungan and teak etc. with a survival rate of 85-90% (*Tamrindus indicus*), Neem (*Azadirecta indica*), Kalli (*Euphorbia* sps.), Echam (*Phoenix syevestris*), Mango (*Mangifera indica*), Palmyra (*Borassus flabelifera*), etc. were dominant species. Presence of large number of Phanerophytes (shrubs and trees) and therophytes (annuals) converted the semiarid region to a tropical vegetative area in the study area. Hemicryptophytes (predominantly grasses and sedges) were found to be significantly grown. CPCB recommend 63,819 trees for plantation and were planted at a cost of Rs. 7 lakh around the industry to adsorb air pollution.
12. Every cement industry has to be doing energy audit due to expansion and retention of heat every 6 months and every three years equipment modification should be carried out by changing the mechanism (Viredra et al. 2015), Energy management system: EnMS: ISO: 50001 Energy performance was 4.9% of total

energy. Cost incurred over improvement was 2060,000 USD to implement total EnMS Rs. 5.4 lakh payback per year on EnMS implementation for 25 years. Total energy saving over improvement period was (GJ) 38,303 GJ, Total $CO_2$ emission reduction over improvement period was 38,371 MT of $CO_2$.

Effluent treatment plant is used to treat the sewage from the cooling tower water through the process of activated sludge process (ASP), secondary treatment process also used and finally passed into the dairy farming, plantation and green belt.

## 4 Reuse of Treated Trade Effluent in Process

Investment of Rs. 3 Lakh as part of the initiative to conserve natural resources, Captive Power Plant effluent (waste water) was used for internal water spray in cement mill by which it was able to reduce the raw water consumption by approximately 80 KL (kilo liter) water per day, leading to raw water conservation.

## 5 Discussion

The Cement plant falls under large scale industry and Red category (Annexure A, in EIA). In Ariyalur in Tamil Nadu, calcium carbonate and gypsum was available in higher quantity. Originally Installed capacity (Clinker), Installed capacity (Cement) 3MTPA, but currently has a rated capacity of Raw mill, Specific Power Consumption—Section wise 15.14 (kWh/Mt), Specific Power Consumption—Section wise 4.36 (kWh/Mt), Klin productivity Average is 4643 TPD, Ordinary pozalanna cement OPC cement grinding 30.20 (kWh/Mt), Cement Dispitch—14.55 LacMT. Cement industry has adopted cleaner technology on process modification but Industry ecology in air pollution control was still using conventional technology at the End process. With Vertical roller mill inlet duct Modification the ROI was (−79.03%) but still conventional technology has a loss but on CBR value of conventional technology is 0.2341 higher that cleaner technology, on energy savings of 518,400 kWh per year, Vertical roller mill Grit Cone Modification has the of ROI (−30.4%) and CBR value (0.762) cleaner technology is higher than conventional technology. Energy saving was 95,900 kWh per year. Vertical roller mill internal modification had a ROI of (−26.5%) loss for Rs. 1 investment and the CBR value 0.76) in cleaner technology, Energy savings 120,000 kWh per year. Power Reduction in OPC Portland cement grinding showed (−67.8%) loss in two technology for Rs. 1 investment and CBR value is 0.337, energy savings 52,680 kWh per year. Variable frequency drive (VFD) ROI was loss in two technologies for Rs. 1 investment and CBR value is higher in cleaner technology 0.564, Energy saving 51,840 kWh per year. False Air Reduction had the ROI which was loss but cost benefit ratio value is higher in cleaner technology 0.1184, energy is gained, 1600 kWh per annum. Compressor Power Reduction ROI was loss in two technologies with CBR value is 0.070 in cleaner technology and Energy savings of 6500 Kwh per year. Coal Mill Sp. Power Reduction by output increased with ROI was a loss but in CBR conventional technology value is higher 0.21 and energy saving was 11,000 kWh per year. Vertical roller mill Fan optimization ROI was loss in two technologies and CBR cleaner technology value is high (0.310), Energy saving was 38,500 kWh per year. To stop the bag filter fan by connection of venting line with ESP inlet ROI was loss for Rs. 1 investment and CBR cleaner technology value is high (0.3555), energy savings 55,2000 kWh per year. Pipe conveyor ROI was 0.97, 426% and the energy savings 55,500 kWh. Air pollution control Conventional technology ROI was (−121.2%) and CBR is 0.0160 but still using conventional technology Energy savings 45,000 kWh per year. Recycling Technology like oil disposal, Waste fuel, fly ash, bed ash, green belt and trade effluent were used. Green rating project (GRP) and Energy management system (Energy audit) EnMS-50001 was used in this industry to calibrate data, analysis (life cycle of energy assessment (LCEA), calculate, document, review, feedback, comparing the efficacy, interpretation and planning per annum is done.

## 6 Conclusion

The present study aimed at analyzing the pros and cons of old pollution control technology against newer cleaner technology in select cement is a most polluting large scale industry. From the results it could be construed that the old technology was not redundant as some are still economically viable than newer ones. Most new technologies are "add on" ones wherein the old technology has been upgraded with new inputs and thereby leading to better environmental protection. Although the total investment and the operation and maintenance cost on cleaner technology was high in all the selected units of study, but the environmental benefits in terms of water and energy saving, besides recycling of water was greater due to the application of cleaner technologies Therefore, cleaner technology will be the future in India and abroad towards sustainable industrial production.

**Acknowledgements** The author thanks the help rendered by Dr. M. Ravichandran, Professor (Retried) Department of Environmental Management, Bharathidasan University, (Tiruchirappalli) in correcting the final version of the paper.

## References

Bapat JD (2001) Application of ESP for gas cleaning in cement industry—with reference to India. J Hazard Mater 81(3):285–308

Chinkal RP, Shah TH, Thakore SB, Desai R (2013) Cleaner production assessment in cement plant using vertical shaft technology. J Pet Gas Explor Res 3(2):26–30

Danielle ML, Patrick M (2010) Industrial strategies to help reduce energy consumption: A holistic approach for cement producers. 2010 IEEE-IAS/PCA 52nd Cement Industry Technical Conference. https://doi.org/10.1109/CITCON.2010.5469768

Emad B, Zahedi G, Shamsaei E, Bahadori A (2013) Global strategies and potentials to curb $CO_2$ emissions in cement industry. J Cleaner Prod 51:142–161

Gupta PK (2013) Indian scenario of co-processing of wastes in cement industries. CMA – IIP International conference on enhanced usage of alternate fuels and raw materials – co-processing in cement plants

Jankovic A, Valery W, Davis E (2004) Cement grinding optimization. Minerals Eng 17(11–12):1075–1081

Kabir G, Abubakar AI, El-Nafaty UA (2010) Energy audit and conservation opportunities for pyro processing unit of a typical dry process cement plant. Energy 35(3):1237–1243

Khurana S, Banerjee R, Gaitonde U (2002) Energy balance and cogeneration for a cement plant. Appl Therm Eng 22(5):485–494

Ministry of Power, Government of India (2012) PAT: Perform Achieve and Trade. http://beeindia.in/content.php?page=schemes/schemes.php?id=8

Phillips PP, Phillips JJ (2006) Return on investment (ROI) basics, American Society for Training and Development

Sengupta B (2016) Status of co-processing of waste in cement plants as per hazardous and other wastes management rules. In: 2016 Presented at 3rd international conference on alternate fuels and raw materials in cement industry–towards realising swachh bharat dream on 23rd and 24th March 2017 at New Delhi, India

Siva Chandran J (2016) Cost-benefit approach of Bt cotton and Non-Bt cotton seeds application in agriculture: a case evidence from Salem and Coimbatore districts of Tamil Nadu, India. p 8

Udara SP, Arachchige R, Kawan D, Tokheim LA, Melaaen MC (2014) Impact of kiln thermal energy demand and false air on cement kiln flue gas CO2 capture. Int J Energ Environ 5(1):45–52

# Analysis of Microbial Isolate from Contaminated Street Food with a Potential to Degrade Food Waste

**S. Swathi Priya, C. Shanmuga Priya, and J. Maria Shyla**

**Abstract** Contaminated food stuff is a potential source of bacteria with interest encompassing mirid attributes prompting their exploitation in assorted applications including environment domain like degradation of waste and discarded food. Street food has bacterial contamination like *Salmonella sp.*, *Staphylococcus aureus, Escherichia coli, Bacillus sp.,* Mostly non-significance bacteria are present in the street foods but still point to environmental contamination, in this work food samples were collected and serially diluted, isolated confirmed using different media. The isolation and identification process were done to know and characterise the bacteria via 16 s rRNA gene sequence and the DNA these further elucidated via BLAST search engine.

**Keywords** Bacteria · DNA sequence · *Bacillus. sp* · Micro-organisms · Street food

## 1 Introduction

Bacteria are ubiquitous in nature. Gram-positive bacteria that are encountered in food samples; *S. aureus, Streptococcus pneumoniae, Staphylococcus epidermidis, Listeria monocytogenes, Bacillus* etc. Gram-negative bacteria; *Escherichia coli, Salmonella, Shigella, and other Enterobacteriaceae, Pseudomonas* etc. (Gdoura-Ben Amor et al. 2018) There are 30–40 million bacterial cells in a gram of food. Genus *Bacillus* is of prominence especially when taken in consideration the significance for its role in the biodegradation of fatty materials in food wastes, as it is armed with an array of important producing capabilities like lipase- and biosurfactant-producing species, also they are capable of creating endospores that empowers them resistant to conflicting; extreme environmental situations, hence they have a special niche in the biological handling of pollutants and xenobiotics. *Bacillus* spp. along with other fat-degrading microorganisms like as *Acinetobacter* spp. and/or *Pseudomonas* spp. have been isolated from an assortment of sources, including contaminated food stuffs

S. Swathi Priya · C. Shanmuga Priya · J. Maria Shyla (✉)
Department of IT, PSGR Krishnammal College for Women, Coimbatore, TamilNadu, India
e-mail: mariatommy07@gmail.com



M. Prashanthi et al. (eds.), *Bioremediation and Green Technologies*, Environmental Science and Engineering, https://doi.org/10.1007/978-3-030-64122-1_16

and have been explored for these features in laboratory-level works (Sayyadifar et al. 2012; Awasthi et al. 2018).

Studies have been carried out to identify various bacteria from different food products. Bakery items, Spoiled foods, Dairy products and Junk foods were collected from the local market of Bagalkot, Karnataka district in India. Samples were confirmed by both phenotypic and genotypic characteristics. The bacterial isolates such as *Pseudomonas* sp. [23.66%], *S. aureus* [22.76%], *Salmonella* sp. [21.87%], *E. coli* [22.32%] and *Klebsiella* sp. [8.92%] were present in large amount (Hemalata and Virupakshaiah 2016).

## 2 Materials and Methods

The material was collected from a bakery which was a packed egg puff and which serves for a good microbial activity. The packed food manufactured on 10/01/2020 with expiry date of 11/01/2020. The sample was taken as 1gm and grind in a pestle into small pieces and further proceeded to serial dilution. A 100 μL of each food suspension was spreader over clearly labelled nutrient agar medium plates from each dilution. The injected Petri plates were incubated at 37°C for 24 h to enable bacterial growth (Liu et al. 2012).

### *2.1 Isolation of Bacteria*

After the bacterial growth, the bacteria were isolated by using specific media like Mannitol Salt Agar, MacConkey Agar, Triple Sugar Iron, SS Agar, Nutrient broth. Mannitol Salt Agar is used to identify the *S. aureus* or Bacillus sp., MacConkey Agar is used to identify the Salmonella sp., Triple Sugar Iron and SS Agar are used to identify the Salmonella- Shigella (Khaneja et al. 2010).

### *2.2 DNA Extraction and PCR Sequencing of Bacterial Sample*

The isolated foodborne pathogens processed to identify the bacteria using molecular biology techniques. The steps to be followed for the extraction of DNA from the bacterial sample in the laboratories are shown in the Fig. 1. Bacterial Genomic DNA Spin-50 kit was used. The solution was prepared with the necessary components to make the RNase solution. About 100 mL of bacterial culture was taken and added with the required components. The DNA concentration was determined by both UV

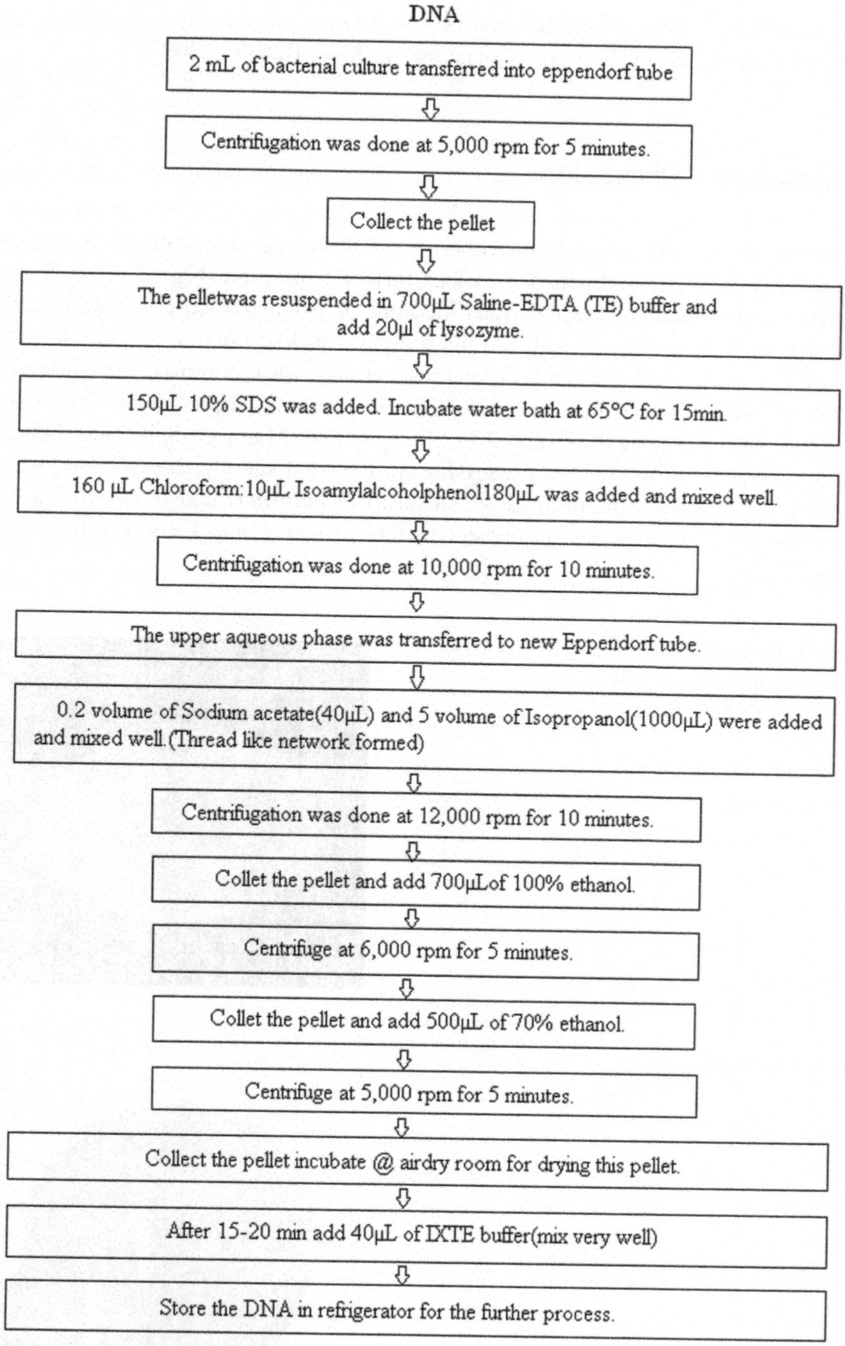


**Fig. 1** Flow of DNA extraction

spectrophotometer and quantitative analysis on Agarose gel. Finally, sequencing was obtained with the aid of the ABI sequencing machine (Babalola 2003).

## 3 Result and Discussion

The result of the estimation of microbial population size in specific dilution was estimated (Fig. 1). The obtained microbes further confirmed (Figs. 2, 3, 4, 5 and 6). The results obtained from various medium of isolated food-borne pathogens shown in the Table 1. The presence of microorganisms like *Bacillus* and *Salmonella* proved via Mannitol salt agar and SS Agar respectively, while comparing the different media for the growth of bacteria; MacConkey Agar and Triple Sugar Iron Agar. Very less bacterial growth observed in SS Agar than Mannitol Salt Agar, hence, this was considered for further processing. In the food sample, *Bacillus sp.*, were identified (asides some gram-negative bacteria) *B. anthracis* and *B. cereus* causes food poisoning. *Bacillus sp.*, is present widely in nature and food (Privitera and Nesci 2015).

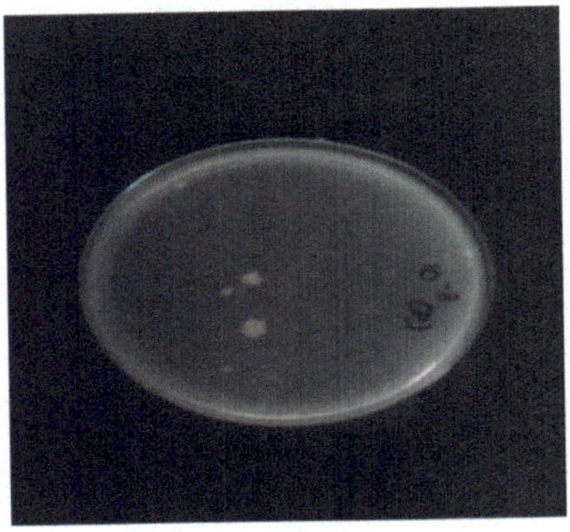

**Fig. 2** Estimation of microbial population size via serial dilution method

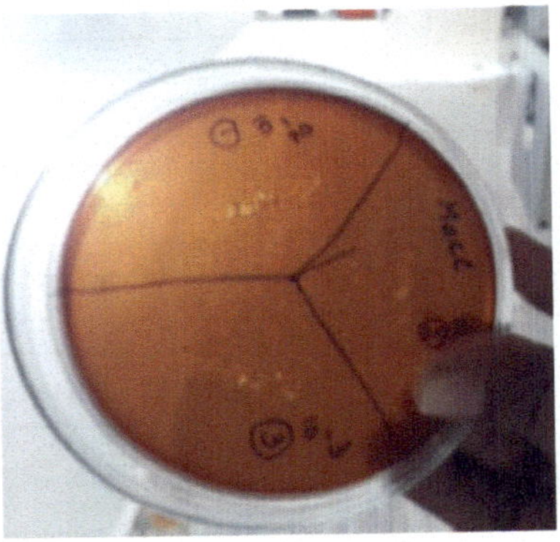

**Fig. 3** Mac Conkey Agar

**Fig. 4** Mannitol agar (*Bacillus* sp.)

**Fig. 5** Triple sugar agar

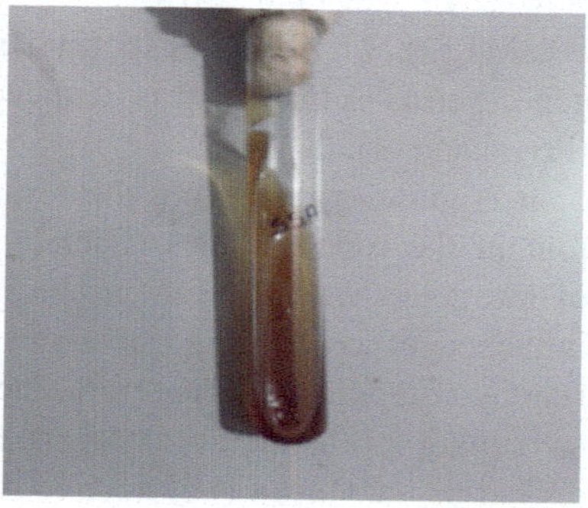

**Fig. 6** SS Agar (*Salmonella*)

**Table 1** Comparison of different medium

| S. No | Medium used | Result | Organisms | Percentage |
|---|---|---|---|---|
| 1. | MacConkey Agar | Absent | – | – |
| 2. | Mannitol salt agar | Present | *Bacillus sp.* | 21. 3% |
| 3. | Triple sugar agar | Absent | – | – |
| 4. | SS Agar | Present | *Salmonella sp.* | 5. 1% |

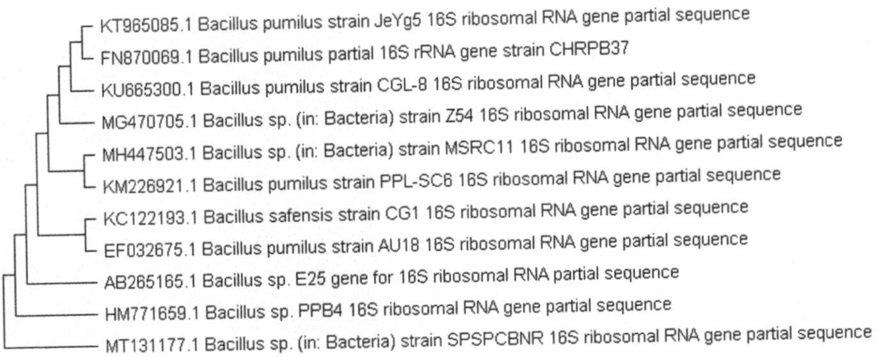


**Fig. 7** Phylogenetic tree

This study was evaluated due to the prevalence of genetic diversity of the *Bacillus sp*.., (in egg puff with 21.3% and *salmonella* with 5.4%). Phylogenetic study also performed to map the molecular level similarity of DNA sequence among the various partial 16 s rDNA. The Gene sequence of *Bacillus species* was taken for the study. This study has been performed to enlarge the molecular level gene similarly identification to enhance the various *Bacillus species*. The sequence is closely related to bacillus were matched under multiple sequence alignment (MSA) using mega tool software. Finally, it was found that our bacteria were under the group of *bacillus* and closely related to uncultured bacterium clone (Fig. 7) (McLauchlin 2007; Pundir et al. 2013; Roy et al. 2018).

>MT131177.1 Bacillus sp. (in: Bacteria) strain SPSPCBNR 16S ribosomal RNA gene, partial sequence was sequenced and a phylogenetic tree was constructed (Priyanka et al. 2016).

It is quite evident from the results the influence of environmental contamination in the food samples (Sanlier et al. 2018; Soni et al. 2016). Many strains of the genus *Bacillus* can produce different enzymes and surfactants under the influence of contrasting substrates including, food wastes (high percentage of starchy components). Therefore, the use of pollutant substances and or its isolates (screening for distinct microorganisms able to produce biosurfactants or other food degrading enzymes), is of notable interest. Others researches investigation also use similar concepts of isolating indigenous bacterial species from contaminated food and employ them for biodegradation of food wastes (Hasan and Zulkahar 2018; Thanigaivel and Anandhan 2015).

## 4 Conclusion

The study focused on the presence of bacteria associated with environmental contamination in street food. The predominate bacteria was found to be *Bacillus sp*. (21.3%). The gene sequence was submitted to NCBI and tree access number was granted

(MT131177.1). Genus *Bacillus* is of prominence especially when taken in consideration the significance for its role in the biodegradation of fatty materials in food wastes, as it is armed with an array of important producing capabilities like lipase- and biosurfactant—producing species, also they are capable of creating endospores that empowers them resistant to conflicting; extreme environmental situations, hence they have a special niche in the biological handling of pollutants and xenobiotics.

**Acknowledgements** We, the authors express our sincere gratitude to the Department of Biotechnology, Ministry of Science and Technology which has financially supported to publish this research work under DBT Star Status grant of our Institution.

## References

Awasthi MK, Wong JWC, Kumar S, Awasthi SK, Wang Q, Wang M et al (2018) Biodegradation of food waste using microbial cultures producing thermostable a-amylase and cellulase under different pH and temperature. Bioresour Technol 248:160–170

Babalola OO (2003) Molecular techniques: an overview of methods for methods for the detection of bacteria. Afr J Biotechnol 2(12):710–713

Gdoura-Ben Amor M, Siala M, Zayani M, Grosset N, Smaoui S, Messadi-Akrout F et al (2018) Isolation, identification, prevalence, and genetic diversity of *Bacillus cereus* group bacteria from different foodstuffs in Tunisia. Front Microbiol 9:447. https://doi.org/10.3389/fmicb.2018.00447

Hasan NA, Zulkahar I (2018) Isolation and identification of bacteria from spoiled fruits. AIP Conf Proc 2020:020073. https://doi.org/10.1063/1.5062699

Hemalata VB, Virupakshaiah DBM (2016) Isolation and identification of food borne pathogens from spoiled food samples. Int J Curr Microbiol App Sci 5(6):1017–1025

Khaneja R, Perez-Fons L, Fakhry S, Baccigalupi L, Steiger S, To E et al (2010) Carotenoids found in *Bacillus*. J App Microbiol 108:1889–1902

Liu W, Wang X, Wu L, Chen M, Tu C, Luo Y et al (2012) Isolation, identification and characterization of *Bacillus amyloliquefaciens* BZ-6, a bacterial isolate for enhancing oil recovery from oily sludge. Chemosphere 87(10):1105–1110

McLauchlin J (2007) Foodborne pathogens: microbiology and molecular biology. J Antimicrob Chemother 60(5):1180

Privitera D, Nesci FS (2015) Golbalization versus local: the role of street food in the urban food system. Proc Econ Financ 22:716–722

Priyanka B, Rajashekhar K, Dawarkanath SA (2016) review on detection methods used for foodborne pathogens. Ind J Med Res 144(3):327–338

Pundir RK, Rana S, Kashyap N, Kaur A (2013) Probiotic potential of lactic acid bacteria isolated from food samples: an in vitro study. J App Pharma Sci 3(03):085–093

Roy S, Pankaz S, Saha T, Torun P (2018) Isolation and identification of bacteria in different street vended foods collected from selected areas of Bangladesh. Asian Australas J Food Saf Secur 2(2):65–70

Sanlier N, Sezgin AC, Sahin G, Yassibas E (2018) A study about the young consumers' consumption behaviors of street foods. Cien Saude Colet 23(5):1647–1656. https://doi.org/10.1590/1413-81232018235.17392016

Sayyadifar F, Morteza H, Munawar T, Jayasimha R (2012) Molecular identification of bacteria isolated from meat sample. Int J App Biol Pharm Technol 3(4):34–39

Soni A, Oey I, Silcock P, Bremer P (2016) *Bacillus* spores in the food industry: a review on resistance and response to novel inactivation technologies. Compr Rev Food Science Food Saf 15:1139–1148. https://doi.org/10.1111/1541-4337.12231

Thanigaivel G, Anandhan AS (2015) Isolation and characterization of microorganisms from raw meat obtained from different market places in and around Chennai. J Pharm Chem Biol Sci 3(2):295–301

# Screening, Isolation and Molecular Characterization of Biosurfactants Producing *Serratia Marcescens* from Petrochemical Exposed Site

**V. Vijayalekshmi, P. Muthukumaran, and Aravind Jeyaseelan**

**Abstract** The present study was taken up to test the soil sample isolated from an automobile workshop located at Coimbatore for the presence of biosurfactant producing bacteria. Ten isolates were initially screened from the soil collected from the petrochemical exposed site. Isolation was done in the minimal salt medium, and this was followed by screening for the biosurfactant production through three different methods, namely emulsification index, oil displacement activity and hemolytic assay. In comparison with all the three screening methods, two best isolates were chosen and was then subjected to mass production in the mineral salt medium for biosurfactants generation. The isolate identified as *Serratia marcescens* through 16S rRNA sequencing. Pinkish red-coloured occurrence was observed during the mass production, which may be possibly due to pigmentation. The extracted biosurfactants were analyzed with Fourier transform infrared spectroscopy spectra. The potent nature of the soil collected from the contaminated petrochemical site for the production of biosurfactants has a remarkable growth prospective, sourcing as a better replacement to the increasing environmental concern associated with the chemical surfactants.

**Keywords** Bacteria · Biosurfactants · Emulsification index · Extraction · Serriatia marcescens

V. Vijayalekshmi · P. Muthukumaran (✉)
Department of Biotechnology, Kumaraguru College of Technology, Coimbatore 641 049, India
e-mail: m.kumaran005@gmail.com

A. Jeyaseelan (✉)
Department of Civil Engineering, Dhirajlal Gandhi College of Technology, Omalur 636309, Tamilnadu, India
e-mail: dr.j.aravind@gmail.com

 
M. Prashanthi et al. (eds.), *Bioremediation and Green Technologies*, Environmental Science and Engineering, https://doi.org/10.1007/978-3-030-64122-1_17

# 1 Introduction

Biosurfactants refer to any biomolecules which exhibit characteristics like surface-active and are found to be amphipathic in nature, which can be produced extracellularly by various microbes including bacteria, fungi and yeasts. They form aggregates at the interfaces amid different fluid polarities, wherein decreasing the surface (Saravanan and Vijayakumar 2012). The unique features of microbial surfactants include biodegradability, emulsifying and demulsifying ability, wetting and penetrating agents, tolerance to temperature, pH, ionic strength, low toxicity and viscosity reducing agents make them most suitable for remediation of environmental contamination, environmental safety (Akbari et al. 2018) and finds application in food, pharmaceutical and allied industries (Shekhar et al. 2015; Sharma et al. 2018). Oil spills and various other hydrocarbon contaminants in the soil and aquatic environment is becoming a significant focus of attention in the today world. Absorbents and synthetic organic products like lime, plastic polymers and cellulose-based materials, polyurethane foams, polypropylene and elastomers were preferred as the commercial sorbents in the oil spill cleanup considering the properties of good hydrophobicity, uptake capacity and oil recovery (Adebajo et al. 2003; Seo et al. 2018), but then it faced the problems of low degradability comparing to the naturally occurring mineral and vegetable products (Teas et al. 2001). Such disadvantages has to be mitigated via microbialy generated surfactants. Microorganisms exposed to oil-contaminated sites, develop the potential to harness hydrocarbons as carbon and energy for its metabolic activities. Biosurfactants increase the substrate bioavailability for microorganisms and interacts with the cell surface, thereby increasing the surface hydrophobicity, thus allowing substrates to accumulate faster within the bacterial cells (Liu et al. 2017). In comparison with its chemical counterparts, biosurfactants are very particular, potent and competent under an extensive range of oil and receptacle environment and may only require meagre quantities. Because of these attributes, they are preferred in different industrial processes and physicochemical phenomena, wherein they attribute enhanced solubility, increased mobility (Pacwa-Płociniczak et al. 2011; Rufino et al. 2014). Microbial biosurfactant with all these unique properties is considered to be a best green alternative potent source when comparing to other methods for clearing oil spills and plays a promising role in bioremediation of hydrocarbons contaminated sites, proving to be better than the chemical surfactants in many aspects including degradability (Sammarco et al. 2013; Matvyeyeva et al. 2014). *Serratia marcescens*, which is a gram-negative bacillus, belonging to Enterobacteriaceae, is better known for the production of red pigment prodigiosin and biosurfactant serrawettin (Muthukumar et al. 2016; Sunaga et al. 2016). At the time when the cell growth rate is sluggish under unfavourable conditions, the pigment biosynthesis has a role of being the protective mechanism (Li et al. 2005). *Serratia* produces chitinolytic enzymes that have a practical part in the biological degradation of chitin (Someya et al. 2000) and its related environmental applications along with environmental protection (Brzezinska et al. 2014). Biosurfactants with the current light of focus are the potent tool for developing a sustainable environment. This study accounts for

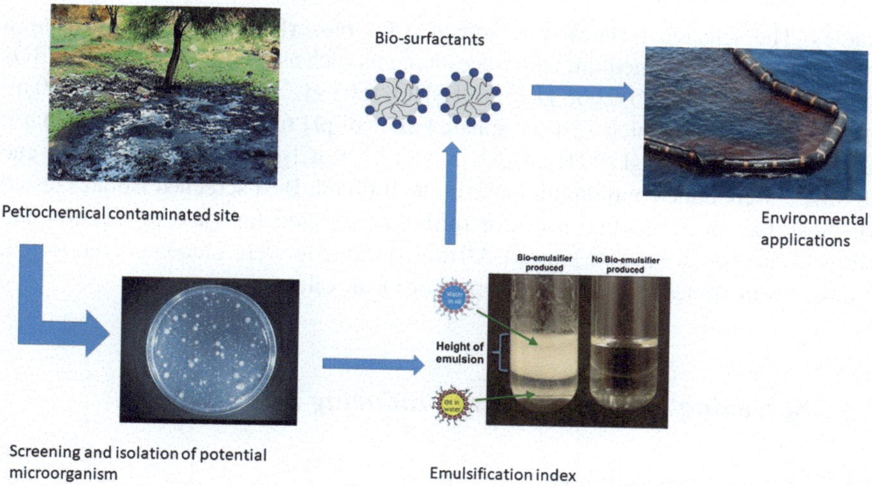


**Fig. 1** Biosurfactant isolation, characterization and possible application

the isolation and screening of biosurfactants producing indigenous bacteria, that has been identified as *Serratia marcescens*, isolated from a soil sample collected from contaminated site identified in Coimbatore district, Tamilnadu, India (Fig. 1).

# 2 Materials and Method

## 2.1 *Sample Collection*

The soil sample was collected from an automobile workshop located at Saravanampatti, Coimbatore district and used for further studies.

## 2.2 *Isolation of Bacterial Colonies*

Soil sample of 5 g was inoculated in 100 ml of Minimal Salt Medium (MSM) with the composition containing 1.5 $gL^{-1}$ of $NaNO_3$, 1 $gL^{-1}$ of $KH_2PO_4$, 1.5 $gL^{-1}$ of $(NH_4)_2SO_4$, 0.5 $gL^{-1}$ of $MgSO_4$, 0.01 $gL^{-1}$ of $FeSO_4$ and 0.002 $gL^{-1}$ of $CaCl_2$ was added along with 3 ml kerosene oil as the carbon source in a 250 ml conical flask. Incubation carried out for 72 h at a temperature of 30 °C. These samples serially diluted up to $10^{-6}$ dilution. From this, 1 ml of each dilution transferred to nutrient agar for spread culture, and these plates were at incubated at 37 °C for about 72 h (Dewaliya and Jasodani 2013). Incubation is followed by single colony isolation and based on the morphology; ten distinct isolates selected for further studies. The selected isolates were screened for biosurfactant production using a modified Mineral salt medium with trace elements such as $FeSO_4$ 0.05 $gL^{-1}$, $H_3BO_3$ 0.5 $gL^{-1}$, $MnSO_4.4H_2O$ 0.008 $gL^{-1}$, $C_2H_7NO_2$ 0.05 $gL^{-1}$ were dissolved in 250 ml of distilled water to which 1 M phosphate buffer of pH 6.8 of 20 $mlL^{-1}$ with major salts such as $KNO_3$ 1 $gL^{-1}$, $H_{14}MgO_{11}S$ 1 $gL^{-1}$, $CaCl_2H_{12}O_6$ 0.1 $gL^{-1}$, kerosene 20 $mlL^{-1}$ were added and finally made up to 1000 ml. Best screened isolates tested for biosurfactant production used for further considered for mass production and subjected to characterization studies. All the experiments were adequately replicated, and the result values expressed in average or mean values.

## 2.3 *Screening of Biosurfactant Producing Bacteria*

### 2.3.1 Emulsification Index

Emulsification activity of the biosurfactant was checked, wherein 2 ml of kerosene oil was added to 1 ml of cellfree extract that had been obtained by centrifugation (Abd-alridha 2014). This mixture was then vortexed well for about 2–3 min at high speed. The emulsion activity was observed after 24 h and calculated as Eq. 1.

$$\text{Emulsion index} = (\text{Total height of the emulsion layer}/\text{height of the aqueous layer})^{*}100 \tag{1}$$

Oil displacement activity is a measure of the ability of the biosurfactants to alter the contact angle of the oil, which is studied on oil-water interphase. Also, it is a measure of the reduction of surface tension by the biosurfactant (Freitas et al. 2016). The oil displacement activity was checked by adding 20 μl of kerosene oil to the surface of the Petri plate containing 50 ml of distilled water. Upon this 20 μl of culture supernatant was added to check for the formation of clear zones formed in the presence of the biosurfactants.

## 2.4 *Blood Hemolysis Activity*

The single isolated colonies were streaked on the blood agar plates and incubated for about 48–72 h under the temperature of 37 °C (Shanks et al. 2012), and then observed for the type of clear zone based on which the presence of biosurfactant producing bacteria was determined (Pacwa-Płociniczak et al. 2014).

## 2.5 *Extraction of Biosurfactant*

The extraction of the biosurfactant was done with the initial step of bacterial cell removal by centrifugation at 12,000 rpm for 20 min, the cultured supernatant adjusted to pH 2.0 with the aid of 0.1 M HCl. The biosurfactants extracted with a solvent mixture containing chloroform and methanol in the ratio 2:1 v/v in a separation funnel. After shaken vigorously, and allowed for phase separation, the lower organic phase concentrated, and anhydrous sodium sulfate added to remove water, and the residue was dried to obtain biosurfactants (Pacwa-Płociniczak et al. 2011).

## 2.6 *Molecular Characterization of Biosurfactant Producing Bacteria–16S RRNA Gene Sequencing*

The DNA was isolated from the bacterial culture and was used in PCR to amplify the bacterial 16 s rRNA using 16 s rRNA PCR kit (800). Using the primers from the kit, amplification of 800 bp amplicon done, and no amplicon was visible in the negative control. The expected sized amplicon of 800 bp observed in the positive control. The test amplicon of 800 bp was purified using magnetic beads, and the product sequenced by Sanger's method of DNA sequencing. The sequencing results were assembled and compared with the NCBI database.

## 2.7 *Characterization of Biosurfactants from Serratia Marcescens Using Thin Layer Chromatography*

Silica gel plate prepared on the ratio 1:2, with 1 mm thickness, i.e., 10 g in 20 ml distilled water which was then allowed it to stay for 30 min. The silica gel plate was kept in the hot-air oven for 1 h at 100 °C to activate the absorption. Chloroform-methanol of ratio 85:15 used as the developing solvent. 20 μl of biosurfactant spotted on the silica gel plate (Sivagamasundari and Jeyakumar 2016), and Ninhydrin reagent sprayed. The retention factor Rf (eq. 2), which defined as the distance travelled by the compound to the distance travelled by the solvent calculated.

$$\text{Rf} = (\text{Distance travelled by the compound}/\text{Distance moved by the solvent front}) \quad (2)$$

### 2.8 Fourier Transform Infrared Spectroscopy

For identifying types of chemical bonds (functional groups), Fourier transform infrared spectroscopy (FT-IR) analysis used to elucidate some components of an unknown mixture (Suryawanshi et al. 2014). One milligram of partially purified biosurfactant was wholly dried and then analyzed in an FTIR (Shimadzu), device that obtains the spectrum with high S/N ratio 30,000:1, 1-minute accumulation and a maximum resolution of 0.5 $cm^{-1}$. All data corrected for the background spectrum.

## 3 Results and Discussion

Soil samples inoculated in minimal salt medium and serial diluted to obtain bacterial isolates and subcultured for the isolation of the single colonies, as shown in the Fig. 2.

### 3.1 Screening of Biosurfactant Producing Bacteria

All the ten isolated strains screened for biosurfactant production. The bacterial isolates named as Petrochemical Contaminated Bacteria (PCB); PCB1, PCB2, PCB3, PCB4, PCB5, PCB6, PCB7, PCB8, PCB9 and PCB10. These were subjected to the following screening assays for the selecting the biosurfactant producing bacterial isolates.

### 3.2 Emulsification Index

When kerosene used as the oil source for testing the emulsification activity of isolated bacteria, positive results were noted and tabulated in Table 1. The emulsification activity was noted after 24 h and emulsification index was calculated to screen for the

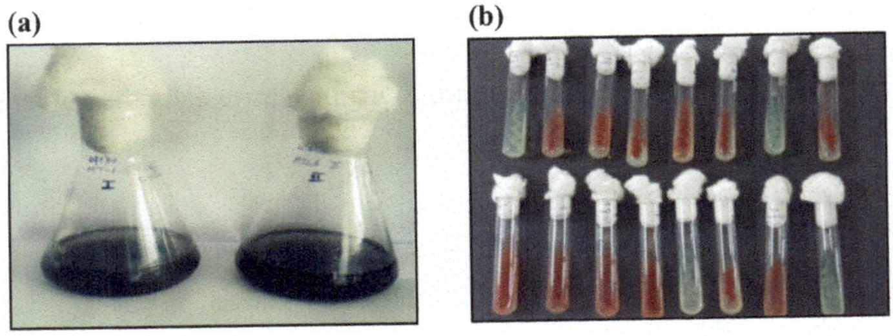


**Fig. 2** **a** Soil samples from petrochemical contaminated site **b** Single colony isolation

**Table 1** Screening results for biosurfactant producing bacterial isolates

| Sr. No. | Bacterial Isolates | Emulsion height (cm) | Total liquid height (cm) | Emulsification index (%) | Oil displacement activity | Color observed (Hemolysis) | Type of hemolytic activity |
|---|---|---|---|---|---|---|---|
| 1 | PCB*1 | 0.20 | 1.20 | 16.67 | + | Green | α Hemolysis |
| 2 | PCB2 | 0.30 | 1.30 | 23.00 | + | Yellow | β Hemolysis |
| 3 | PCB3 | 0.25 | 1.15 | 21.74 | ++ | Green | α Hemolysis |
| 4 | PCB4 | 0.25 | 1.20 | 20.83 | + | Yellow | β Hemolysis |
| 5 | PCB5 | 0.30 | 1.20 | 25.00 | + | Green | α Hemolysis |
| 6 | PCB6 | 0.20 | 1.25 | 16.00 | + | Green | α Hemolysis |
| 7 | *Serratia marcescens (This study)* | 0.30 | 1.40 | 21.43 | ++ | Green | α Hemolysis |
| 8 | PCB8 | 0.30 | 1.20 | 25.00 | + | Yellow | β Hemolysis |
| 9 | PCB9 | 0.25 | 1.35 | 18.52 | + | Yellow | β Hemolysis |
| 10 | PCB10 | 0.30 | 1.40 | 21.74 | ++ | Yellow | β Hemolysis |

*PCB- Petrol chemical bacteria

biosurfactant producing bacteria among the ten bacterial isolates. Among the studied strains, the highest emulsification index of 25 obtained in the two isolates, namely PCB5 and PCB8. It was also noted that the bacterial isolates PCB2, PCB3, PCB7 and PCB10 had emulsification index greater than 20. Similar range of emulsification index reported in the literature, when tested for seven different isolates of *Bacillus sp.*, and *Actinomyces sp.*, according to Bento et al. (2005) and Wang et al. (2014).

### 3.3 Oil Displacement Activity

Oil displacement activity helps in testing the ability of the biosurfactant to alter the contact angle at the oil-water interface when the supernatant added to the kerosene oil. Among the obtained results, the isolates namely PCB3, PCB7 and PCB10 showed good oil displacement activity indicated by ++ within the ten isolated tested for oil

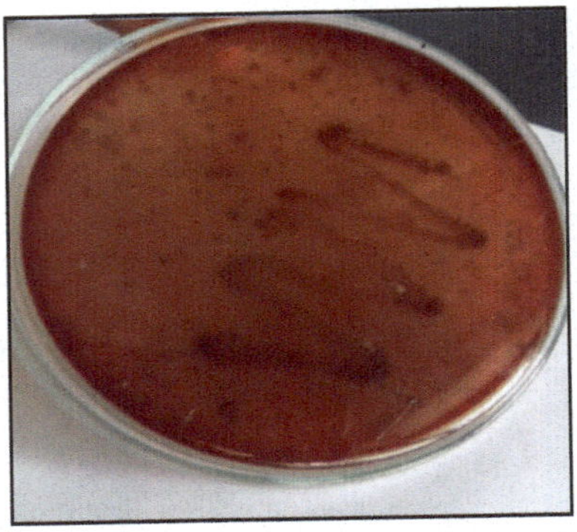

**Fig. 3** Hemolysis activity of *Serratia marcescens*

displacement activity. Rest of the strains only showed partial displacement activity reported by + as noted in Table 1.

## 3.4 Blood Hemolysis Activity

Blood hemolysis activity is tested for checking the ability of bacterial colonies to induce hemolysis when it was grown on blood agar. When the agar under the colony turns darkish green, $\alpha$ hemolysis is said to occur, and when it turns light yellow with transparent nature it is noted to be $\beta$ hemolysis, and $\gamma$ hemolysis observed with no change in the agar. All the ten isolates showed positive hemolytic activity in reference to the results obtained in Table 1. The presence of hemolytic activity in *Serratia marcescens* demonstrated via Fig. 3.

Based on the overall screening results of biosurfactant producing bacterial isolates, *Serratia marcescens* was selected among the ten isolates to narrow down the research for the mass production of biosurfactants with reference to emulsification activity, oil displacement and hemolytic activity. The screened isolates of *Serratia marcescens* and PCB10 were chosen for the mass production of biosurfactants in the mineral salt medium where the biosurfactants were formed as a separate layer with pigmentation. Confirmation test for emulsification activity produced good results (Fig. 4).

## 3.5 Extraction of Biosurfactant

The biosurfactants extracted with a solvent mixture containing chloroform and methanol in the ratio 2:1v/v in a separation funnel (Sukirtha and Usharani 2013; Yap et al. 2016). The lower organic phase concentrated, and anhydrous sodium sulphate added to remove water and dried to obtain biosurfactants as shown in the Fig. 5.

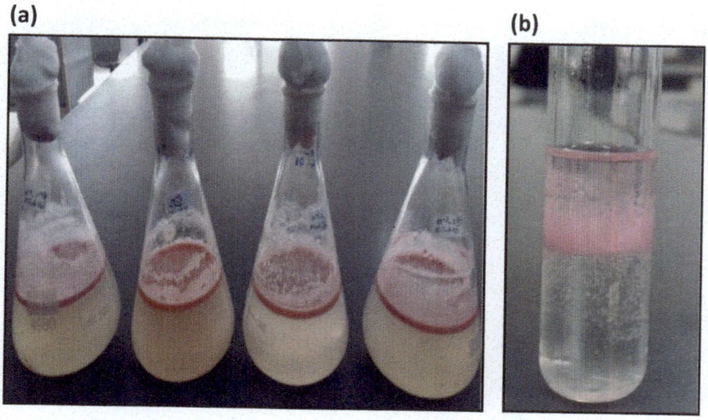


**Fig. 4** **a** Mass production in mineral salt medium **b** Emulsification activity: *Serratia marcescens*

**Fig. 5** **a** Extraction of biosurfactants with the solvents chloroform and methanol, **b** Dried biosurfactants of *Serratia marcescens*

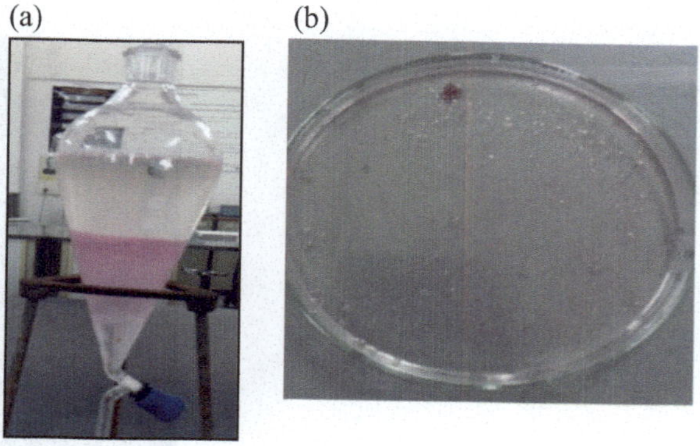


## 3.6 16S RRNA Gene Sequencing

The DNA isolated from the bacterial culture was used in PCR to amplify the bacterial 16 s rRNA. The molecular analysis for the isolated bacterial strain showed 99% identity to the sequence of *Serratia marcescens*. The phylogenetic tree generated for the isolated bacteria shown in Fig. 5. and 99% identity to the sequence of *Serratia marcescens*. Based upon the sequence analyses, the isolate is likely to be *Serratia marcescens*. The phylogenetic tree generated for the isolated bacteria provided in Fig. 6.

### 3.6.1 Thin Layer Chromatography

The retention factor for the compounds separated, calculated using the Formula (3),

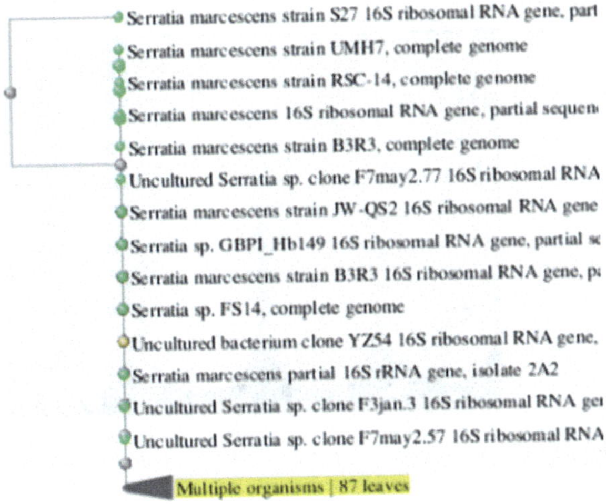


**Fig. 6** Sequence analysis for the bacterial isolate *Serratia marcescens*

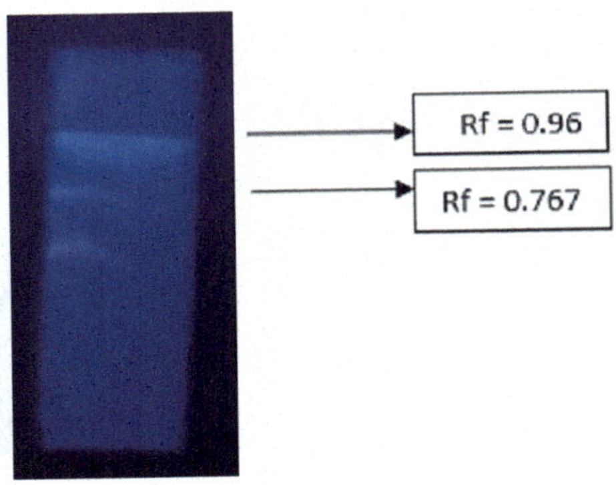


**Fig. 7** TLC plate visualized under UV to detect the separation of compounds for the biosurfactants obtained from *Serratia marcescens*

$$\text{Rf} = \text{Distance travelled by the compound/Distance travelled by the solvent front} \tag{3}$$

The results of TLC shown in Fig. 7.

## 3.7 Thin Layer Chromatography

The Rf value of biosurfactant obtained from *Serratia marcescens* as shown in Fig. 6 established to be 0.76 and 0.96 which was similar to the results reported with a red pigment isolated from *Serratia marcescens* (Vora et al. 2014). The quantitative analysis of the biosurfactant from *Serratia marcescens* is yet to be done.

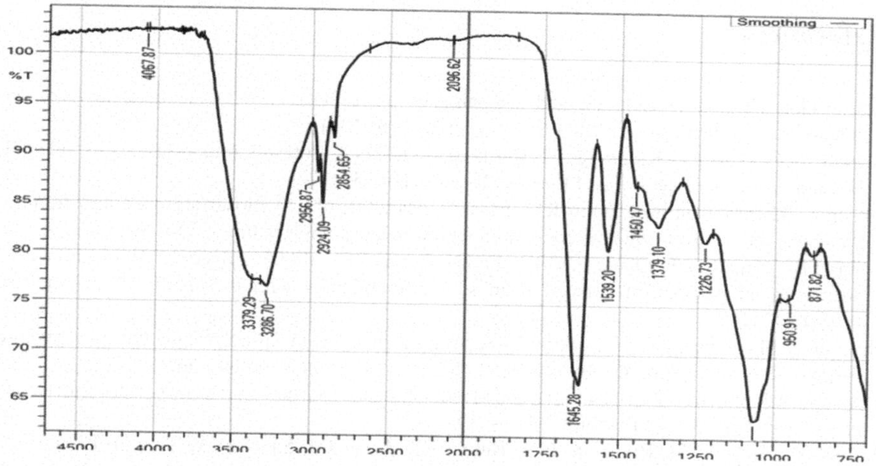


**Fig. 8** FTIR spectrum for the partially purified biosurfactant obtained from PCB7 identified as *Serratia marcescens*

### 3.8 *Fourier Transform Infrared Spectroscopy*

The FTIR studies for the extracted biosurfactant obtained from the bacterial isolate Serratia *marcescens* showed C–H stretching bands of –CH2 and –CH3 groups observed in the region 3000–2700 $cm^{-1}$ in Fig. 8. The deformation vibrations at 1467 and 1379 $cm^{-1}$ also confirmed the presence of alkyl groups. The wavenumber 1066.64 $cm^{-1}$ indicated the presence of C-O bonds. The peaks observed at 1066.64 $cm^{-1}$, 1226.73 $cm^{-1}$, 1539.2 $cm^{-1}$, 1645.28 $cm^{-1}$, 2096.62 $cm^{-1}$, 2956.87 $cm^{-1}$, 3286.7 $cm^{-1}$, 3379.29 $cm^{-1}$ indicated the presence lipid moieties. In this, the absorbance of N–H stretching bond at 3132–3379 $cm^{-1}$ was also observed (Vigneshwaran et al. 2018).

## 4 Conclusion

Among the ten bacterial isolates, *Serratia marcescens*, which was identified through 16 s rRNA sequencing, isolated from the soil sample that was collected from an automobile workshop located in Coimbatore district showed good activity in oil displacement, emulsification index proving to be a potent source for biosurfactant production. This could be further analyzed with the pigmentation studies and could be used for the potential application of environmental remediation and protection.

## References

Abd-alridha FT (2014) Isolation and screening of biosurfactant producing bacteria from oil contaminated soils in Iraq. IOSR J Pharm Biol Sci 9(6):10–15

Adebajo MO, Frost RL, Kloprogge JT, Carmody O, Kokot S (2003) Porous materials for oil spill cleanup: a review of synthesis. J Porous Mater 10(10):159–170

Akbari S, Abdurahman NH, Yunus RM, Fayaz F, Alara OR (2018) Biosurfactants-a new frontier for social and environmental safety: a mini review. Biotechnol Res Innov 2:81–90

Bento FM, Camargo FAO, Okeke BC, Frankenberger WT (2005) Diversity of biosurfactant prodcuing microorganism isolated from soils contaminated with diesel oil. Microbiol Res 160(3):249–255

Brzezinska MS, Jankiewicz U, Burkowska A, Walczak M (2014) Chitinolytic microorganisms and their possible application in environmental protection. Current Microbiol 68:71–81

Dewaliya V, Jasodani R (2013) Isolation and identification of *Bacillus licheniformis* for biosurfactant production. CIBTech J Microbiol 2(4):14–19

Freitas BG, Brito JGM, Brasileiro PPF, Rufino RD, Luna JM, Santos VA, Sarubbo LA (2016) Formulation of a commercial biosurfactant for application as a dispersant of petroleum and by-products spilled in Oceans. Front Microbiol 7:1–5

Li H, Tanikawa T, Sato Y, Nakagawa Y, Matsuyama T (2005) *Serratia marcescens* gene required for surfactant serrawettin W1 production encodes putative aminolipid synthetase belonging to nonribosomal peptide synthetase family. Microbiol Immunol 49(4):303–310

Liu Z, Li Z, Zhong H, Zeng G, Liang Y, Chen M, Wu Z, Zhou Y, Yu M, Shao B (2017) Recent advances in the environmental applications of biosurfactant saponins: a review. J Environ Chem Eng 5:6030–6038

Matvyeyeva OL, Vasylchenko OA, Aliieva OR (2014) Microbial biosurfactants role in oil products biodegradation. Int J Environ Bioremed Biodegrad 2(2):69–74

Muthukumar A, Pradeep P, Thigale I, Mohanasrinivasan V, Jemimah NS, Subathra Devi C (2016) Exploring the bioactive potential of Serriatia marcescens VITAPI isolated from soil. Front Biol (Beijing) 11(6):476–480

Pacwa-Płociniczak M, Płaza GA, Piotrowska-seget Z (2011) Environmental applications of biosurfactants: recent advances. Int J Mol Sci 12(1):633–654

Pacwa-Płociniczak M, Płaza GA, Poliwoda A, Piotrowska-Seget Z (2014) Characterization of hydrocarbon degrading and biosurfactant-producing *Pseudomonas sp.* P-1 strain as a potential tool for bioremediation of petroleum-contaminated soil. Environ Sci Pollut Res 21(15):9385–9395

Rufino RD, de Luna JM, Takaki GMC, Sarubbo LA (2014) Characterization and properties of the biosurfactant produced by *Candida lipolytica* UCP 0988. Electron J Biotechnol 17:34–38

Sammarco PW, Kolian SR, Warby RAF, Bouldin JL, Subra WA, Porter SA (2013) Distribution and concentrations of petroleum hydrocarbons associated with the BP/Deepwater Horizon Oil Spill. Gulf of Mexico Mar Pollut Bull 73(1):129–143

Saravanan V, Vijayakumar S (2012) Isolation and screening of biosurfactant producing microorganisms. J Acad Ind Res 1(5):1–5

Seo S, Mastiani M, Mosavati B, Peters DM, Mandin P, Kim M (2018) Performance evaluation of environmentally benign nonionic biosurfactant for enhanced oil recovery. Fuel 234:48–55

Shanks RMQ, Stella NA, Lahr RM (2012) Serratamolide is a hemolytic factor produced by *Serratia marcescens*. PLoS ONE 7(5):1–8

Sharma R, Singh J, Verma N (2018) Production, Characterization and environmental applications of biosurfactants from *Bacillus amyloliquefaciens* and *Bacillus subtilis*. Biocatal Agric Biotechnol 16:132–139

Shekhar S, Sundaramanickam A, Balasubramanian T (2015) Biosurfactant producing microbes and their potential applications: a review. Crit Rev Environ Sci Technol 45(14):1522–1554

Sivagamasundari T, Jeyakumar N (2016) Biosurfactant production using diesel oil degrading bacteria. Int J Adv Res Sci Eng Technol 3(12):3039–3046

Someya N, Kataoka N, Komagata T, Hirayae K, Hibi T, Akutsu K (2000) Biological control of cyclamen soilborne diseases by *Serratia marcescens* strain B2. Plant Dis 84(3):334–340

Sukirtha T, Usharani MV (2013) Production and qualitative analysis of biosurfactant and biodegradation of the organophosphate by *Nocardia mediterranie*. J Bioreme Biodegrad 4(6):4–11

Sunaga S, Li H, Sato Y, Nakagawa Y, Matsuyama T (2016) Identification and characterization of the pswP gene required for the parallel production of prodigiosin and serrawettin W1 in *Serratia marcescens*. Microbiol Immunol 48(10):723–728

Suryawanshi RK, Patil CD, Borase HP, Salunke BK, Patil SV (2014) Studies on production and biological potential of prodigiosin by *Serratia marcescens*. Appl Biochem Biotechnol 173(5):1209–1221

Teas Ch, Kalligeros S, Zanikos F, Stournas S, Lois E, Anastopoulos G (2001) Investigation of the effectiveness of absorbent materials in oil spills clean up. Desalination 140(3):259–264

Vigneshwaran C, Sivasubramanian V, Vasantharaj K, Krishnanand N, Jerold M (2018) Potential of *Brevibacillus* sp. AVN 13 isolated from crude oil contaminated soil for biosurfactant production and its optimization studies. J Environ Chem Eng 6:4347–4356

Vora JU, Jain NK, Modi HA (2014) Extraction, Characterization and Application studies of red pigment of halophile *Serratia marcescens* KH1R KM035849 isolated from Kharaghoda soil. Int J Pure Appl Biosci 2(6):160–168

Wang W, Cai B, Shao, Z (2014) Oil degradation and biosurfactant production by the deep sea bacterium *Dietzia maris* As-13–3. Front Microbial 1-11

Yap AC, Chan KG, Choo YM (2016) Isolation and identification of metabolites from the gram-negative proteobacteria of Burkholderia cenocepacia and Serratia marcescens. Sains Malaysiana 45(7):1073–1077

# Alternative Approaches and Suggested Studies for Prospective Environmental Remediation

# Sustainable Packaging from Waste Material: A Review on Innovative Solutions for Cleaner Environment

**Paneti Rajesh and V. Subhashini**

**Abstract** Packaging plays a very important role in our daily life especially when buying commodities from a retail or a wholesale store. The importance of packaging is when items are shipped around the world, which in recent years is on a daily basis. Though packaging ensures the safety and durability of the merchandise and carries the brand name of the seller, it leads to accumulated waste where most of these conventional packing materials cannot be recycled. With the large amount of companies transporting goods from factories to warehouses and then to retailers, as well as the enormous online orders placed daily, there is a massive amount of plastic waste such as Styrofoam, cardboard, and paper that are utilized in the process. However, sustainable packaging is a boon to this problem, which is referred to as Green packaging; it offers an environmental friendly alternative playing a great role in protecting products, preventing waste and enabling efficient business conduct promoting environmental sustainability. Sustainable packaging is made by recycling materials thereby reducing the waste during production or raw material processing. The manufacturing process of such packaging materials also tends to be more efficient, further minimizing resources needed and reducing the negative impact that the business has on the environment. Sustainable packaging is produced in an environmentally friendly manner through the use and reuse of biodegradable and recyclable materials and is considered energy efficient. As a part of Corporate Social Responsibility, sustainable packaging is a relatively new addition to the environmental considerations, where, reduced as well as ecofriendly material for packaging provide an attractive opportunity to promote environmental sustainability. Industries, Promoters and Companies using such packaging material not only reduce their carbon footprint but campaign the use of recycled materials minimizing waste generation. Through this, it is indirectly aimed at preserving the world's ecosystems, improving human life quality and viability for a longer period. Moreover, sustainable packaging is economically viable

P. Rajesh (✉)
Sanitation and Environment Secretary, Municipal Corporation of Nellore, Andhra Pradesh, India
e-mail: panetirajesh@gmail.com

V. Subhashini
Department of Environmental Sciences, Acharya Nagarjuna University, Guntur, Andhra Pradesh, India



M. Prashanthi et al. (eds.), *Bioremediation and Green Technologies*, Environmental Science and Engineering, https://doi.org/10.1007/978-3-030-64122-1_18

for both consumers and manufacturers additionally ensuring a cleaner environment for the future generations.

**Keywords** Biodegradable packing · Sustainability · Biopolymers · Energy efficiency

## 1 Introduction

Packaging waste is a significant part of solid waste generated increasing the environmental concerns, which has led to the strengthening of EU Regulations to reduce amounts of packaging waste. Among other raw materials, a wide range of oil-based polymers are currently used in packaging material manufacturing which are non-biodegradable and difficult to recycle or reuse due to mixed levels of contamination and complex composites. Due to surge of plastic industry and the durability of the material, most traditional methods of packaging has long been forgotten. Use of leaf and leaf composites, cloth and paper packing and tin or tin foils were used for food and beverage industries and cardboards were preferred for other merchandises.

Globally more than 200 million tonnes of plastic waste is generated annually with the APAC region reporting the highest consumption, followed by the EU and North America. Plastic is a major global concern, accounting for 12% of the world's solid waste contributing to a major environmental impact on carbon footprint, global warming and blue water footprint. At the same time, the rapid increase in urban population with higher disposable incomes has contributed to increased food consumption, higher rates of food waste and relative food loss. Food packaging and supply companies and basic material producers are now investing in sustainability plans and innovations to improve the management of paper and plastic waste and at the same time to extending shelf-life to ensure food quality and reduce food waste without compromising the quality.

The food and beverage industry has contributed extensively to the production of excess plastic due to the high consumption rate of increasing urban population, which in turn has led to high environmental implications. On the other hand, consumers life style today has changed as they are more concerned about the brands they consume through packaging however fortunately, they are more aware of the environmental impact of the products they consume.

One of the greatest consumers of the packaging industry are the food packing industry and in the recent years and due to the availability of mobile and online facility, the increase in food packing and delivery has reached its highest limits. Earlier the commonly used packaging material were wood, paper, glass, metals and tins, plastics and composites. However the recent decades have seen the increased use of plastics and Styrofoam in the packaging industry. Due to its superior properties like of non permeability, inertness to the environment, durability, lesser weight, stability and availability, plastic are preferred. These properties of plastic also make them last in the environment forever and accumulate as solid waste, if not recycled properly (Andrady and Neal 2009; Ashok et al. 2016a). Additives, plasticizers and colorants in plastic make serious environmental issues during disposal of the same (Andrady and Neal 2009; Thompson et al. 2009). Packaging plays a significant role in bringing products to consumers in a safe and wholesome manner without compromising product quality (Rhim and Ng 2007;Youssef et al. 2012). Interactions between packaging materials and the packed product might contribute in changing the product quality. This paper discusses the biodegradable packaging materials that are most suitable for single use disposable discussing the composting issues. Establishment of appropriate collection, recycling and composting facilities are considered crucial to the success widespread applications of sustainable biodegradable packaging materials.

Most merchandise leave the production facility with three levels of packaging i.e., Primary, Secondary, Tertiary. Primary packaging is the one which is directly handled at the consumers end. Secondary packaging is the packaging where the individual units are packed together for transportation, or for multipacks sold to the end user. Secondary packaging is mainly used for physical protection of the product. Secondary package provides easy handling during storage or distribution and safety against mechanical damage. Tertiary packaging is used for pallets, trays, cartons. This type of packaging mainly aims to protect the product from the physical and mechanical damage due to transport and weather conditions (Brown 1992; Satheesh Kumar and Yaakob 2011).

## 2 Sustainable Packaging and Materials Used

It is well known that packaging protects the material and economic investment in products and contributes to socio economic development by facilitating the distribution and delivery of products to the consumer. However, once used, packaging contributes to municipal solid waste that is managed at the community level. Effective management of this waste is a challenge in many regional communities and especially in developing or underdeveloped economies. In order to obtain a sustainable materials management, it is essential to create a economically viable, closed loop systems for the recovery of packaging materials. Such a sustainable and efficient strategy supports individuals and communities by offering gainful employment, developing infrastructure for material recovery, conserving resources and reducing environmental impacts. In addition, sustainable approaches such as corporate social responsibility, accountability, and equitable wages are all part of creating a more environmental friendly and safe system.

The use of sustainable packaging ensures the elimination of the contaminant and chemicals that are harmful to the environment, which can be achieved through biodegradable and recyclable packaging. Sustainable packaging also has a lower packaging content, and alternative energy sources such as biofuels, wind, and solar energy are used in the production and transport process. The materials such as Biodegradable plastics, Plant-based plastics, Recycled products, Alternative energy sources, Post-consumer recycled polyethylene bags made from recycled waste, recycled molded packaging for eggs etc. are preferred for sustainable packaging.

The Four Main Types of Sustainable packaging Solutions

Looking at sustainable packaging in general terms, there are four main types of sustainable packaging Solutions.

- Renewable resource-based packaging. Starch-derived PLA (lactic acid) is a material that is 90% biodegraded within nine months when placed in an industrial composting facility. PLA can be by processed from sugar cane, potatoes, corn or other starchy foods.
- Packaging with additives: The packaging market has two types of additives such as OXO and BIO additives which are added to make the packaging degradable.
- Packaging optimization: Packaging reduction, is a prime area where many companies focus on while retaining product protection
- Recycled materials: Recycling materials is a prime and staple process within the sustainable packaging industry.

## 3 Biodegradable Packaging Material

In recent years, the research and development of biodegradable packaging materials from renewable natural resources has received increasing attention, particularly in the developed countries. Researchers have made significant progress in producing biodegradable materials with similar functionality to that of the oil-based synthetic polymers. As the raw materials are sourced from renewable resources and are biodegradable, they would contribute to sustainable development and reduce the environmental impact upon disposal.

Owing to the type of packaging and the content they carry, in the recent years biopolymer are suggested for sustainable packing material with preferable properties such as permeability (gas and vapour), sealing and resistance to chemicals, UV and light, transparency, mechanical properties and machinability (Ashok et al. 2016b). Research shows that many physical and thermal properties like Atomic Layer Deposition, Moisture Bearing property, Biodegradability, Elasticity and Gas barrier property, edible Film coating play an important role in refining biopolymers(Ashok et al. 2016b).

Sustainable packaging is the development of packaging film by utilizing recyclable materials and it involves the use of lifecycle assessments and lifecycle inventories to minimize the ecological footprint and environmental impact of the packaging. The Sustainable Packaging Coalition (http://www.sustainablepackaging.org) defines sustainable packaging as "Beneficial, safe and healthy for individuals and communities throughout its lifecycle; Meeting market criteria for both performance and cost; Sourced, manufactured, transported and recycled using renewable energy; Optimizing the use of renewable or recycled source materials; Manufactured using clean production technologies and best practices; Produced from materials that are safe in all probable end of life scenarios; Physically designed to optimize materials and energy; Effectively recovered and used in biological and/ or industrial closed-loop cycles". Valdes et al. (2014) reported that packaging waste accounted for 29.5% of the total municipal solid waste (MSW) in 2009 in the USA and 25% of the total MSW in Europe in 2006. Various laws have been implemented by governments globally to limit the use of plastics and to reduce the amount of packaging waste that end up In the landfill. In Europe, the Packaging and Packaging Waste Directive (94/62/EC amended by2004/12/EC) is used to reduce packaging waste and to encourage the recovery and recycling of the materials (Guzman et al. 2011).

Packaging waste is currently disposed by landfilling, recycling, incinerating and composting; therefore, there is still much work has to be done to significantly reduce packaging waste present in MSW. The contaminations and their impacts associated with synthetic food packaging and their related concerns causes a negative environmental impact of plastic packaging materials derived from petroleum. This has significantly routed the interests of academia and industry to natural and biodegradable materials (Fama et al. 2010; Bonilla et al. 2013; Chang-Bravo and López-Córdoba 2014; Gonzalez et al. 2016). Apart from direct food wastes, other sectors such as foods, beverages and consumer goods packaging generate more non-eco-friendly plastic wastes and this has resulted into huge impact on the environment. With insufficient prioritization source reduction, recyclability, compostability, recycled content and recycling policies in packaging industry (MacKerron and Hoover 2015), wastes are likely to increase in the coming years. It is estimated that less than 14% of plastic packaging materials are recyclable (MacKerron and Hoover 2015), and as plastic commands the greatest proportion of food packaging industry, the need to design biobased material is a priority. Furthermore, consumers have increased awareness on the consequences of plastic consumption and have diverted their attention towards sustainability, which can generally be achieved at different levels.

## 4 Non Plastic Environmental Friendly Packaging Material

Some of the materials on current demand, technological effectiveness, and having a potential for reducing future problems are Glass, paper and textile materials, which already have been used as packaging materials in different ways due to their abundance. Composites are applied to all materials in order to enhance their properties and to reduce their negative points. The development of technology (Tutus et al. 2016; Costa et al. 2019) population growth and urbanization has led packaging technologies into an alternative—plastics.

- Glass is widely used as a packaging material due to the chemical inactivity of silica and also due to its long term reusability, barrier properties, transparency, strength, and ability to mold easily. However, the breakability quality of glass packaging under shocks, heaviness and its vulnerability under temperature shocks are the negative aspects of using glass for packaging (ASTM 2000, 2004)

- Paper is very popular and trendy among most industries as a cheap, widely higher abundance and flexible packaging material as well as rigid packaging applications. The main reasons for the preference of paper packaging are its availability, durability, eco-friendliness, cost effectiveness, long-term reusability, disposal and biodegradability. However, when it comes to surface properties such as moisture and liquid absorptivity, the paper is not the preferred packaging material (Cho et al. 2008; Tutus et al. 2016).
- Textiles and Composites such as Potatoes, tea leaves, coffee, sugar, and dried cereals are used in textile package synthesis. However, plastics and composite material bags are preferred over cloth bags and their utility has increased within the past few decades as they have enhanced barrier and strengthened property (Costa et al. 2019). Ideally, packaging containing composite materials are used by the beverage and dairy industries to pack soft drinks, fruit juices and dairy products.
- Agro Fiber-Based Plastics: Material scientists are carrying out research with the intention of inventing biodegradable plastics to deal with non degradable conventional plastics. Some properties such as matrix breakdown from sunlight exposure have already been achieved with starch plastic combination. Yet 100% achievement in terms of breakdown under the sunlight is to be established (Sarker et al. 2012).
- Bacteria-Based Plastics: Bacteria-based plastics where PHA polymer chains which are synthesized inside bacteria cells are used to make films. In these the mechanical properties can be changed depending on the intention and the type of bacteria and growth condition can influence the polymer outcome. PHA enhances the biodegradability, thermal resistance property and mechanical flexibility of a packaging material (Chen and Patel 2011). The use of genetically modernized bacteria to produce biodegradable plastics via cell metabolic reactions is being recently researched however, it is deemed to be cost consuming for the industries.
- Lignin-Based Plastics Lignin-based plastic is another low cost renewable biodegradable plastic option and it enhances the compatibility of the matrix polymer and UV stability (Hadad et al. 2005, Fabbri et al. 2005 and Cazón et al. 2017). Lignin has resulted in high modulus value from the reinforcing action and thermal stability. It exhibits higher thermal insulation values than the other materials.
- Soy-Based Plastics Soy-based plastic, an alternative for biodegradable plastics has a limited number of proteins with fats and oils, which allow soy to become molded into films and plastic material. It has commonly been used in food coatings and vehicle applications but it is unpopular due to its low availability (Chua et al. 1999).
- Natural Fiber Reinforced Plastics Nowadays, natural fiber reinforced plastics are used due to their availability, low cost, high modulus value, abundance, low density, the high number of reactive ends on the surface, ease of the process and high biodegradability. Bamboo, flax, hemp, pineapple, banana fibre and kenaf are added as reinforcing agents for the polymer matrix showing increased heat resistivity however, they exhibit a brittle property for the tensile test. Also, high

flame retardance is enhanced by adding phenol novolac and aluminum hydroxide in PLA. Carbon dioxide absorption ability of natural fibers (Ramos et al. 2016; Sarker et al. 2012; Chen and Patel 2011); non-renewable raw material usage and accumulation of non-biodegradable materials causes an environmental problem.

## 5 Biopolymer Based Packaging Materials

There is a great demand for plant-based materials as an alternative to conventional single-use plastic in food packaging and are known to reduce the effect of the environmental footprint on the packaging industry. Bio-based plastics from organic polymers, paper and pulp are gaining importance as alternatives to petroleum-based packaging. Just like their counterparts like PHB, bioplastic materials made from biomass waste, oleochemicals waste and like-plastic plant-based materials, the Polylactic acid (PLA) is an excellent substitute for plastic made from fermented plants and are proven to be biodegradable. Polysaccharide-based polymers are recently found to be more suitable for use in coatings films and composites—such as edible films or coatings for fresh food packaging to increase the product shelf-life. However, cellulose-based packaging is gaining importance in food packaging due to its excellent properties and cost-effective solutions as well as reducing environment footprint compared to fossil-based polymers.

Several biopolymer sources have been researched and experimented as reviewed by Siracusa and Rosa (2018). Through the use of recycled and renewable raw materials and energy efficient production process and minimal waste disposal, sustainability can be achieved. The end product should be biodegradable and/or compostable thereby contributing to reduced municipal solid waste problem. Biodegradable polymers are polymers that undergo decomposition into $CO_2$, $CH_4$, $H_2O$, inorganic compounds or biomass through predominantly the enzymatic action of microorganisms. Some of these polymers can also be compostable, where decomposition takes place in a compost site at a rate consistent to known compostable materials (Siracusa et al. 2008; Song et al. 2009). The most prominent polymers are extracted directly from biomass such as polysaccharides obtained from starches of potatoes, rice, corn, maize, and wheat, from hemicelluloses of barley; from gums of guar, alginate, carrageenan and pectin, and from chitosan and chitin. Additionally polymers extracted from animal proteins (such as casein, whey, collagen, and casein) and from plant proteins (such as zein, soy, and gluten) are also in use. Synthesized monomers from renewable resources (named bioderived monomers) such as polylactic acid, bio-polyethylene terephthalate, and bio-polyolefins such as bio-polyethylene are also used as polymers. Some polymer are directly obtained from microorganisms such as the family of polyhydroxyalkanoates e.g., polyhydroxybutyrate, polyhydroxyvalerate, and polyhydroxybutyrate-co-valerate copolymers. Biodegradable materials obtained from monomers of petroleum resources such as polybutylene adipate and its copolymers with polyethylene terephthalate, polybutylene succinate and its copolymers with polybutylene adipate, polycaprolactone, polyglycolic acid, and polypropylene carbonate can considered as the fourth category (Robertson 2013; Siracusa 2016). Recently biopolymers stemming from the waste streams generated during agro-industrial processing have emerged as suitable candidates for playing the role of potential carrier. They also present a double benefit through the promotion of sustainable raw materials but also limit the health impact of synthetic plastics as color additives, BPA entry into the food, thereby limiting the use, disposal and accumulation of conventional plastics in the environment (Jose et al. 2019). Nanocellulosic fibers are promising biopolymers which are advantages in all aspects of sustainability. These materials are mostly suitable for food packaging, medical and pharmaceutical packaging, industrial packaging and other potential areas.

Sustainable packaging materials from biopolymers are effective in minimal waste generation (Davis and Song 2006; Lavoine et al. 2014), maximum functionality (Moon et al. 2011) and are proven to be cost effective (Khalil et al. 2012). These materials exhibit efficiency in maximizing product packaging ratio and materials used and are environmental friendly in terms of recyclability (Abou-Yousef et al. 2005; Davis and Song 2006; Khalil et al. 2014), reusability and biodegradability (Moon et al. 2011). The biopolymer based packaging materials minimize airborne and waterborne emission to the environment, minimize green house gas emission and reduce toxicity and litter impacts (Khalil et al. 2014).

## 6 Composting and Disposal of Biodegradable Packaging Materials

The compositions of sustainable packaging material are very different from conventional one and are designed to degrade after use hence they are generally considered unsuitable for the conventional plastics recycling methods. These are however more suitable for composting advantageously as a form of materials recovery (European Plastics News 2001). The various compositions of packing material derived from biopolymers have various stages of biodegradability or recyclability and these varies from one material to another. Though all biopolymers are generally unsuitable for recycling once biodegradation has been triggered as these materials vary significantly in biodegradability and they may or may not be readily compostable. Another issue is that the biodegradable packaging is generally unsuitable for landfill (due to the requirements of Article 5 of the EU Landfill Directive seeking to reduce biodegradable materials sent to landfill because of their propensity to release methane under anaerobic conditions) or conventional recycling. Though there is no clear reason as to why biodegradable packaging materials cannot be collected with other plastic packaging for incineration, the growth in biodegradable plastics complicates the identification and sorting process due to its similaylty with conventional plastic. These materials contain reasonable GCV and do not produce hazardous emissions.

To ensure material separation, strong clear labelling is required so that such products can be easily identified.

Biodegradable packaging are most suitable for domestic and municipal composting and should be separated from other non-biodegradable packaging and collected as organic waste at household level for composting. In order to facilitate composting, an infrastructure must be established to classify biodegradable packaging materials, separate them and to treat them as organic wastes. By using local or regional composting facilities, the total waste that reaches the landfill can be reduced, in addition to the reduction in transport cost and associated emissions.

In summary, biodegradable packaging broaden the range of waste management treatment option over traditional plastics and this is supported by Life Cycle Assessment (Murphy and Bartle 2004). The most favoured end-of-life disposal options for these materials are domestic and municipal compositing. Methods that are used recycling of conventional polymers are generally not suitable for biodegradable polymers but the value of the materials may be recovered in the form of useful compost. Biodegradable polymers can thus make significant contributions to material recovery, reduction of landfill and utilisation of renewable resources. Widespread public awareness of the material and effective infrastructure for stringent control of certification, collection, separation and compositing are crucial to materialize the benefits.

The recent technological innovations has led to intelligent packaging where the packaging system is capable of carrying out intelligent functions such as detecting, sensing, recording, tracing, communicating and applying scientific logic. This helps in the decision making regarding the extension of shelf life, enhancement of safety and quality, provision of information and trouble shooting (Yam and Takhistov 2005; Kerry et al. 2006). Of late, active and smart packaging such as QR-codes, real-time quality sensors, time-temperature tags and labels or atmospheric gas indicators are used to extend shelf-life of the product and thereby reduce food waste. This innovations seems to substantially improve the food inventory management and in turn increases sustainable business operations.

Smart packaging is the new way to address key challenges in food consumption, environmental sustainability in terms of food and packaging waste and to reduce the environmental footprint of the packaging industry. As new digital technologies coupled with the existing packaging methods in terms of its functionality, traceability, consumer communication, Smart packing technology must coexist towards reaching the sustainability goal.

In an attempt to reduce paper and plastic waste especially in the ready to eat joints, edible packaging has become a new trend. However, its creation and use, storage and shelf life are a great concern specifically in terms of hygiene and logistical issues. Spoons, cups and plates made of edible material like wheat based dough or biscuits are in vogue.

Innovations in sustainable packaging aims to address food waste and food material loss by extending the shelf-life, preserving food quality, as well as meeting food safety standards issues by preventing health issues.

## 7 Conclusion

Reduction in the amount of packaging and use of eco-friendly packaging material provide an attractive opportunity to promote environmental sustainability. Sustainable packaging from the raw material, to process and its biodegradability is an environmentally sound advice. It not only is cost effective for both the consumers and manufacturers but it also helps to ensure a cleaner environment for the future generations.

An LCA of the packaging industry shows that climate change, cumulative energy demand, minerals and fuels consumption, Photochemical oxidation, Eutrophication, Land use water use and soild waste are the environmental impacts caused. According to Silvenius et al. (2011), 2–5% of the total environmental impact could be attributed to the packaging industry to be responsible for greenhouse emission, eutrophication, and acidification impacted categories Though several issues discourage the implementation of sustainable packaging usage, recent awareness and consumer attitude has increased the demand for such biodegradable packages. Biopolymers were developed in an environmental perspective and they exhibit excellent competitive properties (temperature resistance, thermal conductance, oxygen transmission rate tensile strength and water vapor transmission rate) when compared to other materials. Biodegradable packaging has reduced the shelf life of plastics and enhanced plastic recycling programs. Additionally, the bulk material abundance, renewability, and recyclability provide better options to the biopolymer industry. Green Packaging and Sustainable packaging has increased in popularity in the last few years. This shift in wave of popularity is due the increased awareness on environmental sustainability, move towards combating climate change and protection of vulnerable ecosystems. Though many packing firms now operate in support of sustainable packaging in the trend of circular economy, more innovations and sustainable packing solutions are expected to boom in the coming decades.

## References

Abou-Yousef H, El-Sakhawy M, Kamel S (2005) Multi-stage bagasse pulping by using alkali/Caro's acid treatment. Ind Crops Prod 21(3):337–341

Andrady AL, Neal MA (2009) Applications and societal benefits of plastics. Philos Trans R Soc B: Biol Sci 364(1526):1977–1984

Ashok A, Rejeesh CR, Renjith R (2016a) Biodegradable Polymers for Sustainable Packaging Applications: A Review. Int J Bionics Biomater 2(2):1–11

Ashok A, Mathew M, Rejeesh CR (2016b) Innovative value chain development of modified starch for a sustainable environment: a review. Int J Polym Sci Eng. 2(1):20–32

ASTM (2000) Standard test methods for water vapor transmission of materials. ASTM Annual Book of ASTM Standards 2000:907–914

ASTM C 1048 (2004) Standard specification for heat-treated flat glass-kind HS. Kind FT Coated and Uncoated Glass

Bonilla J, Talón E, Atarés L, Vargas M, Chiralt A (2013) Effect of the incorporation of antioxidants on physicochemical and antioxidant properties of wheat starch–chitosan films. J Food Eng 118 (3):271–278

Brown WE (1992) Plastics in food packaging: properties, design and fabrication. Marcel Dekker, New York, pp 66–102

Cazón P et al (2017) Polysaccharide-based films and coatings for food packaging: A review. Food Hydrocolloids 68:136–148. https://doi.org/10.1016/j.foodhyd.2016.09.009

Chang-Bravo L, López-Córdoba A (2014) Martino M, Biopolymeric matrices made of carrageenan and corn starch for the antioxidant extracts delivery of Cuban red propolis and yerba mate. Reactive Funct Polym 85:11–19

Cho B, Ryu J, Song B (2008) Pilot study on the manufacture of kraft paper from OCC pilot study on the manufacture of kraft paper from OCC. J Korea Tech Assoc Pulp and Paper Indus 40(5):27–35

Chen G, Patel MK (2011) Plastics derived from biological sources: present and future: a technical and environmental review. Chem Rev 112(4):2082–2099. https://doi.org/10.1021/cr200162d

Chua H, Yu PHF, Ma CK (1999) Accumulation of biopolymers in activated sludge biomass. Appl Biochem Biotechnol 78(1–3):389–399. https://doi.org/10.1385/abab:78:1-3:389

Costa JL, Legrand V, Freour S (2019) Durability of composite materials under severe temperature conditions: influence of moisture content and prediction of thermo-mechanical properties during a fire. J Compos Sci 3:55. https://doi.org/10.3390/jcs3020055

Davis G, Song J (2006) Biodegradable packaging based on raw materials from crops and their impact on waste management. Ind crops and Prod 23(2):147–161

European Plastics News (2001) Nat Select 28(5):37–38

Fabbri C, Bietti M, Lanzalunga O, Chimica D (2005) Generation and reactivity of ketyl radicals with lignin related structures. on the importance of the ketyl pathway in the photoyellowing of lignin containing pulps and papers. J Organ Chem 70(7):2720–2728. https://doi.org/10.1021/jo047826u

Fama L, Bittante AMBQ, Sobral PJA, Goyanes S, Gerschenson LN (2010) Garlic powder and wheat bran as fillers: their effect on the physicochemical properties of edible biocomposites. Mat Sci Eng: C 30(6):853–859

Gonzalez Seligra P, Medina Jaramillo C, Famá L, Goyanes S, (2016) Biodegradable and non-retrogradable eco-films based on starch–glycerol with citric acid as crosslinking agent. Carbohydrate Polym 138:66–74

Guzman A, Gnutek N, Janik H (2011) Biodegradable polymers for food packing- factors influencing their degradation and certification types-a comprehensive review. Chem Chem Technol 5(1):115–123

Hadad D, Geresh S, Sivan A (2005) Biodegradation of polyethylene by the thermophilic bacterium Brevibacillus borstelensis. J Appl Microbiol 98(5):1093–1100. https://doi.org/10.1111/j.1365-2672.2005.02553.x

Kerry JP, O'Grady MN, Hogan SA (2006) Past, current and potential utilisation of active and intelligent packaging systems for meat and muscle-based products. A Rev Meat Sci 74 (1):113–130

Khalil HPSA et al (2014) Production and modification of nano fibrillated cellulose using various mechanical processes : a review. Carbo hydr Polym 99:649–665

Khalil HPSA et al (2012) Bamboo fibre reinforced biocomposites :a review. Mater Des 42:353–368

Lavoine N et al (2014) Elaboration of a new anti bacterial bio-nano-material for food-packaging by synergistic action of cyclo dextrin and micro fibrillated cellulose. Innov Food Sci Emerg Technol 26:330–340

MacKerron CB, Hoover D (2015) Waste and opportunity, environmental progress and challenges in food, beverage, and consumer goods packaging report. The Natural Resources Defense Council (NRDC), New York City

Moon RJ et al (2011) Cellulose nano materials review: structure, properties and nanocomposites. ChemSocRev 40(7):3941–3994

Murphy R, Bartle I (2004) Summary report, biodegradable polymers and sustainability: insight from life cycle assessment. National Non Food Crops Centre, UK

Robertson GL (2013) Edible, biobased and biodegradable food packaging materials. In: Food packaging principles and practice, 3rd ed, CRC Press, Taylor and Francis Group, Boca Raton, London, New York (Chapter 3), pp 49–90

Ramos M, Valdes A, Beltran A, Garrigós M (2016) Gelatin-based films and coatings for food packaging applications. Coatings 6(4):41. https://doi.org/10.3390/coatings6040041

Rhim J-W, Ng PK (2007) Natural biopolymer-based nanocomposite films for packaging applications .Crit Rev Food Sci Nutr 47(4):411–433

Sarker M et al (2012) Chemical products produced from high-density polyethylene (HDPE) waste plastic. Int J Mat Sci 2(3):56–62

Satheesh Kumar MN, Yaakob Z (2011) Siddaramaiah, Biobased materials in food packaging applications, In: Pilla S (ed) Handbook of bioplastics and biocomposites engineering applications, pp 121–159

Silvenius S, Katajajuuri JM, Gronman K, Soukka R, Koivupuro HK, Virtanen Y (2011) Role of packaging in LCA of food products. Towards Life Cycle Sustainability Management. Springer, Netherlands, pp 359–370

Siracusa V, Rocculi P, Romani S, Dalla Rosa M (2008) Biodegradable polymers for food packaging. Trends Food Sci Technol 19(12):634–643

Siracusa V (2016) Packaging material in the food industry. In: Barros-Velazquez J (ed) Antimicrobial Food Packaging. Academic Press, Elsevier, Cambridge, MA, pp 95–106 (Chapter 7)

Siracusa V, Rosa MD (2018) Sustainable packaging in sustainable food systems from agriculture to industry improving production and processing. Acad Press, Elsevier, pp 275–307

Song JH, Murphy RJ, Narayan R, Davies GBH (2009) Biodegradable and compostable alternatives to conventional plastics. Philosop Trans Royal Soc B 364(2009):2127–2139

Thompson RC, Swan SH, Moore CJ et al (2009) Our plastic age. Philos Trans R Soc B: Biol Sci 364:1973–1976

Tutus A, Karatas B, Cicekler M (2016) Pulp and paper production from hemp by modified kraft method. In: Presented at the 1st international mediterranian science and engineering congress, Adana, Turkey

Valdés A et al (2014) Natural additives and agricultural wastes in biopolymer formulations for food packaging. Front Chem 2:6

Wicochea-Rodríguez Jose Daniel, Chalier Pascale, Ruiz Thierry, Gastaldi Emmanuelle (2019) Active food packaging based on biopolymers and aroma compounds: how to design and control the release. Frontiers in Chemistry 7(398):1–16

Yam KL, Takhistov PT (2005) Miltz J, Intelligent packaging: concepts and applications. J Food Sci 70(1):R1–R10

Youssef AM, El-Samahy MA, Rehim MHA (2012) Preparation of conductive paper composites based on natural cellulosic fibers for packaging applications. Carbo hydr Polym 89(4):1027–1032

# Influence of Sugar Mill Effluent on Germination and Growth of Fenugreek (*Trigonella foenumgraecum*) and Mustard (*Brassica nigra*)

**A. Arivoli, Sathiamoorthi Thangavelu, and X. Agnello J. Naveen**

**Abstract** The study is to understand the physico-chemical characteristics and different applications of sugar mill effluent considered compared with (permissible limits) standards given by the Tamilnadu pollution control board (TNSPCB) towards treated effluent water for agricultural irrigation purpose from a point source. The effluent was used in the application of seed germination and seedling growth on fenugreek *(Trigonella foenumgraecum)* and mustard (*Brassica nigra*) in five different concentrations (10, 25, 50, 75, and 100%). The seeds of fenugreek and mustard shows maximum germination of 95 and 80% growth in effluent irrigated soil. The result shows water and the soil irrigated with 10% effluent were found most suitable for germination. The morphological parameters such as germination percentage, shoot length, root length, fresh weight, and dry weight of seedlings and seed vigor index were calculated. The seed germination and growth parameters increased in lower (10%) concentration of sugar mill effluent and these morphological parameters gradually decreased with increasing effluent concentration. The less amount (10%) concentration of sugar mill effluent may be used for irrigation purposes.

**Keywords** Phytoremediation · Vigor index · *Trigonella foenumgraecum* · *Brassica nigra* · Seed germination · Sugar mill effluent

## 1 Introduction

Sugar cane is a seasonal crop and widely grown in India, considering the production of the crop the economic value is high and there are 550 sugar mills with a

A. Arivoli (✉)
Department of Environmental Science, Government Arts College, Ariyalur, India
e-mail: arivoli82@gmail.com

S. Thangavelu
Department of Microbiology, Alagappa University, Karaikudi, India

X. A. J. Naveen
Department of Environmental Science, Thanthai Roever Institute of Agriculture and Rural Development (TRIARD), Perambalur, India



M. Prashanthi et al. (eds.), *Bioremediation and Green Technologies*, Environmental Science and Engineering, https://doi.org/10.1007/978-3-030-64122-1_19

production capacity of 13.5 million tons cane per year and total sugarcane production 220 million tons per year (Kaur et al. 2010) fully operational for 150–210 days in a year (November to May) (Kolhe et al. 2009).And other additional materials like molasses, bagasse, and chemicals are produced for alcohol production. The impact of the sugar industry effluents is causing odor during decomposition. And contains high Biological Oxygen Demand (BOD) and Chemical Oxygen Demand (COD) which reduce the available oxygen supply into the water and adversely affect the aquatic ecosystems (Jagannathan et al. 2014). Due to this, it has been added in the Ultra-red category (high source of water-polluting industry) by central pollution control board. The factors are a High amount of organic and inorganic compounds, like a high amount of suspended solids and dissolved solids that reduce the penetration capability of light into the water (Singh and Bhati 2005). The organic compounds percent in the sugar effluent water has beneficial nutrients add with press muds or filter muds gives fertilizer to promote the development of plant-like cane cutting and also used composting material (good organic fertilizer) in enhancing agricultural viability (Kumar and Chopra 2012). Sugar effluent by-products were considered as a waste which can be recycled (Almodares and Sharif 2007). The effluent directly from the sugar industry cannot be used for agricultural irrigation uses. And farmers (stakeholders) are the real sources of raw material like sugar cane, so every industry has to do a different kind of research and development scheme to enhance the sufficient supply of raw material source like soil fertility for better germination, Most crops gives higher potential yield with less wastewater irrigation, reduce the usage of chemical fertilizers, resulting in net cost savings to farmers, giving cane cutting, etc., The focus of the research problem is a mutual concern between farmers and sugar industry top-level management to satisfy the farmers by enhancing the yield of the crop in utilizing the organic matter from the effluent utilizing concept like bioremediation, phytoremediation, press muds, filter cake, cane cutting, treatment of effluent for irrigation there is a proper application(methods) to reconcile in practices (Vinod et al. 2010). The review of past studies gives a systematic idea on usage of Phytoremediation than mechanical and technological process Siddiqui and Waseem (2012) studied the effect of treated and untreated sugar industry water on bajra, bram, maize, and wheat. And the same type of study has been done by other (Damodharan and Reddy 2012) for Cuddalore in *Arachis hypogaea* (Peanut), *Vigna radiata* (Green gram) (Santhi and Pandian 2012), *Vigna angularis*, Vigna cylindrical, and *Sorghum cernum* (Doke et al. 2011), wheat, garden pea, black gram, and mustard (Nath et al. 2007), paddy (Ramkrishan et al. 2001; Samuel and Muthukkaruppan 2011). *Trigonella foenumgraecum* (fenugreek) is an annual herb of the Leguminosae family. The effluents which are approaching from the sugar industry are directly used for agricultural practices, i.e., irrigation purpose will affect the soil fertility as well as in plant growth and seed germination. The present study deals with the analyses of sugar mill effluent and its efficacy on seed germination and growth of fenugreek and mustard in an organic way.

# 2 Materials and Methods

## 2.1 Analysis of Sugar Mill Effluent

The effluent sample was collected in precleaned, sterilized plastic containers from the outlet of the sugar industry Cuddalore District, (Cuddalore sugar mills) during January 2018 and March 2019 were stored at 4 °C till further investigation. Physico-chemical parameters, such as colour, odour, pH, temperature, electrical conductivity (EC), total solids (TS), total dissolved solids (TDS), total suspended solids (TSS), chemical oxygen demand (COD), biochemical oxygen demand (BOD), total hardness (TH), total alkalinity (TA) and chloride ($Cl^-$) were measured using standard methods (APHA 2005).

## 2.2 Preparation of Different Concentrations of Effluent

The sugar mill effluent sample collected from the outlet of sugar mill industry was treated as 100% raw effluent. The different concentrations (10, 25, 50, 75 and 100%) of sugar mill effluent were prepared freshly using distilled water whenever necessary (Lakshmi and Sundaramoorthy 2000). The concentrations were used for germination studies.

## 2.3 Germination Studies

The healthy and uniform sized *Trigonella foenum-graecum L. (Fenugreek) and Brassica nigra* (Mustard) seeds were selected and surface sterilized with 0.1% $HgCl_2$ for 2 min and then thoroughly washed with tap water. Twenty seeds were placed equidistantly in plastic trays filled with 100 g sterilized soil [clay + sand (1:1)]. The seeds were irrigated with equal quantity of different concentrations of effluent and the seeds irrigated with distilled water were treated as control. Five replicates were preserved for each treatment including control. The seed germination percentage, shoot length, root length, seedling fresh weight and seedling dry weight (Milner and Hughes 1968; Sundaramoorthy et al. 2006; Sajani and Muthukkaruppan 2011) were taken and seed vigor index, documented.

## 2.4 Germination

The number of seeds germinated in each concentration was analysed on the 15th day and the germination percentage was calculated using the following formula.

$$\text{Germination percentage} = \frac{\text{Number of seeds germinated}}{\text{Total numbers of seeds sown}} \times 100 \quad (1)$$

***Shoot and root length***

Five seedlings were taken from each treatment and their shoot length and root length were determined by using a cm scale and these values were recorded.

***Fresh weight***

Five seedlings were collected from each treatment and their fresh weights were considered with the help of an electrical single pan balance.

***Dry weight***

The same seedlings used for fresh weight were kept in hot air oven at 80 °C for 24 h. The seedlings were taken from the oven and kept in desiccators for some time. The dry weights were taken using an electrical single pan balance.

***Vigor index***

Vigor index of the seedling were calculated by using the formula proposed by (Aery 2010) (Fig. 1).

$$\text{Vigo index} = \text{Germination percentage} \times \text{Length of seedling} \quad (2)$$

**Fig. 1** **a** Different concentration of sugar mill effluent in control **b** Fenugreek seed germination **c** Mustard seed germination **d** After 15th day of seedling grown

## 3 Results and Discussion

The Physico-chemical analysis of sugar mill effluent is given in Table 1. Some of the parameters of the effluent were found to exceed by the TNSPCB given standard towards irrigated water. The analyses of sugar mill effluent showed that it is acidic with a white color. It showed a high value of Biological Oxygen Demand (BOD) and Chemical Oxygen Demand (COD). The presence of considerable amounts of calcium, chloride, and sulfate, were also observed in the effluent (Lakshmi and Sundaramoorthy 2001; Ayyasamy et al. 2008; Rath et al. 2010, Sajani and Muthukkaruppan 2011, Sarala Thambavani and Sabitha 2012; Hussain et al. 2013; Saurabh and Shailja 2014; Elayaraj 2014). The pollution load of the effluent depends upon the nature of raw materials, chemicals used, the processes involved in the factory, and also the methods of treatments given to the effluent before they are discharged from the factory.

The pH was comparatively low due to the use of phosphoric acid and sulfur dioxide during the clarification of sugar cane juice (Manivasakam 1987; Palharyal et al. 1993). The pH is an important factor in the formation of algal blooms that

**Table 1** Physico-chemical characteristics of sugar industry effluent

| Parameters | Effluent | Different concentration (%) | | | | | | TNPCB (2009) |
|---|---|---|---|---|---|---|---|---|
| | | 10 | 25 | 50 | 75 | 100 | Control | |
| Temperature (°C) | 28.2 | 28.5 | 22.8 | 22.8 | 28.8 | 28.1 | 28.2 | 40 |
| Colour | Brownish | Yellowish | Brown | Brown | Brown | Brown | – | Colourless |
| pH | 6.5 | 7.2 | 7.37 | 7.43 | 7.55 | 7.72 | 6.99 | 6.5–8.5 |
| EC (mS) | 2.21 | 1.63 | 1.86 | 2.17 | 2.48 | 3.26 | 1.56 | – |
| TS (mg/L) | 3900 | 2000 | 2300 | 2500 | 2800 | 3800 | 2000 | 2100 |
| TDS (mg/L) | 2700 | 1000 | 1000 | 3000 | 2000 | 3000 | 1000 | 200 |
| TSS(mg/L) | 1200 | 1000 | 13000 | 500 | 800 | 800 | 1000 | 200 |
| Chloride (mg/L) | 325 | 280 | 300 | 310 | 310 | 320 | 260 | 200 |
| Hardness (mg/L) | 380 | 260 | 340 | 350 | 355 | 375 | 180 | 600 |
| Ca (mg/L) | 160 | 160 | 180 | 200 | 200 | 220 | 130 | – |
| Mg (mg/L) | 120 | 100 | 160 | 150 | 155 | 150 | 50 | – |
| Alkalinity (mg/L) | 200 | 75 | 125 | 200 | 375 | 550 | 100 | – |
| $PO_4$(mg/L) | 28 | 18 | 20.6 | 22.5 | 24 | 26 | 12.5 | 10 |
| COD (mg/L) | 780 | 320 | 420 | 480 | 580 | 620 | 180 | 250 |
| DO(mg/L) | 4.52 | 3.24 | 3.64 | 3.64 | 3.64 | 3.7 | 1.21 | – |
| BOD (mg/L) | 450 | 180 | 220 | 240 | 280 | 380 | 80 | 30 |

makes the water unfit for irrigation; if the same water is used for irrigation over a large area, the soil becomes acidic resulting in poor crop growth and yield. Similarly, the effluent had a very high TDS, which was in agreement with the previous report of sugar factory effluent (Abdul and Sirajudeen 2006). The COD value of the effluent was 1330 mg/L, while the recommended level set by BIS is 250 mg/L; the measured COD indicates the high organic load. The remaining measured parameters are within the permissible limits.

Seed germination and seedling growth are vital for maintenance of plant life and they are particularly vulnerable to environmental stress. Since germination is the first physiological process, several growth parameters such as percentage of germination (Mishra and Pandey 2002). In the present investigation, the effects of different concentrations of sugar mill effluent on seed germination of fenugreek and mustard were reported. The highest values of seed germination percentage (90 and 80%) were recorded at 10% of sugar mill effluent concentration and the lowest values of germination percentage (45 and 50%) were recorded at 100% of sugar mill effluent concentration (Table 2). The increase in germination percentage over control at lower concentrations (10%) indicates the due to increase usage of effluent treatment process as reported by (Vinod 2014; Vaithiyanathan et al. 2014; Suresh et al. 2014). The lower concentration of effluent has many nutrients values like nitrogen, phosphorous, etc., which might promotes in enhancement of plants growth as suggested by (Augusthy and Annsherin 2001). At the same time, the higher concentrations of sugar mill effluent inhibited the germination of fenugreek. Large amount of organic and inorganic substances are presented in higher concentration of sugar mill effluent which adversely affects the seed germination process because of the higher salt content which causes changes in osmotic pressure outside of the seedling. It also decreases the water absorption ability of the seed and inhibited the seed germination process (Adriano et al. 1973).

The fenugreek seed grown under different concentration of sugar mill effluent of lowest concentration 10% of shoot length (37.98 cm/seedling), root length (6.12 cm/seedling), fresh weight (0.865 gm/seedling) and dry weight (0.172 mg/seedling) were recorded. The concentration (100%) of sugar mill effluent

**Table 2** Seed germination percentage of fenugreek and mustard grown under different concentrations of sugar mill effluent

| Effluent concentration (%) | Seed germination percentage (%) | | Germination (Hour) | |
|---|---|---|---|---|
| | Fenugreek | Mustard | Fenugreek | Mustard |
| Control | 90 | 65 | 24 | 24 |
| 10 | 95 | 80 | 24 | 24 |
| 25 | 80 | 70 | 24 | 24 |
| 50 | 70 | 75 | 24 | 24 |
| 75 | 85 | 70 | 24 | 24 |
| 100 | 45 | 50 | 24 | 24 |

**Table 3** Shoot length and root length of fenugreek and mustard grown under different concentration of sugar mill effluent (Mean ± dev)

| Effluent concentrations (%) | Fenugreek | | Mustard | |
|---|---|---|---|---|
| | Root length (cm) | Shoot length (cm) | Root length (cm) | Shoot length (cm) |
| Control | 5.98 ± 1.05 | 32.14 ± 1.16 | 9.72 ± 1.13 | 36.06 ± 0.71 |
| 10% | 6.12 ± 0.56 | 37.98 ± 4.94 | 9.82 ± 0.64 | 43.75 ± 1.15 |
| 25% | 5.64 ± 0.92 | 30.46 ± 0.58 | 9.71 ± 1.04 | 39.98 ± 1.02 |
| 50% | 5.18 ± 0.78 | 26.96 ± 1.45 | 9.62 ± 0.69 | 35.12 ± 0.93 |
| 75% | 3.78 ± 1.26 | 25.68 ± 1.18 | 10.1 ± 0.57 | 32.64 ± 1.14 |
| 100% | 1.04 ± 0.58 | 1.52 ± 0.19 | 8.4 ± 0.27 | 0.95 ± 0.14 |

**Table 4** Fresh weight and Dry weight of fenugreek and mustard grown under different concentration of sugar mill effluent

| Effluent concentrations (%) | Fenugreek | | Mustard | |
|---|---|---|---|---|
| | Fresh weight (gm) | Dry weight (gm) | Fresh weight (gm) | Dry Weight (gm) |
| Control | 0.831 | 0.164 | 0.112 | 0.050 |
| 10% | 0.865 | 0.172 | 0.164 | 0.080 |
| 25% | 0.745 | 0.145 | 0.081 | 0.090 |
| 50% | 0.658 | 0.132 | 0.123 | 0.070 |
| 75% | 0.625 | 0.157 | 0.101 | 0.060 |
| 100% | 0.595 | 0.128 | 0.120 | 0.050 |

and the lowest shoot length (1.04 cm/seedling), root length (1.52 cm/seedling), fresh weight (0.595 mg/seedling) and dry weight (0.128 mg/seedling) were observed in higher concentration (100%) of sugar mill effluent (Tables 3 and 4).

The mustard seed is grown under different absorption of sugar mill effluent of lowest concentration 10% of shoot length (36.06 cm/seedling), root length (9.72 cm/seedling), fresh weight (0.164 gm/seedling), and dry weight (0.080 mg/seedling) were observed 10% concentration. The lowest of shoot length (1.04 cm/seedling), root length (1.52 cm/seedling), fresh weight (0.595 mg/seedling) and dry weight (0.128 mg/seedling) were observed in higher concentration (100%) of sugar mill effluent. The presence of various pollutants in lower concentrations of sugar mill effluent (at 10%) has increased the growth and development (Lakshmi and Sundaramoorthy 2000; Saxena and Madan 2012; Ali et al. 2012; Malik et al. 2014). The high concentrations of sugar mill effluent inhibited the root and shoot length of seedlings. It limits a higher amount of dissolved solids and suspended solids to percolate, interfere, and reduced the absorption of some other nutrients. The interference of heavy metals decreases the root and shoot length of the plants. And might be due to the involvement of physiological processes of plant and it also occupies in inhibition of enzyme activities. And affects the nutrition, water difference and

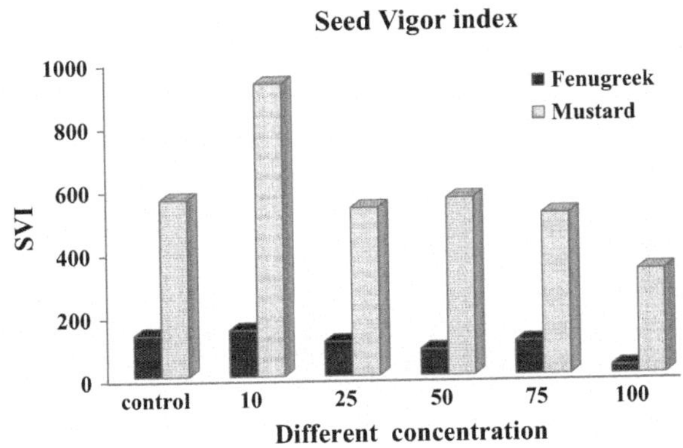


**Fig. 2** Impact of sugar mill effluent of different concentration on seedling vigor index

alters the different hormonal status in membrane permeability (Sharma and Dubey 2005). The fresh and dry weight of seedlings improved in 10% of sugar mill effluent concentration while decreasing at a higher concentration of sugar mill effluent. Some amount of nutrients and trace elements may be needed for seeds. The lower concentration of sugar mill effluent contained requires a high amount of nutrients which develops the growth of seedlings as well as fresh and dry weight of the seedlings. The required amount of various chemicals presented in the lower concentration of sugar mill effluent promotes plant growth (Lakshmi and Sundaramoorthy 2000; Siva and Suja 2012). The higher concentrations of sugar mill effluent reduced the fresh weight and dry weight of seedlings. The decrease of seedling weight may be due to the poor growth of seedlings under the higher concentrations of effluent irrigation.

The highest vigor index (925) and (147) of fenugreek and mustard seedlings were observed in 10% of effluent treated seedling and lowest vigor index (328) and (24.5) were observed in 100% of effluent (Fig. 2). The presence of different pollutants in lower concentrations of sugar mill effluent (at 10%) increased the growth and development. These observations were conformed to (Lakshmi and Sundaramoorthy 2000; Saxena and Madan 2012; Ali et al. 2012; Malik et al. 2014).

## 4 Conclusion

The Physico-chemical parameters such as TDS, BOD, and COD were proved to be higher in sugar mill effluent and it severely affects plant growth. From this study, it concluded that Physico-chemical parameters like pH, electrical conductivity, total solids, dissolved solids, suspended solids, chloride, hardness, calcium, magnesium, COD, BOD, sulphate, and TDS were relatively higher in the sugar industry effluent and severely affected seed germination. There was a gradual decrease in the percentage of seed germination and seedling growth in vigor index value with Sugar mill effluent concentration. The untreated effluent could lead to soil fertility deterioration like NPK value and lower the productivity of yielding. The results show a lower concentration of sugar mill effluent of (10%) stimulates the seed germination percentage, vegetative growth, and biochemical in fenugreek and mustard seedlings and higher concentrations of sugar mill effluent shows the seed germination percentage, vegetative growth, and vigor index of seedlings is compared to the control results. It can also be concluded that toxic metals are essential for seedling and have a higher concentration of nutrients value for using has an organic fertilizer due to high toxic nutrient the lower survivability and sustainable towards the potential of seedling growth, the germination process is seen by diluting to low the concentration of effluent like thickness, moisture, classification of the particle (nano), breaking of nutrients in structure is calculated and given to the seedlings holding potency or capability. And have higher nutrient value and avoid risk factors on the germination of seedling to enhance the potentiality of organic fertilizer. However, In lower concentration of effluent can be utilized for agricultural fertilizer after suitable treatment with proper dilution.

## References


Abdul JA, Sirajudeen J (2006) Risk assessment of physicochemical contaminants in ground water of Pettavaithalai area, Thiruchirappalli, TamilNadu, India. Environ Monitor Assessment 123:299–312

Adriano DC, Chang AC, Pratt PE, Sharpness R (1973) Effect of application of dairy manure on germination and emergence of some selected crops. J Environ Quality 2:396–399

Aery NC (2010) Manual of environmental analysis. Anne Books pvt, Ltd

Ali HM, Khamis MH, Hassan FA (2012) Growth, chemical composition and soil properties of *Tipuana speciosa* (Benth.) Kuntze seedlings irrigated with sewage effluent. Appl Water Sci 2:101–108

Almodares A, Sharif ME (2007) Effects of irrigation water qualities on biomass and sugar contents of sugar beet and sweet sorghum cultivars. J Environ Biol 28(2):213–218

APHA (2005). In: Standard methods for the examination of water and wastewater. American Public Health Association (APHA), American Water Works Association (AWWA) andWater Environment Federation (WEF), 21st edn, Washington, DC, pp 1368

Augusthy PO, Annsherin M (2001) Effect of factory effluent on seed germination and seedling growth of *Vigna radiatus* (L.). J Environ Res 22(2):137–139

Ayyasamy PM, Yasodha R, Rajakumar S, Lakshmana perumalsamy P, Rahman PKSM, Lee S (2008) Impact of sugar factory effluent on the growth and biochemical characteristics of terrestrial and aquatic plants. Bullet Environ Contamination Toxicol 81:449–454

Damodharan U, Reddy MV (2012) Impact of sugar industrial treated effluent on the growth factor in Sugarcane-Cuddalore. India J Sustain Bioenergy Syst 2:43–48

Doke KM, Ejazuddin MK, Joseph R, Asif S (2011) Physico-chemical analysis of sugar industry effluent and its effect on seed germination of *Vigna angularis, vigna cylindrical* and *sorghum cernum*. Annuls Environ Sci 5:7–11

Elayaraj B (2014). Physico-chemical analysis of sugar factory effluent stress on seedling growth of black gram (*Vigna mungo* (L.) Hepper) varieties. Int Lett Natural Sci 12(1):85–93

Hussain I, Iqbal M, Nawaz M, Rasheed R, Perveen A, Mahmood S, Yasmeen A, Wahid A (2013) Effect of sugar mill effluent on growth and antioxidative potential of maize seedling. Int J Agriculture Biol 15:1227–1235

Jagannathan R, Venkatraman K, Vasuki R (2014) Bioremediation of sugar mill effluent by immobilized Bacterial consortium. Int J Pharm Technol 6(3):7107–7114

Kaur A, Vats S, Rekhi S, Bhardwaj A, Goel J, Tanwar RS, Gaur KK (2010) Physicochemical analysis of the industrial effluents and their impact on the soil microflora. Proc Environ Sci 595–599

Kolhe AS, Sarode AG, AndIngale SR (2009) Study of effluent from sugarcane industry. Sodh Samiksha Mulyankan 2:303–306

Kumar V, Chopra AK (2012) Fertigation effect of distillery effluent on agronomical practices of *Trigonella foenum-graecum* L. (Fenugreek). Environ Monitor Assessment 184(3):1207–1219

Lakshmi S, Sundaramoorthy P (2000) Effect of sugar mill effluent on germination, seedling growth and biochemical changes in ragi (*Elusine corocana* Gaertn). Indian J Environ Ecoplann 3(3):501–506

Lakshmi S, Sundaramoorthy P (2001) Screening of paddy cultivars towards tannery effluent pollution. Geobios 28(1):45–48

Malik S, Bhati H, Kumar D, Kumar V (2014) Germination and seedling growth of Vigna radiata L. under sugar mill effluent stress. Int J Pharm Res Biol Sci 3:54–59

Manivasakam N (1987) Industrial effluents origin, characteristics, effects analysis and treatment. Sakthi Publications, Kovai pudur, Coimbatore, India

Milner C, Hughes RE (1968) Methods for the measurement of primary production of grassland. IBP handbook no. 6. Blackwell, Oxford

Mishra V, Pandey SD (2002) Effect of distillery effluent and leachates of industrial sludge on the germination of blackgram (*Cicer arietinum*). Pollution Res 21(4):461–467

Nath K, Singh D, Sharma YK (2007) Combinatorial effects of distillery and sugar factory effluents in crop plants. J Environ Biol 28(3):577–582

Palharyal JP, Siriah VK, Shobana M (1993) In: Environmental impact of sewage and effluent disposal on river systems. Ashish Publishing House, pp 325

Ramkrishan K, Thamizhiniyan P, Sundaramoorthy P, Kumar SD (2001) Effect of sugar mill raw and yeast treated effluent on screed germination and seedling growth of paddy. Asian J Microbiol Biotechnol Environ Sci 3(1–2):37–40

Rath P, Pradhan G, Mishra MK (2010) Effect of sugar factory Distillery Spent Wash (DSW) on the growth pattern of sugarcane (*Saccharum officinarum*) crop. J Phytol 2(5):33–39

Sajani S, Muthukkaruppan SM (2011) Physico-chemical analysis of sugar mill effluent, contaminated soil and its effect on seed germination of paddy (*Oryza sativa* L.). Int J Pharm Biol Achieve 2(5):1469–1472

Samuel S, Muthukkaruppan SM (2011) Physico-chemical analysis of sugar mill effluent, contaminated soil and its effect on seed germination of paddy. Int J Pharm Biol Achieve 2(5):1469–1472

Sarala Thambavani D, Sabitha MA (2012) Kinetic and isotherm studies on Biodegradation of sugar mill waste water using *Mangifera indica* leaf powder. Res J Pharmaceutical, Biol Chem Sci 3(3):247–260

Saurabh S, Shailja P (2014). Physicochemical analysis of sugar mill effluent and their impact on changes of growth of wheat (*Triticum aestivum*) and maize (Zea mays L.). J Environ Sci Toxicol Food Technol 8(4):57–61

Saxena C, Madan S (2012) Growth response of *Solanum melongena* in three different adsorbents irrigated with sugar mill effluent. J Agri Biol Sci 7:325–329

Sharma P, Dubey RS (2005) Lead toxicity in plants. Brazil J Plant Physiol 17(1):35–52

Siddiqui WA, Waseem M (2012) A comparative study of sugar mill treated and untreated effluent-a case study. Oriental J Chem 28(4):1899–1904

Singh G, Bhati M (2005) Growth of *Dalbergia sissoo* in desert regions of western India using municipal effluent and the subsequent changes in soil and plant chemistry. Bioresource Technol 96:1019–1028

Siva Santhi K, Suja Pandian R (2012) Effect of sugar mill effluent on seed germination of peanut (*Arachis hypogaea*) and green gram (*Vigna radiata*). Int J Chem Pharm Sci 1:804–806

Sundaramoorthy P, Chidambaram ALA, Sankar Ganesh K, Rajesh M (2006) Growth response on paddy (*Oryza sativa* L.) for sugar mill effluent irrigation. Pollution Res 25(4):749–752

Suresh B, Abraham K, Damodharam T (2014) Effect of sugar industry effluent on changes of growth and biochemical contents of *Capsicum annuum* L. Adv Appl Sci Res 5(5):305–309

Vaithiyanathan T, Soundari M, Sundaramoorthy P (2014) Response of black gram (*Vigna mungo* L.) to sugar mill effluent treatment. Int J Botany Res 4(1):15–18
Vinod K (2014) Sugar mill effluent utilization in the cultivation of maize (*Zea mays* L.) in two seasons. J Waste Manage 12
Vinod K., Chopra AK, Chakresh P, Sachin P (2010) Agro-potentiality of paper mill effluent on the characteristics of *Trigonella foenum-graecum* L. (Fenugreek). New York Sci J 3(5):68–72

# Quorum Sensing Inhibitors as an Alternate to Antibiotic Against Biotic Pressure Induced Bacterial Contamination in Aquaculture

**Arumugam Veera Ravi, Sivasubramanian Santhakumari, Issac Abraham Sybiya Vasantha Packiavathy, and Jeyaraj Godfred Ponraj**

**Abstract** Aquaculture is a fastest growing, effervescent food producing sector with an annual growth rate of about 10% globally. At present, fin fish and shell fish are the major cog of coastal aquaculture and are now considered as an important source of protein available for human consumption. Despite such progressive growth, the production from aquaculture was hampered by emerging bacterial diseases, resulting in immense mortalities and usher to severe economic losses worldwide. Emerging bacterial disease problems have overwhelmed the aquaculture industry in the past and continue to menace it. Intensification of aquatic animal cultivation has added to the headway of numerous diseases like vibriosis, motile aeromonad septicemia, necrotizing hepatopancreatitis, hemorrhagic septicemia and early mortality syndrome. Subsequently, the unsystematic usage of antimicrobial and other therapeutic agents results in the development of resistance among bacterial pathogens and made the existing treatment measures futile. Increased contamination of the aquaculture with bacterial pathogens has paved the way for efficient approaches which can be applied for environmental restoration. Now, quorum sensing inhibitors are the candidate molecules of choice chosen as the best alternatives in bioremediation process to the existing antimicrobial agents. They act as natural immune enhancers, which hassle the development of drug resistance among bacterial pathogens. This review present an overview of the existing major bacterial diseases in aquaculture sector and endow with an information on the deleterious impacts of antibiotics usage and novel

A. V. Ravi (✉) · S. Santhakumari · I. A. S. V. Packiavathy
Department of Biotechnology, Science Campus, Alagappa University, Karaikudi, Tamil Nadu 630004, India
e-mail: aveeraravi@rediffmail.com

S. Santhakumari
Department of Biochemistry and Molecular Biology, School of Life Sciences, Pondicherry University, Puducherry 605014, India

I. A. S. V. Packiavathy
Department of Biotechnology, Karpagam Academy of Higher Education, Coimbatore 641021, India

J. G. Ponraj
TIL Biosciences—Animal Health Division of Tablets (India) Limited, Jhaver Centre, Egmore, Chennai 600008, India

 
M. Prashanthi et al. (eds.), *Bioremediation and Green Technologies*, Environmental Science and Engineering, https://doi.org/10.1007/978-3-030-64122-1_20

alternative strategies for reducing the risk of antibiotic resistance mediated bacterial infections in aquaculture industry.

**Keywords** Aquaculture · Antimicrobial agents · Antibiotic resistance · Bacterial contamination · Bioremediation · Quorum sensing inhibitors

## 1 Introduction

Aquaculture is one of the emerging food producing zones in the global scenario. The economic yield of this sector have increased at an annual average rate of 125 billion US$ worldwide (Jayaprakashvel and Subramani 2019; Hamza et al. 2015; FAO 2013). Cultivation of aquatic animals in marine and fresh water environments has been practiced in many Asian countries for centuries and it remains as an important economical base for several countries especially in Africa and Latin America in order to raise their national economy (Hall et al. 2013). Though, the aquaculture practices takes place in a vast majority of Asian countries, China alone contributes to more than 70% of aquaculture production in Asia. In global figures, China alone represents more than 27% of total production in value, amounting to $17.1 billion in 2011 and $18.2 billion in 2012. In 2013–14 India has become the second largest aquaculture producer in Asia with the total production of 5.68% of global fish production and ranks next in the world after China. Total fish production is 9.58 million metric tonnes with an input of 6.14 million metric tonnes from internal sector and 3.44 million metric tones from aquatic region, respectively (Handbook on Fisheries Statistics 2014).

Aquaculture would be the significant food sector in the near future to balance the global nutritional demand with special emphasis to developing countries, suffering from food shortage and malnutrition. Amongst all aquatic organisms, fish and penaeid shrimp are presently the major source of animal protein, contributing more than 25% of the total animal protein supply for about millions of people worldwide. Nevertheless, the availability of fish seeds and the emerging microbial diseases hinders the aquaculture production. The extreme release of bacterial pathogens from animal waste into the aquaculture environment has become a major concern for the aquaculture industry (Stabili et al. 2010).

## 2 Role of Bacterial Diseases in Aquaculture

Owing to escalation and commercialization of culturing practices, this sector is facing sequence of struggles. In addition to this, the main hindrance in aquaculture is the anomalous outburst of microbial diseases, especially diseases caused by bacteria is considered as an important impediment for the development of this sector worldwide (Zhao et al. 2015). The bacterial pathogens are considered as major hitch as it causes up to 100% mortalities in aquaculture with harsh economic losses worldwide

(Austin and Zhang 2006). Among these, disease outbreaks caused by pathogenic or opportunistic bacteria especially *Vibrio* spp., remains as a major challenging concern in the intensive rearing of aquatic animals such as molluscs, fish, crab, oysters and shrimp. Vibrios are Gram-negative, rod or curve shaped bacteria, pervasive in marine and estuarine ecosystems as well as aquaculture farms and encompass one of the major microbiota of these ecosystems. In which, *V. harveyi*, *V. splendidus*, *V. penaecida*, *V. anguillarum*, *V. parahaemolyticus*, *V. alginolyticus* and *V. vulnificus* are the major pathogens mostly associated with fish, shrimp and other aquatic diseases. *V. harveyi* is the causative agent of luminescent vibriosis in penaeid shrimp larval rearing systems and is responsible for mass mortality in *P. monodon*. Also, *V. parahaemolyticus* - a halophilic bacterium scattered in temperate water, causes the emerging and most devastating disease, early mortality syndrome (EMS) in shrimp. EMS typically targets the shrimp post larvae between 20 and 40 days and impose 100% mortality within a week. The syndrome is more often lethal to shrimp than compared to other aquatic animals (de Schryver et al. 2014). This pathogen is also responsible for the most familiar vibrio-associated seafood borne gastroenteritis, and wound infections in humans through consumption of contaminated uncooked seafoods (Matsumoto et al. 2000). Similarly, *V. vulnificus* is also well recognized to cause gastrointestinal infections in humans by consumption of uncooked seafood and it also causes wound or soft tissue infections in marine organisms (Bross et al. 2007). In fishes, the pathogenic vibrios causes' intestinal necrosis, anemia, ascetic fluid, petechial haemorrhages in the muscle wall, fluid accumulation in the air bladder, etc. It has been described that shrimp are also prone to vibriosis and it targets the vital organs such as gill, gut and hepatopancreas, etc. Apart from vibrios, *Pseudomonas* and *Aeromonas* species are also considered as the major aquatic pathogens, which cause black spot syndrome and severe necrosis in infantiles of the penaeid shrimp. The prime mode of infection in fish involves the invasion of the host tissues mainly by the help of chemotactic movement, followed by deployment of siderophores system, resulting in eventual damage to the fish by means of extracellular virulent products, which includes different types of extracellular products such as protease, chitinase and hemolysin. After invasion, vibrios colonize the host tissue of shrimp by crossing the epithelial cells. In human, consumption of raw or undercooked seafood, particularly shellfish with vibriosis may lead to development of acute gastroenteritis, wound and soft-tissue infections characterized by diarrhoea, vertigo, vomiting, abdominal pain and high fever. In 2014, vibriosis is ranked as the foremost infectious cause of human mortality by Centre for Disease Control and Prevention revealing its important status among bacterial pathogens.

## 3 Use of Antibiotics and Other Chemicals to Control Bacterial Diseases

Conventionally, to control bacterial diseases, infected animals will be fed with antibiotics and chemicals medicated supplementary diet. For finfish and shellfish, antibiotics are habitually administered in the feed, either as compounds or as surface-coated feed pellets mixed along with oil as vehicle. Similarly, in the shrimp industry, antibiotics are mostly used as an immersion treatment in the hatchery. In aquaculture, antibiotics at curative levels are habitually administered for short periods of time through the oral course to groups of animals in intensive culture. Generally, antibiotics are used in aquaculture during the production phases, both in the larval and growth forms. The routine use of antibiotics in aquaculture is attendant with environmental and human health concerns, including resistance among bacterial pathogens, importunity of the disease in the marine environment. The accretion of antibiotics and its residues in the palatable tissues of shrimp may also modify human intestinal micro flora and cause food poisoning problems. Globally, oxytetracycline, ciprofloxacin, florfenicol, sarafloxacin, chlortetracycline, quinolones, enrofloxacin, norfloxacin, oxolinic acid, perfloxacin, sulfamethazine, gentamicin and tiamulin are the most frequently used antibiotics for the treatment of bacterial diseases in salmon, catfish, trout, other marketable fish and shrimp in aquaculture (Soto-Rodríguez et al. 2006).

In aquaculture, antimicrobial drugs are frequently mixed along with the feed or may be added directly to the water. These procedures result in the selective pressure to the bacteria in the exposed environments. The use of antimicrobials in aquaculture may involve broad environmental application that affects a wide variety of bacteria. Due to the erratic effects, various antibiotics are prohibited for aquatic animals in the food producing sector.

## 4 Development of Antibiotic Resistance

In aquaculture, the unsystematic usage of antimicrobial compounds has escort the development of antibiotic resistance among pathogens, creating serious challenges to human health and environmental security. A number of mechanisms of antimicrobial resistance are readily spread to a variety of bacterial genera. Therefore, resistance development in disease-causing bacteria, renders the life-saving antimicrobial compounds became ineffective (Sahu et al. 2008; Defoirdt et al. 2011). Currently, antibiotics are only moderately successful due to the emergence of resistant bacteria and the therapeutic treatments may have limited success in controlling infectious bacterial diseases. As a consequence of this problem, spread of antibiotic resistance from aquaculture settings to the natural environment is increasing. Thus, due to the overusage, inappropriate dosing of antibiotics and the prevalence of selective pressures, vibrios are now challenging with resistance to several broad spectrum antibiotics. About 70% of the vibrios isolated from aquaculture settings in Mexico are multi-drug resistant (Verschuere et al. 2000). In addition, the antibiotic residues from the excessive usage of antibiotics in aquaculture eventually accumulate in the host tissues of farmed fin fishes and shellfish, thus causing adverse health problems to humans. Hence, antibiotics are no longer considered as an effective measure for treating vibriosis among aquatic animals (Verschuere et al. 2000; Cabello 2006). Therefore, to increase and improve the yield of food production from the aquaculture industry, a novel alternative treatment strategy is urgently required to prevent emerging antibiotic resistant bacterial infections among aquaculture organisms.

## 5 Alternative Strategies—Quorum Sensing (QS) Disruption as a Novel Anti-infective Approach in Aquaculture

The ever increasing antibiotic resistance among bacterial pathogens and the bioaccumulation of antibiotic residues in the edible tissues has become global concern. Therefore, an exigent need is raised for developing alternative approach to antimicrobial compounds to overcome this bacterial infection with similar or enhanced protection to aquatic animals. An alternative approach with considerate on host and environment with deleterious effect to the pathogen will be the most effective method in the long-term to improve aquatic animal health. Recently, vaccination and application of probiotics become ideal methods for prevention of infectious diseases. However, in aquaculture the commercially available vaccines and probiotics are of very limited in success. As limited or no usage of antibiotics are stimulated and chemotherapy has its own limitation in aquaculture production microbial interference including bioremediation for health management and targeting quorum sensing (QS) is fast emerging as an alternative therapy. Among many other alternative strategies, the disruption of QS by inhibiting the virulence and biofilm formation of bacterial pathogens has recently been suggested as a novel alternative strategy to antibiotics for the use in aquaculture (Natrah et al. 2011). QS is the phenomenon of density dependent coordinated gene expression system, which is mediated through small signaling molecules called autoinducers (AIs). It has been well recognized that the QS mediated by autoinducer molecules play a vital role in the regulation of virulence factors production and biofilm formation of aquatic pathogens (Rasch et al. 2004). Therefore, any interference with the autoinducer mediated QS mechanism would be the suitable alternative treatment measure to prevent the emerging aquatic bacterial diseases.

## 6 Quorum Sensing in *Vibrios*

Two general types of QS systems have been reported in *Vibrios*, depending on whether the signals molecules are detected in the cytoplasm or at the cell surface. The former group involves the QS systems that use N-acylhomoserine lactones (AHLs) as signal molecules. These systems are found in many Gram-negative aquatic bacterial pathogens, including *A. hydrophila* and *A. salmonicida*. The latter group of system is multi-lingual QS system as found in vibrios, such as the aquaculture pathogens *V. harveyi*, *V. anguillarum* and *V. vulnificus*.

In vibrios, QS is the phenomenon of multi-channel gene expression system synchronized by three different types of autoinducers. The LuxM/N QS system utilizes AHLs as signal molecules, whereas CqsA/S system utilizes CAI-1 ("Cholera autoinducer -1"), and in the third QS system, the signal molecules are together referred as autoinducer-2 (AI-2) (Milton 2006). Third QS system appears to be shared by many Gram-positive and Gram-negative species and is based on a mixture of inter convertible molecules collectively referred as autoinducer-2 (AI-2). A crucial enzyme in the production of AI-2 is LuxS. LuxS directs the breakdown of S-ribosylhomocysteine to homocysteine and 4,5-dihydroxy-2,3-pentanedione (DPD). Not all QS systems are present in *Vibrio* spp., because most of them contain the AI-2 based QS system. In *Vibrio* spp., AI-2 binds to LuxP, a periplasmic AI-2 receptor that is connected with the LuxQ sensor kinase-phosphatase. However, at higher population density, AI-2 will bind to LuxP and as a result LuxQ will act as a phosphatase, leading to a dephosphorylation of LuxO. Since, dephosphorylated LuxO became quiet, no small regulatory RNAs will be formed and the LuxR mRNA remains stable, resulting in the production of LuxR and ultimately an altered gene expression pattern (Fig. 1; Table 1).

AI-2 based QS play an important role in regulating the secretion of several virulence factors, biofilm formation and stress responses in several *Vibrio* spp. In aquatic bacterial pathogens, the QS circuits, like AI-1 and AI-2 control the genes responsible for bioluminescence, conjugation, motility, sporulation, biocorrosion, antibiotic production, most importantly biofilm formation and the expression of virulence factors such as lytic enzymes, toxins, siderophores and adhesion molecules (de Kievit and Iglewski 2000; de Windt et al. 2003).

Moreover, in vibrios, QS mediated phenotypic factors including virulence factors production and biofilm formation plays a vital role in causing diseases in animals (Randall et al. 2004). These signal molecules are known to coordinate the production of virulence factors and biofilm formation in response to the cell density of the surrounding bacterial population (You et al. 2007; Yildiz and Visick 2009). The biofilm forming potential of *Vibrio* spp. is responsible for their survival, virulence and stress resistance (Packiavathy et al. 2013). Such biofilms are the preferable lifestyle for bacteria as they colonize and enhance growth and survival by affording the free entry to nutrients and develop resistant to antibiotics. It has also been found that bacteria remains vegetative within the biofilm often up to 1000 fold more tougher to antibiotics, chemicals and heavy metals than free swimming cells. It was reported

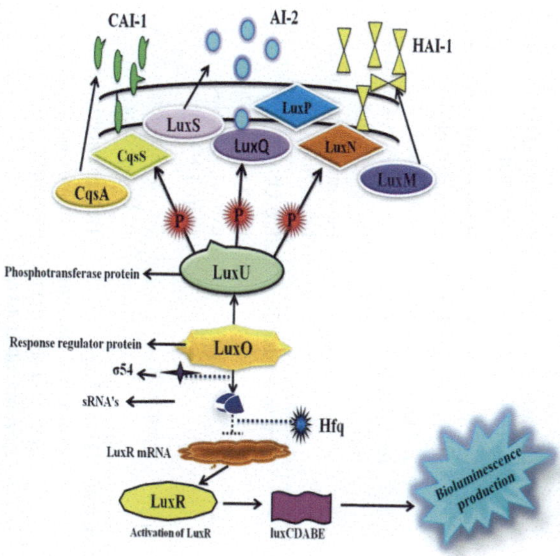


**Fig. 1** Pictographic representation of multi-lingual QS system in *V. harveyi*

that, biofilms are the route cause for more than 80% bacterial infections (Brooun et al. 2000). Hence, interfering with such QS mechanisms by targeting these signal molecules and to inhibit their virulence enzyme production and biofilm formation will pave the way to combat bacterial infections especially vibriosis in the aquaculture industry. It is also envisaged that this new approach has potential in fighting against aquatic bacterial infections and might become part of novel non-antibiotic based line of attack to trounce high mortalities in the industrial production of animals.

## 7 Application of Quorum Sensing Inhibitors (QSIs) from Natural Resources in Aquaculture

The disruption of AI mediated QS activity in aquatic bacterial pathogens could be achieved by three different strategies, which includes (i) inhibition of AIs synthesis, (ii) degradation of AIs by QS inhibitory enzymes and (iii) competitive binding of AIs to receptor proteins by structural antagonistic compounds (Fig. 2).

Natural resources are the hub of biologically vital compounds that are employed for the treatment of infectious diseases. These QSI compounds or enzymes have been identified from a wide range of natural sources such as medicinal plants, common spices, edible fruits, marine resources, etc. Hitherto, a broad spread occurrence of QSI agents from various natural sources that act against signaling molecules dependent QS have been identified. Current studies on QS in aquaculture organisms are very much limited but pointed out with most exciting results. For examples, algal metabolites isolated from red marine algae, components from sour orange and also

**Table 1** List of QS systems, signal molecules and identified QSI compounds for aquaculture pathogens (Zhao et al. 2015)

| Pathogens | Signal molecules | QS regulated phenotypic factors | Host organism | Diseases | Reported QSI compounds |
|---|---|---|---|---|---|
| *V. harveyi* | 3-hydroxy-C4-HSL, AI-2 and CAI-1 | Bioluminescence, Biofilm formation, Extracellular toxins and metalloprotease | Shrimp, Abalone, Oyster and Fish | Luminescent vibriosis, Skin ulcer, Septicemia and Gastroenteritis resulting in mass mortality | Halogenated furanones, Cinnamaldehyde, Brominated thiophenone, Curcumin, Boric acid and Flavonoids and citrus limonoids |
| *V. cholerae* | CAI-1 and AI-2 | Biofilm formation | Humans | Rice water diarrhea | Pro-QS molecules,Turmeric extract, AHL lactonase (AiiA) |
| *Vibrio fischeri* | *N*-(3-oxohexanoyl)-HSL (LuxI/LuxR) | Bioluminescence and Colonization | Hawaiian bobtail squid | Symbiosis | Secondary metabolites of Brown alga |
| *Aeromonas hydrophila* | *N*-butanoyl-HSL AhyI/AhyR | Serine protease, metallo protease production Biofilm formation, enzyme production | Juvenile and mature fish and human | Hemorrhagic septicemia, resulting in fin and tail rot and epizootic ulcerative syndrome Intestinal and Wound infection in human | Benzene acetic acid, Palmitic acid and AHL lactonase |
| *Aeromonas salmonicida* | *N*-butanoyl-HSL AsaI/AsaR | Exoprotease, cytotoxic factor | Salmonids | Furunculosis of salmonids and Ulcer disease | - |

(continued)

**Table 1** (continued)

| Pathogens | Signal molecules | QS regulated phenotypic factors | Host organism | Diseases | Reported QSI compounds |
|---|---|---|---|---|---|
| *V. anguillarum* | 3-oxo-C10-HSL and C6-HSL | Extracellular protease, Pigment production, Haemolysin and Biofilm formation | Fish | Haemorrhagic septicaemia | Furanone C-30, Cinnamaldehyde and its derivatives |
| *V. alginolyticus* | AI-2 | Total extracellular protease, Haemolysin, Motility and Biofilm formation | Fish, Abalone and Mussel | Shell disease, Superficial wound Infections, Septicemia, haemorrhaging, dark skin, and ulcers on the skin surface, mortality | Turmeric extract, Palmitic acid and Secondary metabolites from *Bacillus* spp. |
| *V. parahaemolyticus* | LuxO (represses SmcR) OpaR (LuxR homologue) | Cytotoxins and RtxA1 | Shrimp and humans | Early mortality syndrome in shrimp, Mild gastroenteritis to severe debilitating dysentery, eye and ear infections, wound infections of the extremitiesin human | Curcumin, Phenolic and Flavonoids, Cinnamaldehyde and its derivatives, Turmeric extract, Palmitic acid |
| *V. vulnificus* | AI-2 (SmcR -LuxR homologue) | Metalloprotease, Virulence enzymes and Biofilm formation | Fish and Humans | Primary septicemias, Wound infections in fish, and Gastrointestinal Illnessesin human | Curcumin, Phenolic and Flavonoids, Cinnamaldehyde and its derivatives, Turmeric extract |

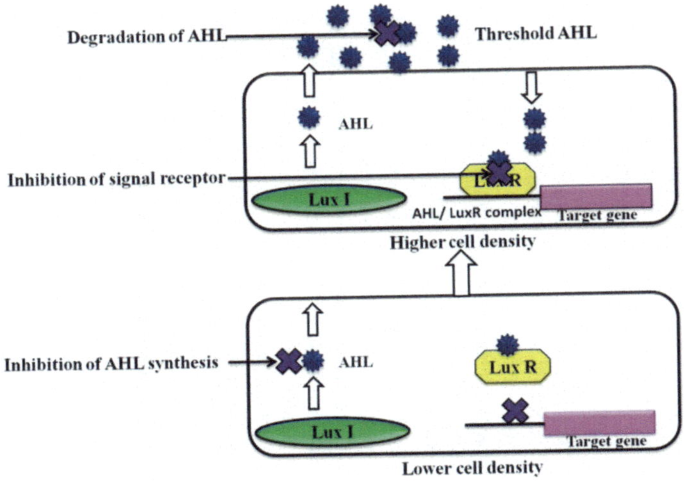


**Fig. 2** General mechanism of QSI strategies

synthetic analogues have been proven to diminish QS regulated gene expression in vibrios and to protect fish and shrimp from vibriosis infections (Rasch et al. 2004; Defoirdt et al. 2006; Brackman et al. 2008; Vikram et al. 2011; Defoirdt et al. 2012). Halogenated furanones, the first discovered and extensively studied group of QS-disrupting compounds, have been extracted from the red marine algae, *Delisea pulchra*. This compounds disrupted the AHL and AI-2 based QS system of *Vibrio* spp. by modulating the DNA binding activity of transcriptional regulators. It also prevented the virulence of several *Vibrio* spp. (including *V. anguillarum*, *V. campbellii*, *V. harveyi* and *V. parahaemolyticus*) in gnotobiotic brine shrimp *Artemia franciscana*.

As well to *D. pulchra*, several other macroalgae, micro-algae, cyanobacteria, mangroves, corals, sponges and associated bacteria from marine resources also secrete compounds which could able to interfere with QS. Marine bacteria can also produce secondary metabolites with QSI activity. For example, marine *Halobacillus salinus* strain isolated from sea grass secretes two phenetylamide metabolites as bioactive compounds that blocked QS-regulated phenotypes in diverse bacteria, including the aquaculture pathogen *V. harveyi*. Similarly, production of QS-inhibitory substances by several epiphytic bacteria associated with marine brown alga *Colpomenia sinuosa* was also reported.

In this view, cinnamaldehyde and cinnamaldehyde derivatives hinder with AI-2 based QS in various *Vibrio* spp., by inhibiting the virulence and biofilm formation of *Vibrio* spp. and protected the gnotobiotic *Artemia* shrimp against virulent *V. harveyi* BB120. Cinnamaldehyde is a non-hazardous synthetic flavouring substance frequently used in food, beverages, chewing gum, and in perfumes. It is generally recognised as safe compound with well-known QSI activity. In addition, the cinnamaldehyde analogs also increased the survival of *Caenorhabditis elegans* nematodes infected with *V. anguillarum*, *V. harveyi* and *V. vulnificus* (Brackman et al. 2008, 2011). The same compound has also enhanced the survival and growth of freshwater prawn larvae when challenged with pathogenic luminescent *V. harveyi* (Pande et al. 2013). In the year 2010, Vikram et al. studied the QSI activity of numerous flavonoids from citrus fruits specifically naringin, quercetin, sinensetin and apigenin. Out of four flavonoids tested, naringin was identified as potent inhibitor of autoinducer-mediated cell–cell signalling by inhibiting the *Escherichia coli* O157:H7 biofilm and *V. harveyi* virulence. Also, citrus limonoids from sour orange, specifically isolimonic acid and ichangin displayed a maximum inhibitory activity against bacterial cell–cell signalling as measured with the *V. harveyi* reporter strain model system (Vikram et al. 2011).

Consequently, in the recent years the marine cyanobacteria is also receiving much attention on their role as source of QSIs. Tumonoic acids isolated from cyanobacteria, *Blennothrix cantharidosmum* exhibited anti-QS activity by inhibiting QS dependent bioluminescence in *V. harveyi* without affecting the bacterial growth (Clark et al. 2008). Malyngolide and lyngbyoic acid from the cyanobacterium *Lyngbya majuscula* inhibited the QS dependent elastase and virulence factors production in *P. aeruginosa* (Dobretsov et al. 2010; Kwan et al. 2011). The ability of cinnamaldehyde and its derivatives to interfere with the AI-2-dependent QS system of *Vibrio* spp. and its ability to rescue *Artemia* shrimp against *V. harveyi* infection has been documented in earlier studies as well (Brackman et al. 2009; Musthafa et al. 2011).

In this outlook, Defoirdt et al. (2012) have studied the effect of synthetic compounds on luminescent QS dependent virulence and biofilm formation of luminescent *V. harveyi*. In this study, brominated thiophenones TF101 and TF310 effectively inhibited the QS-regulated gene expression system in *V. harveyi* by changing the DNA-binding action of the QS master regulator LuxR. Similar to in vitro, in vivo challenge test with gnotobiotic brine shrimp larvae also has the maximum curative index of all QS disrupting compounds tested thus far in our brine shrimp model system against luminescent vibriosis. Another report by Rasch et al. (2004) showed a noticeable decline in transience of rainbow trout by *V. anguillarum* cultures in the presence of furanone C-30 at tested concentration. Based on the second strategy of inhibition, *Bacillus* species were amongst the well-known bacteria accounted to open the lactone ring of AHLs by secreting lactonase enzymes, which inactivate AHLs by degradation. Earlier reports revealed that AHL-degrading bacterial cultures isolated from the intestinal tract of healthy shrimp and fish increased the survival of different aquaculture species, including larvae of turbot (*Scophthalmus maximus*) and the freshwater prawn (Nhan et al. 2010).

In recent years, anti-biofilm and anti-QS molecules have listed out a variety of compounds from diverse sources including natural as well as synthetic compounds against human pathogens such as *Pseudomonas aeruginosa*, *Proteus mirabilis* and *Serratia marcescens* as well as aquatic bacterial pathogens including *V. harveyi*, *V. parahaemolyticus*, *V. alginolyticus*, *V. vulnificus* and *Aeromonas hydrophila*. A few of such compounds, which are known to interfere with the bacterial QS are curcumin from *Curcuma longa* (Packiavathy et al. 2013), phenyl acetic acid from *Streptomyces humidus* (Musthafa et al. 2012), sponge extracts (Annapoorani et al. 2012a), phenolic and flavonoids such as rosmarinic acid, morin, naringin, chlorogenic acid and mangiferin from edible fruits (Annapoorani et al. 2012b). Srinivasan et al. (2016) demonstrated the promising anti-QS and antibiofilm activities of *Piper betel* leaves extract and its active secondary metabolite phytol against *S. marcescens* infections. In another report, Sivaranjani et al. (2016) identified the potential of morin as a promising antibiofilm agent against biofilm associated infections caused by *Listeria monocytogenes*.

The findings on QSIs against aquatic bacterial pathogens includes, the QSI activity of cell free culture supernatants of marine *Bacillus* spp. reported by Nithya and Pandian (2010) against QS dependent biofilm formation, exopolysaccharides production and bacterial attachment of *Vibrio* spp., such as *V. harveyi*, *V. parahaemolyticus*, *V. vulnificus* and *V. alginolyticus*. Similarly, the anti-QS potential of natural compound curcumin from *C. longa* in the prevention of vibriosis caused by *Vibrio* spp. such as *V. harveyi*, *V. parahaemolyticus* and *V. vulnificus* among aquatic organisms were studied by Packiavathy et al. (2013) (Fig. 3).

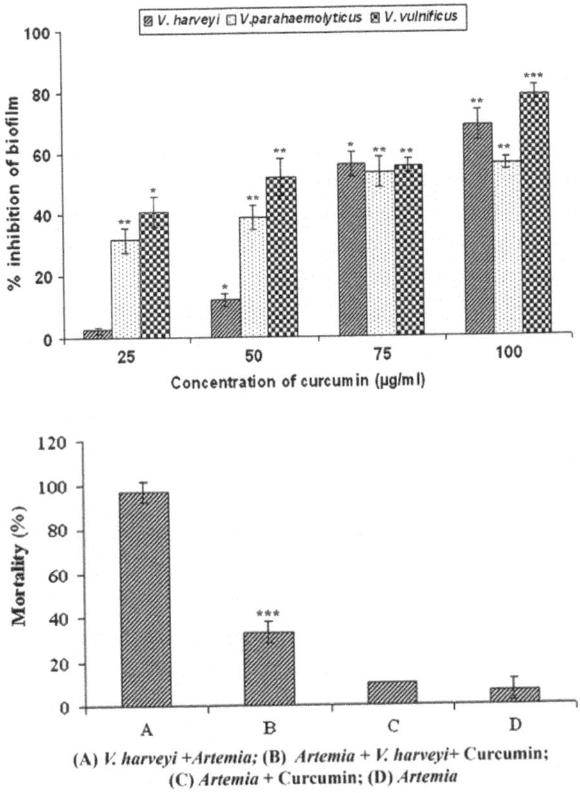


**Fig. 3** Quantitative analysis of biofilm inhibition by curcumin in *Vibrio* spp. and anti-infective potential of curcumin in enhancing the survival of *Artemia* nauplii against *Vibrio harveyi* infection

**Table 2** Molecular docking energy (kcal/mol) of compounds identified from *R. annamalayana* bark extract against receptor proteins of *V. harveyi*

| Against *V. harveyi* LuxP receptor | Docking score |
|---|---|
| Furanosyl borate diester (natural ligand) | −7.484 |
| lH-Purin-6-amine | −5.681 |
| Cycloheptasiloxane | −6.027 |
| Cyclooctasiloxane | −6.074 |
| Cyclononasiloxane | −5.919 |
| Cyclononasiloxane octadecamethyl | −8.459 |
| Cyclodecasiloxane eicosamethyl | −8.175 |
| 1,1,1,5,7,7,7-Heptamethyl-3,3-bis(trimethylsiloxy)tetrasiloxane | −6.186 |

Musthafa et al. (2011) have assessed the anti-QS activity of bark extract obtained from the mangrove plant *Rhizophora annamalayana* against biomarker strain *Chromobacterium violaceum* ATCC 12472 and aquatic bacterial pathogen luminescent *V. harveyi* MTCC 3438. In which, the bark extract inhibited the bioluminescence production upto the level of 99%. Based on the docking score, the compounds cyclononasiloxane octadecamethyl and cyclononasiloxane eicosamethyl were identified as active compounds (Table 2).

Similarly, the in vitro and in vivo effcicacy of phenolic and flavonoids such as rosmarinic acid, naringin, morin, chlorogenic acid and mangiferin present in edible fruits were evaluated against *Vibrio* spp. challenged with *Artemia* nauplii. A recent report of Santhakumari et al. (2016) evidenced that the chloroform/ethyl acetate (1:1) extract of marine cyanobacteria *Synechococcus elongatus* have been proven for their QSI activity against the violacein production in *C. violaceum* (ATCC 12472) and also against the QS mediated virulence and biofilm formation of aquatic pathogens *V. harveyi* and *V. vulnificus*. From GC–MS analysis of cyanobacterial extract, the major peak corresponding to the active compound was identified as hexadecanoic acid using International standard NIST library.

Moreover, Durai et al. (2013) reported that, *C. elegans* based in vivo screening using sponge associated bacteria that inhibits the virulence of *V. alginolyticus* by interrupting the QS pathway. Therefore, focusing the investigation in the discovery of natural resources for QSI would be the best suitable alternative approach to prevent emerging antibiotic resistance bacterial infections.

Similarly, flavonoids from plants are receiving importance due to their biomedical potential. In recent investigation, Rama Devi et al. (2016) explored the effectiveness of rosmarinic acid, is one of the most common flavonoids tested against QS mediated biofilm formation and virulence factor production such as hemolysin, lipase and elastase by *A. hydrophila* isolates and a reference strain.

## 8 Conclusions and Future Perspectives

Exhaustive farming has stimulated the spreading of numerous bacterial diseases, which has propelled the over usage of antimicrobial drugs. The increasing proportions of multi-drug resistant bacteria that persist in residues and farm environments provide a threat to aquaculture. Several alternative approaches to antibiotics have been developed and some of them have been successfully implemented to control bacterial infections in aquaculture facilities.

Overall, this review presented a brief idea on the problems and challenges associated with antibiotic resistance and significance of quorum sensing inhibitors as alternative strategies to conquer the antibiotic resistance and thwart the emerging bacterial infections with less selective pressure among pathogenic bacterial organisms. Hence, usage of QSIs is promising in the development of novel antipathogenic drugs to control vibriosis caused by antibiotic resistant vibrios in aquaculture. Besides, novel therapeutic strategies could be intended based on information garnered from studies of quourum sensing and quorum sensing inhibitors. This research findings and existing positive reports suggest that, quorum sensing could be a thrust area of research in order to combat emerging infectious agents in aquaculture with enormous other practical applications.

**Acknowledgements** The author S. Santhakumari gratefully acknowledges the Department of Biotechnology, Government of India (Grant No. BT/PR4815/AAQ/3/587/2012), for the financial support in the form of DBT- JRF & SRF. The author also thank the Department of Science and Technology, SCIENCE & ENGINEERING RESEARCH BOARD (SERB) (Grant No. PDF/2018/001518), for the financial support in the form of National Post-Doctoral fellowship (N-PDF). Also, the authors gratefully acknowledge the Department of Biotechnology, Governmentof India, for providing Bioinformatics Infrastructure Facility[Grant No. BT/BI/25/012/2012(BIF)] at Alagappa University. The authors also acknowledge DST-FIST [Grant No. SR/FST/LSI-639/2015(C)],UGC-SAP[GrantNo.F.5-1/2018/DRS-II(SAPII)], and DST-PURSE [Grant No. SR/PURSE Phase 2/38(G)]for providing instrumentation facilities.

## References

Annapoorani A, Jabbar AKKA, Musthafa SKS, Pandian SK, Ravi AV (2012a) Inhibition of quorum sensing mediated virulence factors production in urinary pathogen *Serratia marcescens* PS1 by marine sponges. Indian J Microbiol 52:160–166

Annapoorani A, Umamageswaran V, Parameswari R, Pandian SK, Ravi AV (2012b) Computational discovery of putative quorum sensing inhibitors against LasR and RhlR receptor proteins of *Pseudomonas aeruginosa*. J Comput Aided Mol Des 26:1067–1077

Austin B, Zhang XH (2006) *Vibrio harveyi*: a significant pathogen of marine vertebrates and invertebrates. Lett Appl Microbiol 43:119–124

Brackman G, Celen S, Hillaert U, Calenbergh SV, Cos P, Maes L, Nelis HJ, Coenye T (2011) Structure-activity relationship of cinnamaldehyde analogs as inhibitors of AI-2 based quorum sensing and their effect on virulence of *Vibrio* spp. PLoS ONE 6:1–10

Brackman G, Defoirdt T, Miyamoto C, Bossier P, Van Calenbergh S, Nelis H, Coenye T (2008) Cinnamaldehyde and cinnamaldehyde analogs reduce virulence in *Vibrio* spp. by decreasing the DNA-binding activity of the quorum sensing response regulator LuxR. BMC Microbiol 8:149

Brooun A, Liu S, Lewis K (2000) A dose-response study of antibiotic resistance in *Pseudomonas aeruginosa* biofilms. Antimicrob Agents Chemother 44:640–646

Bross MH, Soch K, Morales R, Mitchell RB (2007) *Vibrio vulnificus* infection: diagnosis and treatment. Am Fam Physician 76:539–544

Cabello FC (2006) Heavy use of prophylactic antibiotics in aquaculture: a growing problem for human and animal health and for the environment. Environ Microbiol 8:1137–1144

Clark BR, Engene N, Teasdale ME, Rowley DC, Matainaho T, Valeriote FA, Gerwick WH (2008) Natural products chemistry and taxonomy of the marine cyanobacterium *Blennothrix cantharidosmum.* J Nat Prod 71:1530–1537

de Kievit TR, Iglewski BH (2000) Bacterial quorum sensing in pathogenic relationships. Infect Immun 68:4839–4849

de Windt W, Boon N, Siciliano SD, Verstraete W (2003) Cell density related H-2 consumption in relation to anoxic Fe(0) corrosion and precipitation of corrosion products by *Shewanella oneidensis* MR-1. Environ Microbiol 5:1192–1202

de Schryver P, Defoirdt T, Sorgeloos P (2014) Early mortality syndrome outbreaks: a microbial management issue in shrimp farming. PLoS Pathog 10(4):e1003919. https://doi.org/10.1371/journal.ppat.1003919

Defoirdt T, Sorgeloos P, Bossier P (2011) Alternatives to antibiotics for the control of bacterial disease in aquaculture. Curr Opin Microbiol 14:251–258

Defoirdt T, Benneche T, Brackman G, Coenye T, Sorgeloos P (2012) A quorum sensing-disrupting brominated thiophenone with promising therapeutic potential to treat luminescent vibriosis. PLoS ONE 7:e41788

Dobretsov S, Teplitski M, Alagely A, Gunasekera SP, Paul VJ (2010) Malyngolide from the cyanobacterium *Lyngbya majuscula* interferes with quorum sensing circuitry. Environ Microbiol Rep 2:739–744

Durai S, Vigneshwari L, Balamurugan K (2013) *Caenorhabditis elegans* based *in vivo* screening of bioactives from marine sponge associated bacteria against *Vibrio alginolyticus.* J Appl Microbiol 115:1329–1342

FAO Fisheries and Aquaculture Report (2013) No. 1053. Hanoi, Vietnam, pp 126–130

FAO statistical yearbook (2013), pp 146–150

Handbook on Fisheries Statistics (2014) Department of Animal Husbandry, Dairying & Fisheries, MINISTRY OF AGRICULTURE, Govt of India, New Delhi

Hall SJ, Hilborn R, Andrew NL, Allison EH (2013) Innovations in capture fisheries are an imperative for nutrition security in the developing world. Proc Natl Acad Sci USA 110:8393–8398

Hamza F, Kumar AR, Zinjarde S (2015) Biotechnological applications of quorum-sensing inhibitors in aquacultures. Springer, India. https://doi.org/10.1007/978-81-322-1982-8_30

Jayaprakashvel M, Subramani R (2019) Implications of quorum sensing and quorum quenching in aquaculture health management. In: Implication of quorum sensing and biofilm formation in medicine, agriculture and food industry. Springer, Singapore, pp 299–312. ISBN 978-981-32-9408-0

Kwan JC, Meickle T, Ladwa D, Teplitski M, Paul V, Luesch H (2011) Lyngbyoic acid, a "tagged" fatty acid from a marine cyanobacterium disrupts quorum sensing in *Pseudomonas aeruginosa.* Mol Biosyst 7:1205–1216

Manefield M, Harris L, Rice SA, de Nys R, Kjelleberg S (2000) Inhibition of luminescence and virulence in the black tiger prawn (*Penaeus monodon*) pathogen *Vibrio harveyi* by intercellular signal antagonists. Appl Environ Microbiol 66:2079–2084

Matsumoto C, Okuda J, Ishibashi M, Iwanaga M, Garg P, Rammamurthy T, Wong HC, Depaola A, Kim YB, Albert MJ, Nishibuchi M (2000) Pandemic spread of an O3:K6 clone of *Vibrio parahaemolyticus* and emergence of related strains evidenced by arbitrarily primed PCR and toxRS sequence analyses. J Clin Microbiol 38:578–585

Milton DL (2006) Quorum sensing in vibrios: complexity for diversification. Int J Med Microbiol 296:61–71

Musthafa KS (2011) Anti-quorum sensing and antimicrobial potential of marine sediment bacteria. Ph.D Thesis, Alagappa University, India pp 230

Musthafa KS, Sivamaruthi BS, Pandian SK, Ravi AV (2012) Quorum Sensing Inhibition in Pseudomonas aeruginosa PAO1 by Antagonistic Compound Phenylacetic Acid. Current Microbiol 65(5):475–480

Natrah FMI, Ruwandeepika HAD, Pawar S, Karunasagar I, Sorgeloos P, Bossier P, Defoirdt T (2011) Regulation of virulence factors by quorum sensing in *Vibrio harveyi*. Vet Microbiol 154:124–129

Nhan DT, Cam DTV, Wille M, Defoirdt T, Bossier P, Sorgeloos P (2010) Quorum quenching bacteria protect *Macrobrachium rosenbergii* larvae from *Vibrio harveyi* infection. J Appl Microbiol 109:1007–1016

Nithya C, Pandian SK (2010) The *in vitro* antibiofilm activity of selected marine bacterial culture supernatants against *Vibrio* spp. Arch Microbiol 192:843–854

Packiavathy IASV, Sasikumar P, Pandian SK, Ravi AV (2013) Prevention of quorum sensing mediated biofilm development and virulence factors production in *Vibrio* spp. by curcumin. Appl Microbiol Biotechnol 97:10177–10187

Pande GS, Natrah FM, Sorgeloos P, Bossier P, Defoirdt T (2013) The *Vibrio campbellii* quorum sensing signals have a different impact on virulence of the bacterium towards different crustacean hosts. Vet Microbiol 167:540–545

Rama Devi K, Srinivasan R, Kannappan A, Santhakumari S, Bhuvaneswari M, Rajasekar P, Prabhu NM, Veera Ravi A (2016) *In vitro* and *in vivo* efficacy of rosmarinic acid on quorum sensing mediated biofilm formation and virulence factor production in *Aeromonas hydrophila*. Biofouling 32:1171–1183

Randall LP, Cooles SW, Piddock LJ, Woodward MJ (2004) Effect of triclosan or a phenolic farm disinfectant on the selection of antibiotic resistant *Salmonella enterica*. J Antimicrob Chemother 54:621–627

Rasch M, Buch C, Austin B, Slierendrecht WJ, Ekmann KS, Larsen JL, Johansen C, Riedel K, Eberl L, Givskov M, Gram L (2004) An inhibitor of bacterial quorum sensing reduces mortalities caused by vibriosis in rainbow trout (*Oncorhynchus mykiss*, Walbaum). Syst Appl Microbiol 27:350–359

Santhakumari S, Kannappan AR, Pandian SK, Thajuddin N, Rajendran RB, Ravi AV (2015) Inhibitory effect of marine cyanobacterial extract on biofilm formation and virulence factor production of bacterial pathogens causing vibriosis in aquaculture. J Appl Phycol 28:313–324

Sahu MK, Swarnakumar NS, Sivakumar K (2008) Probiotics in aquaculture: importance and future perspectives. Indian J Microbiol 48:299–308

Sivaranjani M, Gowrishankar S, Kamaladevi A, Pandian SK, Balamurugan K, Ravi AV (2016) Morin inhibits biofilm production and reduces the virulence of *Listeria monocytogenes* - An *in vitro* and *in vivo* approach. Int J Food Microbiol 237:73–82

Soto-Rodríguez S, Armenta M, Gomez-Gil B (2006) Effects of enrofloxacin and florfenicol on survival and bacterial population in an experimental infection with luminescent *Vibrio campbellii* in shrimp larvae of *Litopenaeus vannamei*. Aquaculture 255:48–54

Srinivasan R, Rama Devi K, Kannappan A, Pandian SK, Ravi AV (2016) *Piper betle* and its bioactive metabolite phytol mitigates quorum sensing mediated virulence factors and biofilm of nosocomial pathogen *Serratia marcescens in vitro*. J Ethnopharmacol 193:592–603

Stabili L, Schirosi R, Licciano M, Mola E, Giangrande A (2010) Bioremediation of bacteria in aquaculture waste using the polychaete *Sabella spallanzanii*. N Biotechnol 27:774–781

Verschuere L, Rombaut G, Sorgeloos P, Verstraete W (2000) Probiotic bacteria as biological control agents in aquaculture. Microbiol Mol Biol Rev 64:655–671

Vikram A, Jesudhasan PR, Jayaprakasha GK, Pillai BS, Patil BS (2010) Grapefruit bioactive limonoids modulate *E. coli* O157:H7 TTSS and biofilm. Int J Food Microbiol 140:109–116

Vikram A, Jesudhasan PR, Jayaprakasha GK, Pillai SD, Patil BS (2011) Citrus limonoids interfere with *Vibrio harveyi* cell-cell signalling and biofilm formation by modulating the response regulator LuxO. Microbiol 157:99–110

You JL, Xue XL, Cao LX, Lu X, Wang J, Zhang LX, Zhou SN (2007) Inhibition of *Vibrio* biofilm formation by a marine actinomycete strain A66. Appl Microbiol Biotechnol 76:1137–1144

Zhao J, Chen M, Quan CS, Fan SD (2015) Mechanisms of quorum sensing and strategies for quorum sensing disruption in aquaculture pathogens. J Fish Dis 38:771–786

# Ecofriendly Approach for Bioethanol Production from Microalgae

**Pandian Prabakaran, Virumandi Pradeepa, Nagasundaram Rashiya, Sundaram Ravikumar, Sathiamoorthi Thangavelu, and Gopal Selvakumar**

**Abstract** Cultivation of microalgae biomass for bioethanol production appeared as one of the potential solution to drive green and sustainable fuel production. Microalgae are photosynthetic microorganism that can grow rapidly compared to the terrestrial plants and able to accumulate high content of carbohydrate within their cells. The carbohydrate is usually stored in the form of starch, which is easier to breakdown to simple reducing sugar than lignocellulosic biomass. In the present chapter, the process route to produce bioethanol from microalgae biomass is discussed, including the cultivation strategies to enhance microalgae carbohydrate productivity, biomass pre-treatment methods, hydrolysis and fermentation process.

**Keywords** Microalgae · Bioethanol · Carbohydrate · Renewable energy · Fuel

## 1 Introduction

As reviewed from 2014, the global energy demand was increasing with a slower rate than in 2013 (1.1% vs. 2.5%), which recorded a world total primary energy supply (in megatonnes oil equivalent, Mtoe) of 13,700 Mtoe (IEA 2016). Figure 1 illustrated the distribution of Total Primary Energy Supply (TPES) according to different energy sources (Fig. 1a), in which 13.8% or 1894 Mtoe was contributed by renewable energy (Fig. 1b) in the year of 2014 (IEA 2016). Figure 1a shows that fossil fuels remained to be the largest energy consumption, which in turn resulted to the increment of greenhouse gases (GHGs) emission to the atmosphere. Thus, utilization of renewable energy is encouraged to fulfill the requirement of COP21

P. Prabakaran · S. Ravikumar · G. Selvakumar
Department of Biomedical Sciences, Science Campus, Alagappa University, Karaikudi, India

V. Pradeepa · G. Selvakumar
Department of Zoology, Seethalakshmi Achi College for Women, Pallathur, India

N. Rashiya · S. Thangavelu · G. Selvakumar (✉)
Department of Microbiology, Science Campus, Alagappa University, Karaikudi, India
e-mail: gselvakumar75@gmail.com; skumarg@alagappauniversity.ac.in



M. Prashanthi et al. (eds.), *Bioremediation and Green Technologies*, Environmental Science and Engineering, https://doi.org/10.1007/978-3-030-64122-1_21

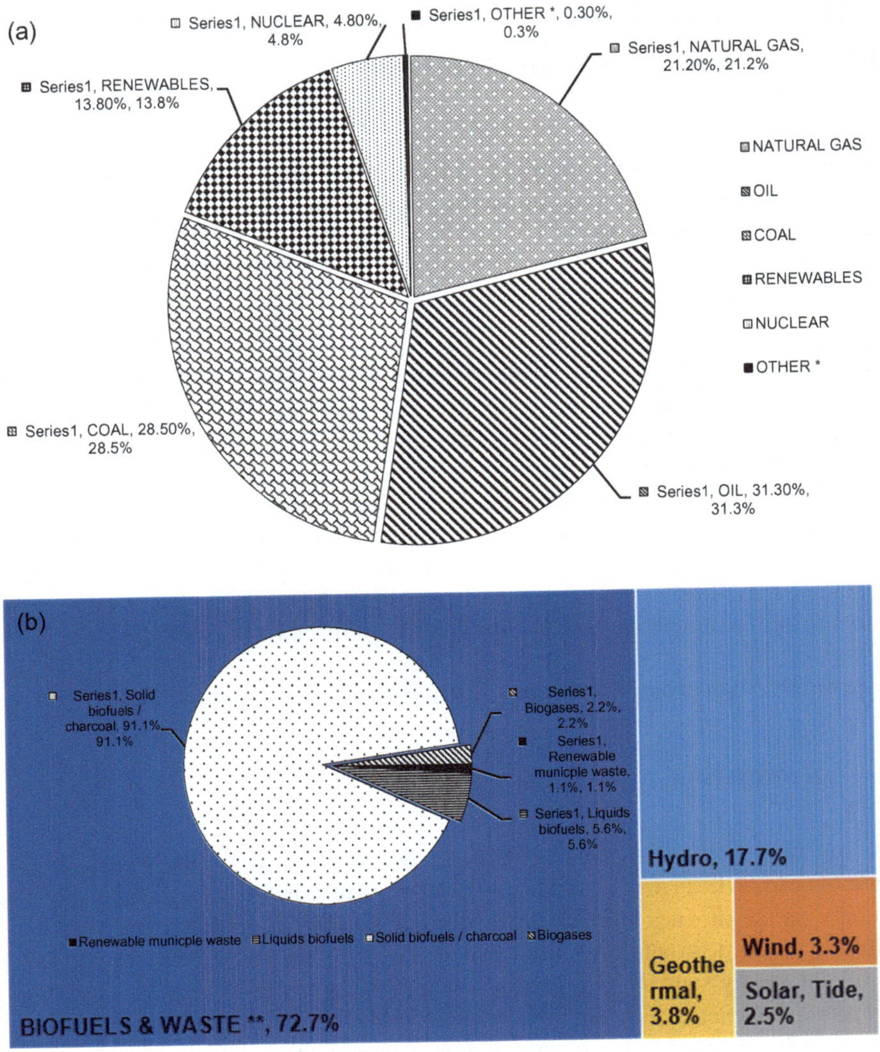


**Fig. 1** Distribution of energy supply by 2014: **a** Fuel shares in world total primary energy supply; and **b** Product shares in world renewable energy supply (IEA 2016). Note: * Others include energy sources such as non-renewable wastes, peat, oil shale and chemical heat. ** Details distribution of Biofuels & Waste energy supply according to categories is represented by the pie chart

(an international environmental agreement that subject to keep the global warming below 2 °C (Ra and Kim 2014; Climate action 2015).

## 2 Bioethanol from Microalgae

Bioethanol emerged as a green and sustainable transportation fuel, as burning this fuel emitted less GHGs than conventional gasoline (RFA 2016; USDOE 2016). However, using sugarcane, coarse grains, wheat and molasses as the main feedstock for bioethanol production could attributed to the two major controversy issues (OECD-FAO 2016): (i) the imbalance of worldwide food markets due to the "food versus fuel" scenario (Lim and Teong 2010); (ii) the use of large agricultural land (Kirkels 2016). Therefore, microalgae with simple cell structure and those that can grow rapidly are potentially employed for sustainable bioethanol production. Furthermore, from the environmental perspectives, cultivation of microalgae coupled with $CO_2$ fixation and bio-treatment of wastewater, evolved to be a potential green energy producer (Maity et al. 2014).

Microalgae are unicellular with size ranging from a few to hundred micrometers (μm). They are commonly found in freshwater or marine water, exists either in the form of individual or groups and chains (Suganya et al. 2016). They are photosynthetic microorganisms that utilized $CO_2$ as carbon source and simultaneously release $O_2$ into the atmosphere (Andersen 2013). The photoautotrophical process (or light-driven redox reaction) convert the absorbed $CO_2$ into carbohydrates and lipid within the microalgae cells (Masojídek et al. 2013).

### *2.1 Effect of Cultivation Conditions to Improve Carbohydrates Production*

Besides screening and choosing carbohydrate, rich microalgae species, inducing specific cultivation conditions is an alternative way to further enhance carbohydrate productivity in the microalgae. In fact, this strategy is also applicable to low carbohydrate content microalgae species, but grows easily and able to withstand extreme environment. The following sections discussed several cultivation strategies to increase the carbohydrate content in microalgae cells.

### *2.2 Effect of Sulfur*

Sulfur is an important element is all living cells, as it is needed to synthesis amino acids (methionine and cysteine), vitamins and sulfolipids (Becker 1994). The strategy to deprive sulfur concentration in microalgae cultivation was previously initiated to produce biohydrogen from *Chlamydomonas reinhardtii* through direct biophotolysis (Melis et al. 2000). When microalgae are cultivated under limited sulfur source, synthesis of the PS II D1 polypeptide chain (32 kDa) is impeded and thus, interfere the repair of PS II from photooxidative damage (Melis 2007 and Wykoff et al. 1998).

As a result, the microalgae cells can maintain high respiration activity that exceeds the rate of photosynthesis, rendering an anaerobic condition in the cultivation system that allows biohydrogen production for 4 to 5 days (Melis 2007).

During biohydrogen production, it was found that the protein and starch content in *Chlamydomonas reinhardtii* were increased 40% and 60%, respectively, from their initial value for the first 24 h and 30 h of sulfur deprivation, but gradually declined thereafter (Zhang et al. 2002). This observation clearly shows that substantial protein and starch catabolism occurs under sulfur limitation condition, and subsequently utilization of these endogenous substrates by microalgae to sustain the production of biohydrogen. For the interest of bioethanol production, once the microalgae cells have accumulated optimum starch concentration under sulfur deprivation, the microalgae need to be immediately harvested instead of extending the cultivation for biohydrogen production.

Apart from that, sulfur deprivation method was found to give more promising result than deprivation of other macronutrients, such as nitrogen and phosphorus (Brányiková et al. 2011). In a study done by Brányiková et al. (2011), when *Chlorella vulgaris* was cultivated with limited nitrogen or phosphorus nutrients, the starch accumulation in the microalgae cells could reached to 55% and 38%, respectively; however, after 20 h of cultivation, the microalgae stopped to synthesize starch and its content began to decrease (Brányiková et al. 2011). On the other hand, under sulfur limited condition, the *Chlorella vulgaris* was able to maintain 60% of starch content even after 32 h that allowed it to attain optimum cell density (stationary phase). In addition, the study also revealed plausible result with sulfur limitation method in a scaled-up open door thin-layer solar photobioreactor (working volume was 250 L). Under outdoor condition, *Chlorella vulgaris* successfully accumulated 50% of starch content, although the cultivation was only exposed to sunlight for 11 h with inconsistent of surrounding temperature.

## 2.3 Effect of Nitrogen

Nitrogen is an essential element of photosynthesis components and usually participating in DNA, proteins, amino acids, enzymes and pigments (Markou et al. 2012). Hence, sufficient nitrogen concentration is always necessary to accelerate the growth of microalgae and to enhance biomass productivity. However, recent studies have shown that when microalgae are cultivated under nitrogen deficient condition, higher accumulation of lipid or carbohydrate could be attained. This is because protein synthesis pathway is diverted to form lipid or carbohydrate as the main energy reserve components in microalgae cells. However, the main drawback of this cultivation method is low biomass yield, predominantly due to the decrease in the synthesis of photosynthesis pigments (chlorophyll) that subsequently inhibit the growth of microalgae (Berges et al. 1996). Thus, two-stage cultivation method has been recommended to enhance microalgae biomass productivity with high lipid or carbohydrate content (Widjaja et al. 2009). In the first stage, microalgae are cultivated in nutrients-rich medium until sufficient biomass is produced, and then transferred to nitrogen-deficient medium to trigger the lipid or carbohydrate accumulation.

Apparently, the competition between lipid and carbohydrate accumulation in microalgae cells under nitrogen deficient condition is still unclear as these two compounds shared similar synthesis pathway (Li et al. 2008). However, some recent studies have reported that cultivation time played a significant role in manipulating the accumulation of lipid or carbohydrate in microalgae cells. For example, the carbohydrate content in *Scenedesmus obliquus* was increased from 20.90% to 49.36% after two days of nitrogen starvation, while the protein content decreased from 50.43% to 27.38% (Ho et al. 2013). Nevertheless, for the subsequent days of cultivation, the carbohydrate content was decreased to 46.98% whereas the lipid content was started to increase from 11.72% to 13.66%. Similar results are also observed in *Chlorella vulgaris* and *Chlorella zofingiensis*.

## 2.4 Effect of Phosphorus

Phosphorus is essential macronutrient that regulates microalgae growth and metabolism. It is mainly incorporated into nucleic acid and phospholipids, an important modifier of protein function, and involved in generating metabolic energy (Yao et al. 2013). Up to now, the effect of phosphorus on carbohydrate accumulation in microalgae cells is not well understood and it is strongly believed to be species-specific (González-Fernández and Ballesteros 2012). Nevertheless, it is well established that carbohydrate accumulation started to increase when intracellular phosphorus content decreased to a critical level (Cade-Menun and Paytan 2010). In a recent study by Brányiková et al. (2011), *Chlorella vulgaris* tended to accumulate higher starch content (55%) under phosphorus starvation condition instead of nitrogen starvation condition (38%) (Rodolfi et al. 2009). In addition, an added advantage of inducing phosphorus starvation instead of nitrogen starvation to microalgae cultivation is that higher biomass productivity can be attained, that subsequently resulted to higher carbohydrate yield. However, in some studies, phosphorus starvation only caused lipid accumulation with no significant effect on carbohydrate accumulation. For example, cultivation of *Tetraselmis subcordiformis* under phosphorus starvation did not increase the carbohydrate content in the microalgae but the concentration remained constant throughout 8 days of cultivation (Ji et al. 2011).

## 2.5 Effect of Carbon Source

***Inorganic carbon***

It is a common practice to cultivate phototrophic microalgae by supplying $CO_2$ as the main carbon source. Through photosynthesis process, $CO_2$ is absorbed by microalgae cells to support their growth by converting the carbon to build proteins, nucleic acids, carbohydrate and lipids (Beer et al. 2009). However, it is interesting to note that the carbohydrate content in microalgae cells usually does not increase linearly with the $CO_2$ concentration. Based on several recent studies, the increment of carbohydrate content in *Chlorella, Scenesdusmus obliquus* and *Chlamydomonas reinhardtii* was not significantly increased although high $CO_2$ concentration was supplied (Thyssen et al. 2001). In a study carried out by Izumo et al. (2007), the carbohydrate content in *Chlorella* was reduced when switching the $CO_2$ supply from 0.04% (atmospheric air) to 3%. It was observed that when supplying low concentration of $CO_2$, the location of starch was shifted from stroma to pyrenoid. As a result, the expression of granule-bound starch synthase (GBSS) was increased, leading to higher accumulation of starch content under low concentration of $CO_2$. However, it is well established that supplying high concentration of $CO_2$ to microalgae cultivation could accelerate the biomass productivity and thus, shorten the overall cultivation time (Kumar et al. 2010). Subsequently, by inducing appropriate stress conditions to the microalgae cultivation, such as nitrogen depletion, the overall carbohydrate yield could be further increased as high biomass productivity were already attained previously (Ho et al. 2013).

***Organic carbon***

Besides $CO_2$, specific microalgae species could utilize organic substrate (e.g. glucose, acetate, glycerol and etc.) as carbon and energy source to grow (Mata et al. 2010). Usually, this is referred as heterotrophic cultivation (independent of light). The consumption of organic substrates by microalgae is achieved through several mechanisms, such as phosphorylation (glucose), simple diffusion (glycerol) or using membrane transporter proteins (organic acids) (Markou et al. 2012). Up to now, several microalgae species such as *Chlorella protothecoide* (Cheng et al. 2009), *Chlorella vulgaris* (Liang et al. 2009), *Crypthecodinium cohnii* (Couto et al. 2010), *Neochloris oleoabundans* (Morales-Sánchez et al. 2013) and *Schizochytrium limacinum (*Johnson and Wen 2009) have been studied under heterotrophic growth conditions to achieve high biomass productivity.

However, most heterotrophic cultivation studies are focused on lipid productivity and studies related to carbohydrate production is rarely reported in the literature. This is because the commonly used organic carbon source in growing heterotrophic microalgae is glucose, which is a derivative of carbohydrate after subjected to hydrolysis process. In other words, the net production of carbohydrate in heterotrophic cultivation could be significantly lower than phototrophic cultivation. In a recent study by Morales-Sánchez et al. (2013), 54 wt % of carbohydrate content (comprised of 90% glucose) was observed when *Neochloris oleoabundans* was cultivated under fed-batch heterotrophic condition using glucose as carbon source. Nevertheless, in term of glucose balance, the overall glucose produced by *Neochloris oleoabundans* was estimated to be 7.67 g/L (based on biomass yield of 14.2 g/L), while the total glucose consumed by the microalgae to grow was 50 g/L. Thus, it is clear that high carbohydrate content attained in microalgae under heterotrophic cultivation does not directly indicates the potential of this approach, but the overall mass balance of carbohydrate should be taken into consideration.

### 2.6 *Light Intensity*

Light is the main energy provider for photosynthetic microalgae. Providing suitable light intensity does not only improve biomass productivity, but also can change the biochemical compositions of microalgae (Hu 2007). Through illumination with high light intensity, the carbohydrate and lipid content in microalgae cells could be further enhanced (Lv et al. 2010). Recent study have shown that supplying light intensity between 200 and 420 μmol $m^{-2}$ $s^{-1}$ could enhance the carbohydrate content of *Scenedesmus obliquus* (Ho et al., 2012). At a light intensity of 420 μmol $m^{-2}$ $s^{-1}$, *Scenedesmus obliquus* accumulated the highest carbohydrate content of 38%; which was about 2.3 times higher compared to cultivation supplied with a light intensity of 60 μmol $m^{-2}$ $s^{-1}$. The same observation was also reported by Brányiková et al. (2011), in which the starch content in *Chlorella* was increased from 8.5% to 40% when the light intensity was switched from 215 μmol $m^{-2}$ $s^{-1}$ to 330 μmol $m^{-2}$ $s^{-1}$. However, it should be noted that providing extremely high light intensity could reduce the biomass productivity of microalgae, which is generally known as photo-inhibition (Chisti 2008). The carbohydrate productivity in *Scenedesmus obliquus* was reported to reduce by about 45% when high intensity was increased from 420 to 540 μmol $m^{-2}$ $s^{-1}$ (Ho et al. 2012).

## 3 Carbohydrate in Microalgae

Microalgae cells are composed by different polysaccharides (polymeric carbohydrate molecules that bonded by glycosidic linkages of long chains of monosaccharides unit) that can be broken down into monosaccharides through pre-treatments and hydrolysis processes. The produced monosaccharides (e.g. glucose) can be converted to bioethanol through fermentation (Barsanti and Gualtieri 2014). Microalgae species such as *Chlorella, Chlamydomonas, Dunaliella,* and *Scenedesmus* consists of high carbohydrates content, which is suitable to be used as the feedstock for bioethanol production (Harun et al. 2014). Cellulose structure has been the research focus in recent reported studies, particularly in bioethanol production.

The types of classes of polysaccharides that found in green microalgae are cellulose, chrysolaminarin, floridean starch, paramylon and starch (Barsanti and Gualtieri 2014). Among the listed polysaccharides, cellulose and starch are the widely studied classes in microalgae research for bioethanol production.

### 3.1 *Cellulose*

The microalgae phyla that have cellulose storage are Chlorophyceae (Chlorophyta), Dinophyta, Phaeophyta (Ochrophyta), Prymnesiophyceae, Rhodophyceae

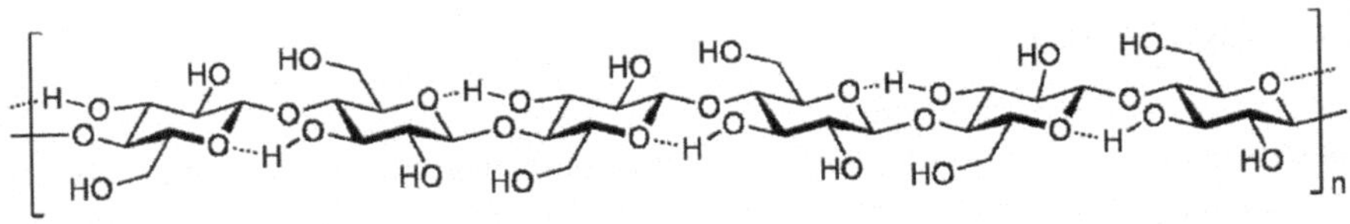

**Fig. 2** Molecular constitution of cellulose (β-1,4-linked polymer of D-glucose) (Stick and Williams 2009)

(Rhodophyta) and Xanthophyceae (Ochrophyta) (Barsanti and Gualtieri 2014). Cellulose is the major component of plant fibres that found in the secondary cell wall, embedded in a matrix of other polysaccharides (hemicelluloses, pectins, and lignins) to form a biocomposite. The molecular constitution of cellulose is β-1,4-linked polymer of D-glucose (Fig. 2). Cellulose is easily to degrade via cellulolytic enzymes. These enzyme complexes (cellulosomes) consist of endo- and exo-cellulase, β-glucosidases and carbohydrate-binding proteins were bind to glycoside hydrolase, are capable to degrade the cellulose for bioethanol production (Stick and Williams 2009).

## *3.2 Starch*

Starch is a complex carbohydrate that found inside cytoplasm as insoluble and semicrystalline granules, but outside of chloroplast, functioned as energy reservoir (Barsanti and Gualtieri 2014). Starch granules comprise of α-glucans that formed by amylose and amylopectin (Fig. 3). Amylose is a linear polymer consists of several thousand of α-1,4-linked D-glucose residues; whereas amylopectin is a branched

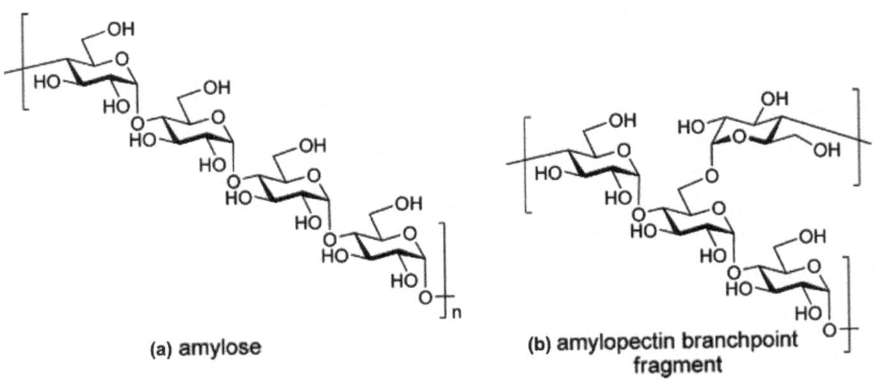


**Fig. 3** Molecular constitution of starch granules: **a** Amylose: linear polymer of α-1,4-linked D-glucose residues; and **b** Amylopectin: branched polymer consists of α-1,4-linked D-glucose residues and α-1,6-linked branch (Stick and Williams 2009)

polymer consists of α-1,4-linked D-glucose residues and α-1,6-linked branch for each 24–30 residues (Stick and Williams 2009).

Similar to cellulose, starches can be degraded by enzymatic hydrolysis to produce bioethanol. Amylases are among the most commonly used enzymes to hydrolyse starch (Simas-Rodrigues et al. 2015).

## 4 Overview of Processes for Microalgae Bioethanol Production

Microalgae based bioethanol can be produced via the following chemical equation, in which the sugar produced from microalgae biomass are fermented into bioethanol:

$$C_nH_{2n}O_n(\text{sugar}) \rightarrow \frac{n}{3}C_2H_5OH + \frac{n}{3}CO_2 + \text{Heat}. \quad (1)$$

The primary carbohydrates ($(C_nH_{2n}O)_n$) source is usually the simple sugar, such as glucose, starch, and cellulose available in microalgae biomass. Generally, bioethanol production from microalgae involved several processes, such as pre-treatment of biomass, hydrolysis or saccharification of complex carbohydrates, fermentation of bioethanol, and bioethanol recovery, as summarized in Fig. 4 (Harun et al. 2014).

Up to date, various researches have been conducted to enhance the microalgae based bioethanol production, especially in the aspect of feedstock pre-treatment and hydrolysis (saccharification) processes. The performance of fermentation process is greatly depending on the operating conditions of the biomass pre-treatment and their respective carbohydrates hydrolysis processes (Harun et al. 2014). This is due to the formation of undesired by-products, such as formic acid, acetic acid, and furanic compounds from the degradation of the hydrolysed carbohydrates during fermentation process (Hargreaves et al. 2013). Basically, the pre-treatment process is functionalized to disrupt the microalgae biomass cell wall and to release the entrapped complex carbohydrates for the subsequent hydrolysis process. The saccharification process is referred to the combination of pre-treatment and enzymatic hydrolysis processes.

## 5 Pre-treatment and Hydrolysis

Since pre-treatment and hydrolysis of microalgae biomass feedstock are playing a vital role in bioethanol production, it is important to optimize the operating conditions prior to the subsequent fermentation process. There are a number of important features to classify the effectiveness of microalgae biomass pre-treatment: (i) quantifying the carbohydrates and sugars contents of the post-filtered pre-treated samples; (ii) choosing the source of bioethanol based on sugar and carbohydrates analyses,

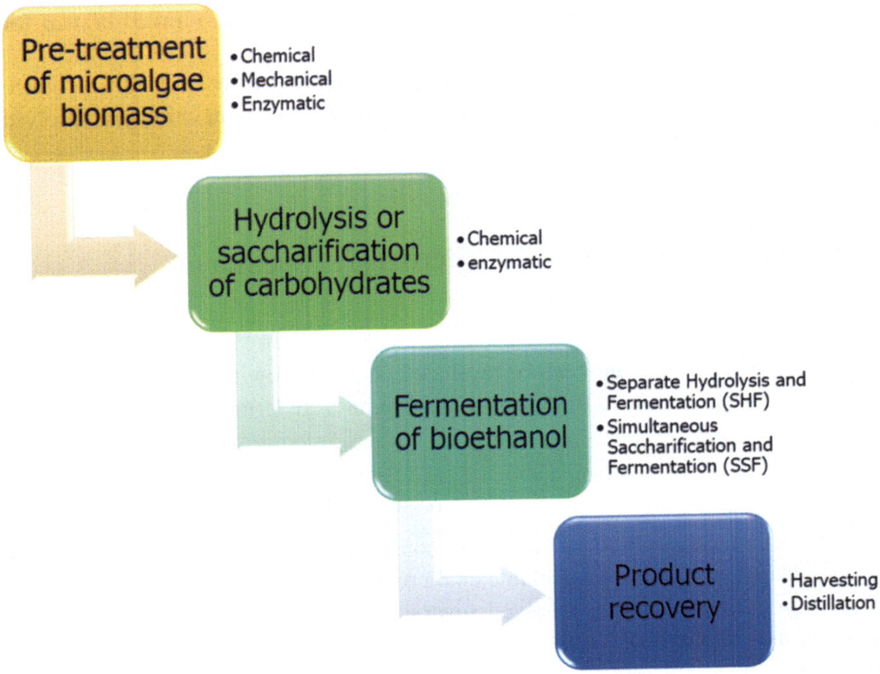


**Fig. 4** Routes for microalgae based bioethanol production

either from liquid hydrolysate or water insoluble solids phase; (iii) investigating the potential inhibitory effects of pre-treat samples and neutralizing the samples subjected to fermentation; and (iv) evaluating the potential value-added products from the pre-treated samples (Agbor et al. 2011). The pre-treatment of microalgae biomass can be categorized into: (i) physical; (ii) chemical; and (iii) enzymatic. The incorporation of the biomass pre-treatment with hydrolysis process, particularly in sugars extraction methods are further discussed in the sub-sections below.

## *5.1 Physical Pre-treatment*

High-pressure homogenization, thermal heating, and ultrasonication can accomplish physical pre-treatments, which aimed to disrupt the microalgae biomass cell wall through mechanical force (Hernández et al. 2015). The ultrasonic employed cavitation process to disrupt the cells, in which the pressures created by bubbles are, collapsed violently on the microalgae cellular walls. During sonication process, excessive heat was dissipated, thus it is important to keep the pre-treatment samples in ice cold condition (Miranda et al. 2012).

Ultrasonic assisted extraction (UAE) was an effective biomass pre-treatment for carbohydrates extraction, as compared with conventional solvent extraction (CSE) and fluidized bed extraction (FBE) methods (Zhao et al. 2013).The glucose extraction by UAE showed an increment of almost fourfold (36.94 g/100 g of DCW biomass), as compared with CSE and FBE methods. In addition, (Hwang et al. 2016) also found that the yeast promoted fermentative bioethanol production can be further enhanced by joint sonication, heat and enzyme (SHE) on biomass pre-treatments, which led to complete destruction of biomass cell walls. Initially, filamentous microalgae were partially disrupted by sonication, whereas complete cells wall (filamentous and cyclotella) destruction was achieved by thermal heating and enzymes.

### 5.2 *Chemical Pre-treatment*

The chemicals hydrolysis method appeared to be a rapid and cheaper method than enzymatic hydrolysis. Usually, acids or alkalis are used for microalgae biomass saccharification, with acids is more commonly employed. Acid hydrolysis is capable to break the intra- and inter-molecular hydrogen bonds across the glycosidic link and thus, further enhance the breakdown of carbohydrates (Simas-Rodrigues et al. 2015). Among the available acids, sulfuric acid (H2$SO_4$) is the most attractive choice for biomass saccharification process. This is due to the advantage of released anions ($SO_4^{2-}$) as nutrient source for yeast fermentation process (Miranda et al. 2012). The optimization parameters of chemical pre-treatment includes: (i) time; (ii) temperature; and (iii) acid or alkaline concentration.

By referring to the acid hydrolysis performed by Miranda et al. (2012), it was found that the optimum operating conditions for *Scenedesmus obliquus* biomass could be achieved by using 2 N of $H_2SO_4$, at temperature of 120 °C for 30 min. The sugar yield attained was 95.6%, indicating that low cost and low energy chemical pre-treatment method could be employed to release the sugar from microalgae biomass efficiently.

Besides, alkaline-peroxide also demonstrated a significantly impact on sugars extraction (maximum 87%) from microalgae biomass cultivated in domestic and pig manure wastewaters. This is due to the enhancement of oxidation of sugars to form organic acids during fermentation process (Martin et al. 2016).

### 5.3 *Enzymatic Pre-treatment*

Enzymatic pre-treatment degrades the cellulose for the subsequent hydrolysis process. As compared with the previous discussed pre-treatment (physical and chemical) methods, enzymatic pre-treatment appeared to be advantages as this method consume less energy, mild conditions and will not incurred corrosion issue to the fermenter. However, the disadvantages of enzymatic pre-treatment are the enzymatic rate is rather too slow for commercial application (10–14 days) (Agbor et al. 2011) and high enzymes specificity towards different microalgae species (Yuan et al. 2016). The current commercial enzymes are cellulase and amylase, which are used to hydrolyze cellulose and starch, respectively (Yuan et al. 2016).

Kim et al. (2014) used cellulase and pectinase to saccharify *Chlorella vulgaris*. From the results obtained, *Chlorella vulgaris* cultivated under nitrogen stress condition were able to increase the carbohydrates content within their cells. By using *Aspergillus* pectinase enzyme, 79% of carbohydrates were successfully extracted from the saccharification process. The study also evaluated the activity of immobilized yeast fermentation and attained 89% of conversion. This can be concluded that immobilized enzymes on support were able to enhance the bioethanol production yield.

## 6 Fermentation Process

Free sugars extracted from microalgae biomass can be fermented into bioethanol through some ethanologen microorganisms (bacteria, filamentous fungi, yeast). There are two different types of fermentation process: (i) separate hydrolysis and fermentation (SHF); and (ii) simultaneous saccharification and fermentation (SSF). The comparison of both SHF and SSF is tabulated in Table 1.

Depending on the microalgae species and cultivation conditions, different hydrolysis and fermentation (SSF and SHF) approaches are introduced for bioethanol production.

**Table 1** Comparison of SHF and SSF process

| Fermentation | Benefit | Drawback |
|---|---|---|
| SHF | • Easier to optimize the hydrolysis and fermentation operating conditions<br>• Simpler equipment design<br>• Shorter residence time<br>• Using cheaper chemicals | • Higher contamination rates<br>• Inhibitory effects |
| SSF | • Higher bioethanol yields<br>• Higher rates of hydrolysis<br>• Less inhibitory effects<br>• Lower operating cost<br>• Lower contamination rate<br>• Required less sterile conditions<br>• Smaller amount of enzyme intake<br>• Shorter residence time | • Only can performed optimal operating conditions for either hydrolysis or fermentation process<br>• Difficulty in process control |

### 6.1 Separate Hydrolysis and Fermentation (SHF)

During SHF, both of the enzymatic hydrolysis pre-treated biomass and fermentation process take place in two different units of reactors. Hence, optimizations of operating conditions for each stage are able to be performed. However, the inhibitory effect from the accumulation of hydrolysis products (especially glucose and cellobiose), will suppress the activity of cellolases and resulted to slower hydrolysis rate (Balat et al. 2008).

### 6.2 Simultaneous Saccharification and Fermentation (SSF)

In SSF, enzymatic hydrolysis and fermentation take place in a single reactor. Therefore, the hydrolysis products (glucose and cellobiose) is consumed by the yeast in fermentation process, which could greatly reduce the inhibitory effects (El-Dalatony et al. 2016). In addition, higher concentration of bioethanol resulted to lower degree of contamination, due to the lower solids contents yields (>15% dry weight basis) (Harun et al. 2011). After all, since both of the enzymatic hydrolysis and fermentation are operated under the same conditions, it would be difficult to attain an optimal temperature for SSF operation.

## 7 Microorganism for Fermentation of Microalgae Biomass

The performance features of fermentation process include: temperature, pH, microorganism growth rate, bioethanol productivity and yield, alcohol tolerance, osmotic tolerance, inhibitor tolerance, and genetic stability (Balat et al. 2008). It is well known that *Saccharomyces cerevisiae* is the most efficient yeast in fermentative bioethanol. It has shown a good alcohol and inhibitor tolerances associated with high bioethanol productivity *Saccharomyces cerevisiae* was utilized to produce bioethanol via SHF and SSF processes (El-Dalatony et al. 2016). The yeast was able to maximize the sugars consumption efficiency (91–98%) and achieving 0.5 g/g or an equivalent of 88.2% (theoretical) of bioethanol yield from SSF. However, energy recovery was higher in SHF mode (85.96%) than SSF (70%).

Other than that, others microorganism such as bacteria are also used in microalgae biomass fermentation process. These bacteria include *Bacillus stearothermophilus*, *Clostridium thermohydrosulfuricum*, *Clostridium thermocellum*, *Escherichia coli*, *Klebsiella oxytoca*, *Thermoanaerobacterium thermosaccharolyticum*, *Thermoanaerobacter ethnolicus*, and *Zymomonas mobilis* (Simas-Rodrigues et al. 2015). Recently, *Z. mobilis* was reported to be an efficient bacterium in sugars fermentation, in which the sugars (e.g. glucose and fructose) uptake rate was increased, high tolerance to ethanol content (~11% v/v) and able to withstand high osmotic pressure caused by higher sugar concentrations (Kalnenieks et al. 2014). Yang et al. (2016) summarized that *Z. mobilis* was able to utilize glucose, sucrose, and fructose to produce bioethanol. For instance, a production of 78 g/L of bioethanol under batch fermentation (pH 6 and 32 °C) with an initial carbon content of 295 g/L; 20.5 g/L of bioethanol from 60 g/L of xylose; 72 g/L of bioethanol from 150 g/L of fructose under continuous fermentation with immobilized cells (pH 5 and 30 °C); and 87.2 g/L of bioethanol from sucrose (350 g/L) under repeated batch fermentation with immobilized cells (pH 4 and 30 °C).

## 8 Products Recovery

The commercial production of bioethanol required a product recovery process (Balat et al. 2008). The bioethanol can be recovered from distillation process, where the bioethanol distillate is concentrated (~37%) below the azeotrope in distillation column, whereas the bottom product is discharged to stripping column to remove excessive water, and the remaining bioethanol was recycled back to the feed. The recovery of bioethanol process can be controlled at 99.6% to minimize the lost. On the other hand, the solids product from the fermentation process was separated by a centrifuge, followed by drying process in rotary dryer. Then, the solids products are recovered in the evaporation process as clean condensate and concentrated syrup (15–20% of total solids weight).

## 9 Conclusion

Microalgae biomass is the third generation feedstock for biofuel production. Besides the accumulation of lipid within the microalgae cells for biodiesel production, the stored carbohydrate also can be further utilized for bioethanol production. It is foreseen that bioethanol derived from microalgae biomass could pave a sustainable route to substitute the gasoline in the global market. However, more researches are still required to justify the commercial feasibility of microalgae bioethanol production, especially the life cycle impact assessment and techno-economic analysis.

**Acknowledgements** We gratefully acknowledge RUSA—Phase 2.0 grant sanctioned vide Letter No.F.24-51/2014-U, Policy (TN Multi-Gen), Dept. of Edn. Govt. of India, Dated 09.10.2018 and financial assistance rendered to Ms. N. Rashiya in the form of a UGC – RGNF by University Grants Commission, New Delhi [No. F1 – 17.1/2014-15/RGNF-2014-15-SC-TAM-75166].

## References

Andersen RA (2013) The microalgal cell. In: Richmond A, Hu Q (eds) Handbook of microalgal culture: applied phycology and biotechnology. Wiley, New York, pp 3–20

Agbor VB et al (2011) Biomass pretreatment: fundamentals toward application. Biotechnol Adv 29:675–685

Balat M et al (2008) Progress in bioethanol processing. Prog Energy Combust Sci 34:551–573

Barsanti L, Gualtieri P (2014) General overview. In: Algae: anatomy, biochemistry, and biotechnology, 2nd ed. CRC Press, New York, pp 1–48

Barsanti L, Gualtieri P (2014) Algae utilization. In: Algae: anatomy, biochemistry, and biotechnology, 2nd ed. CRC Press, New York, pp 267–308

Becker EW (1994) Microalgae: biotechnology and microbiology. Cambridge University Press, Cambridge

Beer LL et al (2009) Engineering algae for biohydrogen and biofuel production. Curr Opin Biotechnol 20:264–271

Berges JA et al (1996) Differential effects of nitrogen limitation on photosynthetic efficiency of photosystems I and II in microalgae. Plant Physiol 110:689–696

Brányiková I et al (2011) Microalgae-novel highly efficient starch producers. Biotechnol Bioeng 108:766–776

Cade-Menun BJ, Paytan A (2010) Nutrient temperature and light stress alter phosphorus and carbon forms in culture-grown algae. Mar Chem 121: 27–36

Cheng Y et al (2009) Biodiesel production from Jerusalem artichoke (Helianthus Tuberosus L.) tuber by heterotrophic microalgae Chlorella protothecoides. J Chem Technol Biotechnol 84:777–781

Chisti Y (2008) Biodiesel from microalgae beats bioethanol. Trends Biotechnol 26:126–131

Climate Action (2015, September 10) What is COP21? Available: https://www.cop21paris.org/about/cop21

Couto RM et al (2010) Supercritical fluid extraction of lipids from the heterotrophic microalga Crypthecodinium cohnii. Eng Life Sci 10:158–164

El-Dalatony MM et al (2016) Long-term production of bioethanol in repeated-batch fermentation of microalgal biomass using immobilized Saccharomyces cerevisiae. Bioresour Technol 219:98–105

Ghirardi ML et al (2009) Photobiological hydrogen-producing systems. Chem Soc Rev 38:52–61

González-Fernández C, Ballesteros M (2012) Linking microalgae and cyanobacteria culture conditions and key-enzymes for carbohydrate accumulation. Biotechnol Adv 30:1655–1661

Hargreaves PI et al (2013) Production of ethanol 3G from Kappaphycus alvarezii: Evaluation of different process strategies. Bioresource Technol 134: 257–263

Harun R, Jason WSY, Cherrington T, Danquah MK (2011) Exploring alkaline pre-treatment of microalgal biomass for bioethanol production. Appl Energy 88:3464–3467

Harun R et al (2011) Analysis of process configurations for bioethanol production from microalgal biomass. In: Shaukat SS (ed) Progress in biomass and bioenergy production. Intech Science, Technology & Medicine, Croatia, pp 978–53

Harun R et al (2014) Algal biomass conversion to bioethanol—a step-by-step assessment. Biotechnol J 9:73–86

Hernández D et al (2015) Saccharification of carbohydrates in microalgal biomass by physical, chemical and enzymatic pre-treatments as a previous step for bioethanol production. Chem Eng J 262:939–945

Ho SH et al (2011) Perspectives on microalgal $CO_2$-emission mitigation systems—A review. Biotechnol Adv 29:189–198

Ho SH et al (2012) Effect of light intensity and nitrogen starvation on $CO_2$ fixation and lipid/carbohydrate production of an indigenous microalga Scenedesmus obliquus CNW-N. Biores Technol 113:244–252

Ho SH et al (2013) Bioethanol production using carbohydrate-rich microalgae biomass as feedstock. Biores Technol 135:191–198

Ho SH et al (2013) Characterization and optimization of carbohydrate production from an indigenous microalga Chlorella vulgaris FSP-E. Biores Technol 135:157–165
Ho S-H et al (2013) Bioprocess development on microalgae-based CO2 fixation and bioethanol production using Scenedesmus obliquus CNW-N. Biores Technol 145:142–149
Ho SH et al (2013) Engineering strategies for improving the $CO_2$ fixation and carbohydrate productivity of Scenedesmus obliquus CNW-N used for bioethanol fermentation. Biores Technol 143:163–171
Hwang J-H et al (2016) Enhancement of continuous fermentative bioethanol production using combined treatment of mixed microalgal biomass. Algal Res 17:14–20
Hu Q (2007) Environmental effects on cell composition. In: Handbook of microalgal culture. Blackwell Publishing Ltd, pp 83–94
International Energy Agency (2016) Excerpt from: renewables information. IEA Publishing, United States of America, Statistics
Izumo A et al (2007) Physicochemical properties of starch in Chlorella change depending on the $CO_2$ concentration during growth: Comparison of structure and properties of pyrenoid and stroma starch. Plant Sci 172:1138–1147
Ji CF et al (2011) Effects of nutrient deprivation on biochemical compositions and photo-hydrogen production of Tetraselmis subcordiformis. Int J Hydrogen Energy 36:5817–5821
Johnson MB, Wen Z (2009) Production of biodiesel fuel from the microalga schizochytrium limacinum by direct transesterification of algal biomass. Energy Fuels 23:5179–5183
Kalnenieks U et al (2014) Modeling of Zymomonas mobilis central metabolism for novel metabolic engineering strategies. Front Microbiol 5(42):2014
Kirkels A (2016) Biomass boom or bubble? A longitudinal study on expectation dynamics . Technol Forecast Soc Chang 103:83–96
Kumar A et al (2010) Enhanced $CO_2$ fixation and biofuel production via microalgae: Recent developments and future directions . Trends Biotechnol 28:371–380
Kim KH et al (2014) Bioethanol production from the nutrient stress-induced microalga Chlorella vulgaris by enzymatic hydrolysis and immobilized yeast fermentation. Biores Technol 153:47–54
Li Y et al (2008) Effects of nitrogen sources on cell growth and lipid accumulation of green alga *Neochloris oleoabundans*. Appl Microbiol Biotechnol 81:629–636
Liang Y et al (2009) Biomass and lipid productivities of *Chlorella vulgaris* under autotrophic, heterotrophic and mixotrophic growth conditions. Biotech Lett 31:1043–1049
Lim S, Teong LK (2010) Recent trends, opportunities and challenges of biodiesel in Malaysia: An overview. Renew Sustain Energy Rev 14:938–954
Lv et al (2010) Enhanced lipid production of Chlorella vulgaris by adjustment of cultivation conditions J. M. Bioresource Technol 101: 6797–6804
Maity JP et al (2014) Microalgae for third generation biofuel production, mitigation of greenhouse gas emissions and wastewater treatment: Present and future perspectives – A mini review, Energy
Martin J et al (2016) Saccharification of microalgae biomass obtained from wastewater treatment by enzymatic hydrolysis. Effect of alkaline-peroxide pretreatment. Bioresour Technol 218:265–271
Masojídek J et al (2013) Photosynthesis in microalgae. In: Richmond A, Hu Q (eds) Handbook of microalgal culture: applied phycology and biotechnology, 2nd ed. Wiley, New York, pp 21–36
Melis A et al (2000) Sustained photobiological hydrogen gas production upon reversible inactivation of oxygen evolution in the green alga Chlamydomonas reinhardtii. Plant Physiol 122:127–135
Melis A (2007) Photosynthetic $H_2$ metabolism in Chlamydomonas reinhardtii (unicellular green algae). Planta 226:1075–1086
Markou G et al (2012) Microalgal carbohydrates: an overview of the factors influencing carbohydrates production, and of main bioconversion technologies for production of biofuels. Appl Microbiol Biotechnol 96:631–645
Mata TM et al (2010) Microalgae for biodiesel production and other applications: A review. Renew Sustain Energy Rev 14:217–232
Morales-Sánchez D et al (2013) Heterotrophic growth of Neochloris oleoabundans using glucose as a carbon source. Biotechnol Biofuels 6

Miranda JR et al (2012) Pre-treatment optimization of Scenedesmus obliquus microalga for bioethanol production. Biores Technol 104:342–348

OECD/FAO (2016) OECD-FAO Agricultural Outlook 2016–2025. OECD Publishing, Paris 04 July 2016

Ra CH, Kim SK (2014) Bioethanol production from macroalgae and microbes. In: Kim S-K, Lee C-G (eds) Marine bioenergy: trends and developments. CRC Press, Taylor & Francis Group, New York, pp 257–271

Renewable Fuels Association (2016) Fueling a high octane future. RFA Publishing, New York

Rodolfi L et al (2009) Microalgae for oil: Strain selection, induction of lipid synthesis and outdoor mass cultivation in a low-cost photobioreactor. Biotechnol Bioeng 102:100–112

Simas-Rodrigues C et al (2015) Microalgae for economic applications: advantages and perspectives for bioethanol. J Exp Bot 66:4097–4108

Stick RV, Williams SJ (2009) Disaccharides, Oligosaccharides and Polysaccharides (Chap. 9). In: Carbohydrates: the essential molecules of life, 2nd ed. Elsevier, Amsterdam, pp 321–342

Suganya T et al (2016) Macroalgae and microalgae as a potential source for commercial applications along with biofuels production: A biorefinery approach. Renew Sustain Energy Rev 55:909–941

Thyssen C et al (2001) The $CO_2$-concentrating mechanism in the physiological context: Lowering the $CO_2$ supply diminishes culture growth and economizes starch utilization in *Chlamydomonas reinhardtii* . Planta 213:629–639

U.S. Department of Energy (2016, March) Ethanol Vehicle Emissions. Available: https://www.afdc.energy.gov/vehicles/flexible_fuel_emissions.html

Widjaja A et al (2009) Study of increasing lipid production from fresh water microalgae *Chlorella vulgaris*. J Taiwan Inst Chem Eng 40:13–20

Wykoff DD et al (1998) The regulation of photosynthetic electron transport during nutrient deprivation in *Chlamydomonas reinhardtii*. Plant Physiology 117:129–139

Yang S et al (2016) Zymomonas mobilis as a model system for production of biofuels and biochemicals. Microbial Biotechnol

Yuan T et al (2016) Microalgae pretreatment with liquid hot water to enhance enzymatic hydrolysis efficiency. Biores Technol 220:530–536

Yao CH et al (2013) Characterization of cell growth and starch production in the marine green microalga Tetraselmis subcordiformis under extracellular phosphorus-deprived and sequentially phosphorus-replete conditions. Appl Microbiol Biotechnol 97:6099–6110

Zhao G et al (2013) Ultrasound assisted extraction of carbohydrates from microalgae as feedstock for yeast fermentation. Biores Technol 128:337–344

Zhang L et al (2002) Biochemical and morphological characterization of sulfur-deprived and $H_2$-producing *Chlamydomonas reinhardtii* (green alga). Planta 214:552–561

# Novel QR Code Tagging System for Campus Vegetation to Promote Ecorestoration

**V. E. Chahana, S. Abirami, S. Karpagavalli, and C. Arunpriya**

**Abstract** This work is an effort to create a means to establish environment sustainability through QR code tagging to trees. In this digital era, with increased usage of smartphones, tablets and the availability of these devices at a lower cost motivated us to adopt QR code tagging for obtaining detailed information about the flora collection available in our college. The objective of this work is to tag each tree in the campus with QR Code and enable the students aware of all the facts and data about the tree from its scientific name to its medicinal and pharmaceutic potential along with other interesting information, by scanning the QR code put up on the tree using their smartphone. These codes can be easily accessed by anyone in the college so that it will help the student, faculty, and others to gain knowledge about the plants available in the college campus.

**Keywords** Matrix barcode · Recording and displaying · Information on trees · Environmental informatics · Environmental modelling

## 1 Introduction

The role and function of Trees are both vital and critical for people and the planet. Numerous studies have proved that the presence of trees and urban nature can improve people's mental and physical health, student's attention and test scores, the property values in a neighbourhood, and beyond (Endreny 2018). Trees are essential for healthy communities and people. Protecting and Investing in trees will result in establishing a sustainable environment and society (Aplin 2008; Turner-Skoff and Cavender 2019). Today most students don't find the time and energy to go to the library and read books, some students who are interested to know about them seek information in textbooks, which may provide a monograph and other statistics. Still, the identification of these plants cumbersome via textbooks. It becomes fascinating

---

V. E. Chahana · S. Abirami · S. Karpagavalli · C. Arunpriya (✉)
Department of Computer Science, PSGR Krishnammal College for Women, Coimbatore, Tamil Nadu, India
e-mail: arunpriya.psgrkcw@yahoo.com



M. Prashanthi et al. (eds.), *Bioremediation and Green Technologies*, Environmental Science and Engineering, https://doi.org/10.1007/978-3-030-64122-1_22

to view a live plant, and this urge to understand these plants in view enables the retention of information, rather than gained through books. Even though there are many dedicated websites available to know more information on plants and trees, there is a huge demand for effective and easy to use the system to provide handy information about the trees (Blicharska and Mikusiński 2014); thus, this motivated us to use QR (Quick Response) code tagging to trees (Agarwal et al. 2006; Wäldchen et al. 2018). The QR codes are better than a traditional barcode since they can accommodate much more data; URL links, geographical coordinates and other information in the form of text. The other unique attribute of QR Codes is that it is not restricted and relied on the hand-held scanner the smartphones can be used to quickly scan the QR Code (Várallyai 2012; Goyal et al. 2016).

A University of Minnesota forestry class engaged QR codes to carter technology to aid nature. Students labelled trees around the St. Paul campus as part of a research project for their arboriculture course, wherein they placed signboard with QR codes on trees, enabling smartphone users to access information about the specific tree by scanning the sign. Arboriculture is the study of how to care for trees and shrubs and what tree most suited for a particular area; thus, the project paved the way for students to enable the public to find out about the trees on their campus (Chuang et al. 2010; QreateTrax 2012). In this context, a project has been done (PB Siddhartha College of Arts and Science, Vijayawada) to create awareness about the conservation of plants among the educated people. Anyone could use their smartphones to scan the QR code hung on the trees to acquire all the information about the plant ranging from its botanical name to its medicinal value at their fingertips (Grimm et al. 2016; Marthi 2019). The Botany department toiled for a month to collate the entire info on trees in the college campus and assigned QR codes to them. They catalogued the trees according to their species and incorporated vital information. The positive comment from the initial set of users was on the ease of using this mode to gain knowledge about the trees (Ballard 2007; Austen et al. 2016).

## 2 Proposed Work

The main objective of the proposed system is to provide information about the trees with the aid of the QR code, employing which the user will be gaining the knowledge by scanning the code via suitable scanner applications. In the proposed system, a QR code will be generated for each plant or tree on the college campus. The code generated with the help of the link would contain the details of the plants that will be stored in the college server. This empowers the user to search for the details of the plants quickly, and also it is user-friendly. User can easily access the code without anyone's help and also saves their time instead of searching it in textbooks (Sun et al. 2017) The proposed work involves many tasks such as tree dataset preparation, web page creation for each tree, generation of QR code, hanging QR code on trees and uploading web page in the webserver (Hagedorn et al. 2010; Joly et al. 2014a, b). The proposed system framework is shown in Fig. 1.

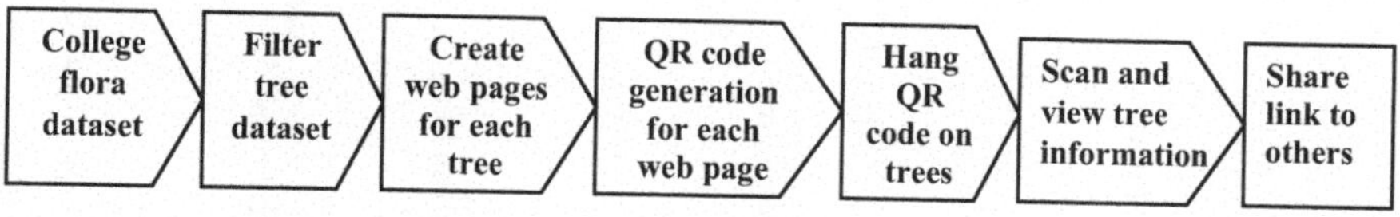


**Fig. 1** Framework of the proposed system

## 3 Design and Development of Web Pages of Trees and QR Code Tree Dataset Preparation

In the current work, detailed information on various plants and trees located inside the campus of "PSGR Krishnammal College for Women", Peelamedu, Coimbatore, were collected involving students of the Botany department. Various information about the plant documented as per family; botanical name; local name; different parts used and their medicinal uses. Among the 1034 plant species belonging to 47 families that frequently used for cultivation, ornamental and therapeutic uses in the campus, and a total of 20 trees taken into consideration for QR code tagging (Farnsworth et al. 2013; Goeau et al. 2016; Endreny 2018). The dataset presented in Table 1.

### *3.1 Web Page Design*

A web page created exclusively for each tree using HTML tags; cascading style sheets and javascript. A sample web page created is shown in Fig. 2. The web page is mainly developed based on the bootstrap framework. Bootstrap is an open-sourced HTML, CSS, and JavaScript framework that enables the UI development simple and can be with pre-built responsive classes and other utilities (Kumar et al. 2012; Pawara et al. 2017).

### *3.2 QR Code Generation*

A QR code for each tree web page generated using an online QR code generator as shown in Figs. 3 and 4. It is printed in sheets and put up on the trees shown in Fig. 5. The benefit of using QR code is that it makes the user very much easy, which can be freely accessed by their smartphones by either using a default scanner or some third-party scanning applications (Picchi et al. 2015).

**Table 1** Sample tree dataset

| Family | Binomial name | Morphological characters | Flowering and fruiting | Local name | Medicinal uses | Ecology |
|---|---|---|---|---|---|---|
| Annonaceae | *Polyalthia longifolia* (Sonn.) Thwaites | Tree, leaves lanceolate, fruit monocarp | Feb–May and Jun-Aug | Nettulingam | Antioxidant and antimicrobial activity | Evergreen tree |
| Fabaceae | Acacia spp. | Trees, Flowers are small, regular and usually bisexual, Spikes are in globular heads | Jan–Jun | Gum tree | Skin disease, mouth ulcer and plaque | Tropical plant |
| Caesalpiniaceae | *Bauhinia tomentosa* L. | Tree, branchlets present, leaflets adnate, terminal raceme | Throughout the year | Iruvaji | Inflammation of the liver, diarrhoea, digestive disorders and cough | Tropical plant |
| Caesalpiniaceae | Delonixregia (Hook.) Raf | Tree, leaves glabrous, pinnate, opposite, corymbose, flowers red | Mar–Aug | Mayflower | Constipation, inflammation, arthritis and dysmenorrheal | Deciduous tree |
| Asteraceae | *Eclipta prostrata* (L.) L. | Herb, rough hairs present, leaves base alternate, solitary flowers, involucral bract | Dec–Mar | Karisalankanni | Liver problems | Cultivated plant |
| Fabaceae | *Samanea saman* (Jacq.) Merr | Tree, leaves bipinnate, tomentose, gland opposite to leaflets, raceme | Mar–Jun and Oct onwards | Raintree | Healing wounds and cuts | Evergreen tree |
| Lythraceae | *Lawsonia inermis* L. | Shrub, leaves elliptic, lanceolate, flowers numerous, terminal panicle | Apr–July | Marudhani | Used for anti-cancer and anti-inflammatory activities | Tropical plant |

(continued)

**Table 1** (continued)

| Family | Binomial name | Morphological characters | Flowering and fruiting | Local name | Medicinal uses | Ecology |
|---|---|---|---|---|---|---|
| Asteraceae | *Balsamorhiza sagittata* (Pursh) Nutt | Leaves triangular to shapeless, hairs are present, terminal or axillary | Dec–Mar | Arrow root balsamoroot | Stomach problems | Cultivated plant |
| Musaceae | *Musa x paradisiaca* L. | Herb, leaves simple, spiral, pinnately parallel, spikelet drooping | Throughout the year | Vaazhai | Cancer, diabetes, diarrhoea and hypertension | Cultivated plant |
| Amaryllidaceae | *Crinum asiaticum* L. | Herb, leaves numerous, perianth white, capsule, limb lobes linear | Jun–Aug | Spider lily | Gastrointestinal disorders, skin disease, rheumatism | Cultivated plant |

**Fig. 2** Screenshot of tree web page

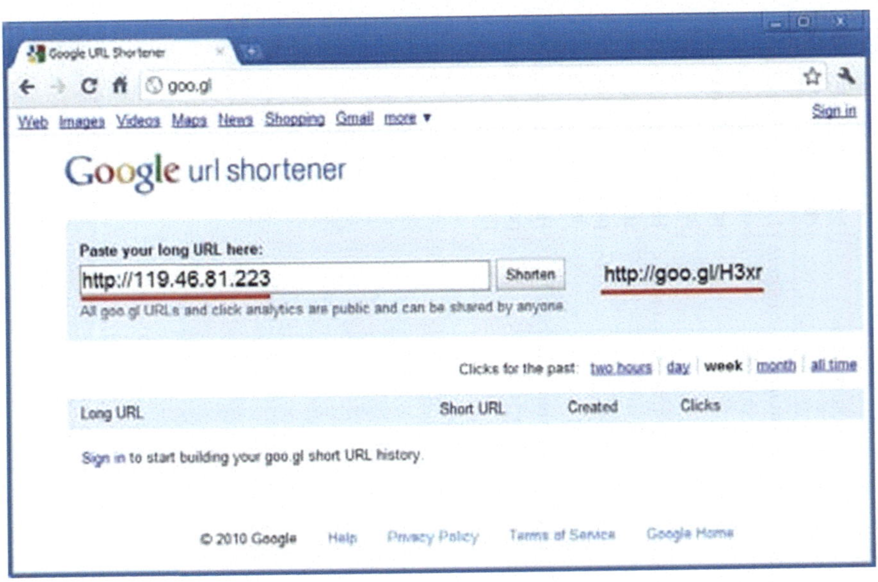


**Fig. 3** Shortening URL for generating QR code

**Fig. 4** Generated QR code

**Fig. 5** QR code hanged to a tree

### *3.3 Scanning the QR Code and Sharing*

Students with their smartphones can scan the QR code signboard placed on the trees on the campus to acquire all the facts about the trees ranging from its botanical name to its medicinal importance etc. The user can install any QR code scanner mobile app which is available for free on the online stores and platforms. Once the QR code on the tree scanned, the mobile interface shown in Fig. 6, appears and enables the user to perform various operations such as linking to the web page loaded in the intranet server, copying the link and sharing the link to others (Cope et al. 2012; Protos 2011; Shengyi 2015).

## 4 Conclusion

This proposed work helps the students to access all the information about the trees ranging from its botanical name to its medicinal value. It saves the time of the students

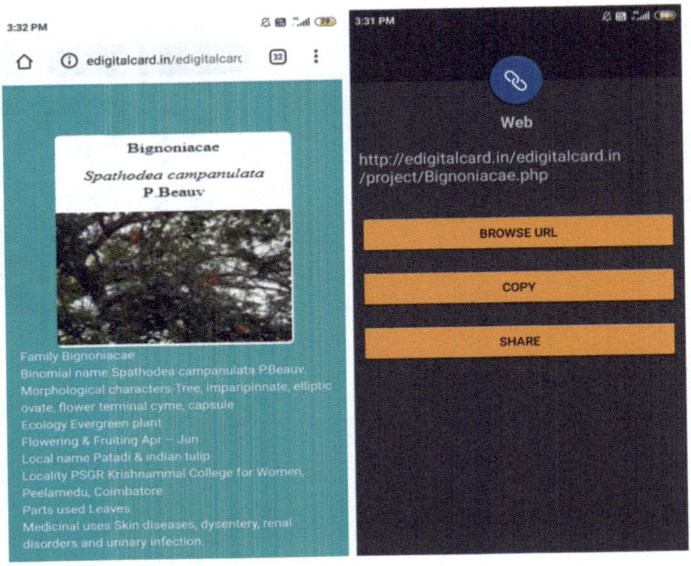


**Fig. 6** Tree information after scanning the QR code and options

and provides quick, handy useful information about the trees of their interest in seconds. This effort may lead to protect the plants and maintain the environmental sustainability. In the future, the work can be extended with the IUCN status of the tree, which indicates whether the tree belongs to endangered species or not. That will motivate the students to conserve such endangered species by growing them in their place. Also, the QR code tagging system will be extended for bio-parks with a large number of species as a service to the society.

**Acknowledgements** This work was financially supported by Department of Bio technology (DBT), Delhi.

## References

Agarwal G, Belhumeur P, Feiner S, Jacobs D, Kress WJ, Ramamoorthi R et al (2006) First steps toward an electronic field guide for plants. Taxon 55(3):597–610

Aplin DM (2008) How useful are botanic gardens for conservation? The Plantsman. R Horticultural Society 7(3):190–193

Austen GE, Bindemann M, Griffiths RA, Roberts DL (2016) Species identification by experts and non-experts: comparing images from field guides. Sci Rep 6:33634

Ballard B (2007) Designing the mobile user experience. Wiley, Chichester, pp 173–174

Blicharska M, Mikusiński G (2014) Incorporating social and cultural significance of large old trees in conservation policy. Conserv Biol 28(6):1558–1567

Chuang J, Hu Y, Ko H (2010) A novel secret sharing technique using QR code. Int J Image Process 4(5):468–475

Cope JS, Corney D, Clark JY, Remagnino P, Wilkin P (2012) Plant species identification using digital morpho-metrics: A review. Expert Syst Appl 39(8):7562–7573

Endreny TA (2018) Strategically growing the urban forest will improve our world. Nat Commun 9(1):1160

Farnsworth EJ, Chu M, Kress WJ, Neill AK, Best JH, Pickering J et al (2013) Next-generation field guides. Bioscience 63(11):891–899

Goeau H, Bonnet P, Joly A (2016) Plant identification in an open-world. In: CLEF 2016 conference and labs of the evaluation forum, pp 428–439

Goyal S, Yadav S, Mathuria M (2016) Exploring concept of QR code and its benefits in digital education system. In: International conference on advances in computing, communications and informatics (ICACCI' 16) Jaipur 2016, pp 1141–1147

Grimm J, Hoffmann M, Stover B, Muller K, Steinhage V (2016) Image-based identification of plant species using a model-free approach and active learning. In: Proceedings of annual german conference on AI, pp 169–176

Hagedorn G, Rambold G, Martellos S (2010) Types of identification keys. In: Tools for identifying biodiversity: progress and problems. EUT Edizioni Università di Trieste

Joly A, Goeau H, Bonnet P, Bakić V, Barbe J, Selmi S et al (2014) Interactive plant identification based on social image data. Ecol Inform 23:22–34

Joly A, Muller H, Goeau H, Glotin H, Spampinato C, Rauber A, et al. (2014b) Lifeclef: Multimedia life species identification. In: Proceedings of ACM workshop on environmental multimedia retrieval, pp 1–7

Kumar N, Belhumeur P, Biswas A, Jacobs D, Kress WJ, Lopez I et al (2012). Leafsnap: a computer vision system for automatic plant species identification. IN: Fitzgibbon A, Lazebnik S, Perona P, Sato Y, Schmid C (eds) Computer Vision–ECCV 2012. Lecture notes in computer science. Springer, Berlin Heidelberg, pp 502–516

Marthi T (2019) In this Vijayawada college, trees have QR codes. https://www.thehindu.com/sci-tech/energy-and-environment/now-even-trees-have-qr-codes/article29938678.ece

Pawara P, Okafor E, Schomaker L, Wiering M (2017) Data augmentation for plant classification. In: Proceedings of international conference advanced concepts for intelligent vision systems. Springer International Publishing, pp 615–626

Picchi G, Kühmaier J, Marqués JDD (2015) Evaluation of RFID UHF tags for electronic marking of standing trees. In: Proceedings of the 48th FORMEC symposium 2015 Oct 4–8, 2015, Linz, Austria

Protos M (2011) Why you should be wary of QR codes. https://gcn.com/articles/2011/09/13/qr-code-vulnerabilities.aspx

QreateTrax (2012) University of Minnesota forestry class uses QR codes on trees to connect with public. https://www.qreatetrax.com/Blog/tabid/235/entryid/9/u-forestry-class-uses-qr-codes.aspx

Shengyi L (2015) Develop PF mobile application for botony catalogue. BE Dissertation, UniversitiTunku Abdul Rahman, Malaysia

Sun Y, Liu Y, Wang G, Zhang H (2017) Deep learning for plant identification in natural environment. Comput Intel Neurosci 2017:7361042

Turner-Skoff JB, Cavender N (2019) The benefits of trees for live able and sustainable communities. Plants People Planet 1(4):23–335

Várallyai L (2012) From barcode to QR code applications. J Agric Inform 3(2):9–17

Wäldchen J, Rzanny M, Seeland M, Mäder P (2018) Automated plant species identification - Trends and future directions. PLoS Comput Biol 14(4):1–19

The manufacturer's authorised representative in the EU is Springer Nature Customer Service Centre GmbH, Europaplatz 3, 69115 Heidelberg, Germany. If you have any concerns regarding our products, please contact ProductSafety@springernature.com

Printed and bound by CPI Group (UK) Ltd, Croydon, CR0 4YY
22/09/2026
02220981-0014